ECONOMICS AND THE ENVIRONMENT

A signalling and incentives approach

Ian Wills

ALLEN & UNWIN

TO BARBARA
For her love and support

First published in 1997 by
Allen & Unwin
9 Atchison Street
St Leonards NSW 2065
Australia
Phone: (61 2) 9901 4088
Fax: (61 2) 9906 2218
E-mail: frontdesk@allen-unwin.com.au
URL: http://www.allen-unwin.com.au

National Library of Australia
Cataloguing-in-Publication entry:

Wills, I. R. (Ian R.).
Economics and the environment: a signalling and incentives approach.

Includes index.
ISBN 1 86448 257 5.

1. Environmental protection—Economic aspects.
2. Externalities (Economics). 3. Economic indicators.
I. Title.

333.72

Set in 10pt Revival by DOCUPRO, Sydney
Printed and bound in Malaysia by SRM Production Sdn Bhd

10 9 8 7 6 5 4 3 2 1

Contents

Preface

The idea that eventually led to this book was planted at my first reading of Nobel Laureate F.A. Hayek's classic paper 'The Use of Knowledge in Society', published in the *American Economic Review* in September 1945. Hayek explains that markets and government planning are alternative systems for coordinating people's use of resources. The effectiveness of coordination depends on the ability of each system to signal accurately information about people's wants and available supplies of resources, and on the incentives each provides for individuals to respond to the desires of others.

Introductory microeconomics textbooks commonly pay little attention to signalling and incentives, or to markets and government as alternative coordination systems. Signalling problems are overcome by assuming that decision makers have perfect information about others' wants and resource supplies. Decision makers are assumed to have appropriate incentives to respond to the desires of others. Given these assumptions, textbooks are able to concentrate on logical explanations of the decisions of consumers and producers in idealised markets, and the resulting market prices and quantities. The choice between market planning and government planning, perhaps the dominant political issue of the century, is generally set aside for later discussion. And this makes logical sense: with signalling and incentive problems assumed away, markets and government planning are equally effective ways of coordinating people's use of resources.

When we apply economics to the environment, this will not do. In core economics courses, there is some justification for introducing important analytical techniques using unrealistic assumptions that yield simple diagrams and equations. The complications due to imperfect knowledge, and the virtues and vices of market and government allocation systems, can be taught in later courses. However, a large proportion of the students in economics courses devoted to the environment take little or even no other economics, and have little interest in economics *per se*. Many of these students come to class perceiving that private property rights and free markets create or exacerbate environmental problems such as air pollution, land degradation, loss of indigenous species and destruction of stratospheric ozone. The limitations of private property and markets, real and supposed, have to be faced up front.

Economists teaching environmental courses have another problem. The economy and the environment are complex interdependent systems whose functioning is imperfectly understood. Continued human progress—and even human survival—is ultimately dependent on life support services of natural ecosystems (such as carbon fixation, waste decomposition and atmospheric screening of the sun's ultraviolet B radiation), and human use of natural resources and discharge of wastes change the environment on which we depend. However, economic analyses usually assume that the life support services of the environment are unaffected by economic activity; this greatly simplifies economic analysis by allowing economists to ignore the wider ecological impacts of human resource use, and their possible consequences for the future of human societies. If this assumption is invalid, as seems increasingly plausible after Chernobyl and overgrazing and desertification of the Sub-Sahara, then the system subject to economic analysis should be the combined economy and environment, not just the economy. Thus a course that uses economics to analyse environmental problems must recognise the complexity of combined economic–environmental systems and the implications of that complexity for the social coordination problem discussed by Hayek.

This book portrays environmental problems as results of failures of coordination between people in their use of natural resources. Its objectives are to clarify how such failures can occur under both market and government signalling and incentive systems, and how signalling and incentives might be improved in order to reduce particular environmental problems. It emphasises that both markets and government are unavoidably imperfect, because information is costly for buyers and sellers, and for government planners and the citizens they serve. It also emphasises our imperfect knowledge of the interactions within and between economic and environmental systems. And, recognising that both market and government decisions have contributed to our present environmental problems, it focuses on the incentives of both private and public decision makers.

This book is designed to teach a way of thinking about environmental problems, based on economic theory. It is not intended as an introductory economics text for students majoring in environmental studies;

for that purpose I use Paul Heyne's *The Economic Way of Thinking*, which is extensively referenced in the early chapters of this book. However, since many students and teachers of environmental courses have little interest in economics as a discipline and no desire to study introductory economics, I have tried hard to make the text accessible to readers with no prior knowledge of economics, without duplicating economics principles texts such as Heyne. Formal exposition of economic theory is confined to the minimum necessary to deal with particular problems. I judge that if readers can grasp a very few basic economic concepts, such as the law of demand and opportunity costs, and can be convinced that information and incentives matter, they will come to adopt the economic way of thinking about private and public decisions affecting the environment.

Consistent with the preceding point about our imperfect knowledge of the economy, the environment and their interactions, this book does not suggest specific answers to environmental problems. The topics covered are chosen to illustrate particular signalling and incentive problems, and for their currency and interest. I have not attempted a comprehensive coverage of important environmental issues, on the grounds that if the student can recognise particular types of signalling and incentive problems, he or she is best placed to apply economic principles in each case.

Acknowledgments

This book has been a long time coming, so I cannot possibly recall all my sources of information and inspiration, either written or verbal. The process began in the late 1970s, when the arguments of my farmer and planner colleagues on the Farm Management Advisory Committee of the former Upper Yarra Valley and Dandenong Ranges Authority first prompted me to think carefully about 'market failure' arguments for regulation of rural land use.

My greatest intellectual debts are to Friedrich Hayek and Ronald Coase. Other economists and political scientists whose writings have had a substantial influence on my thinking include Yoram Barzel, James Buchanan, Stephen Cheung, Carl Dahlman, Douglas North, Elinor Ostrom, Alan Randall, Aaron Wildavsky and Oliver Williamson.

My greatest expository debt is to Paul Heyne, whose style of writing for beginning students I have tried to emulate, I fear most imperfectly.

Much of the reading, thinking and writing leading to the book was accomplished during two periods of study leave, spent at the University of Arizona in Tucson (twice) and the Political Economy Research Center in Bozeman, Montana. Readers familiar with the western United States will agree that the spectacular natural environments of southern Arizona and south-central Montana are ideal locations for developing one's interests in how people use natural resources. I am grateful to Monash University for granting and financing my leaves, and to Jimmye Hillman, Bruce Beattie, Mike Copeland and Terry Anderson and their colleagues and staffs for their hospitality and stimulating working environments during my stays in the USA.

At Monash University, successive chairmen of the Economics Department, in particular Richard Snape and John Freebairn, have been supportive and understanding during the trials of authorship. Ian Thomas and Frank Fisher of the Graduate School of Environmental Science have encouraged me to believe that the book will be useful to non-economists as well as economists. My successive secretaries, Barbara Bergin, Margaret Coates and, especially, Mavis McGill, have done a sterling job of editing, printing and collating the many partial and complete drafts of the manuscript.

At Allen & Unwin, I am indebted to Mark Tredinnick and Joshua Dowse, my publishers, and to Rowena Lennox and Lynne Frolich, my editors. Their patience and advice at all stages of production of the book is greatly appreciated.

My family pitched in. Miranda and Nathaniel advised me on the accessibility of the material to undergraduate students. My wife Barbara edited all of the first drafts, turning literary sows' ears into silk purses.

Many individuals have made helpful comments on the manuscript and the drafts and papers that preceded it. In addition to my family, the students in the graduate program in environmental science at Monash have been a great help in identifying suitable modes of expression and examples of environmental problems. Over the years I have benefited from the comments of Geoff Edwards, Peter Gorringe, Geoff Hogbin, Jeff LaFrance, Alan Moran, Warren Musgrave, Kwang Ng, Ross Parish and Richard Snape. I am especially grateful to Terry Anderson, Jeff Bennett, John Freebairn and Ron Johnson, who took the time to read and comment on complete drafts of the manuscript. Their advice that the task was worthwhile encouraged me to persevere at times of self-doubt.

Last but not least, I am grateful to the taxpayers of Australia, who have underwritten my thinking and writing about economics and the environment. If this book helps to lessen the common impression that economics and the environment are at odds, I think that the taxpayers will be adequately compensated.

I

SOCIAL COORDINATION, THE ECONOMY AND THE ENVIRONMENT

1
Introduction

The economic way of thinking about the environment begins with the recognition that environmental problems are people problems. They occur when some people are unhappy with other people's use of the environment. Such failures of social coordination are unavoidable because people are generally self-interested and imperfectly informed about the environment and other people's wishes. Thus the route to resolution of environmental problems lies through first understanding and then reducing these incentive and informational barriers. But the complexities of human–environment interactions, where people are both dependent on the environment and capable of changing it, mean that the task will be anything but easy.

1.1 What are environmental problems?

Bottles and cans discarded along the highway. Pulp and paper mill effluent polluting rivers. Chlorofluorocarbons (CFCs) from refrigeration and air-conditioning appliances breaking down Earth's protective layer of stratospheric ozone. Illegal trapping and smuggling of Australia's rare parrots. Clearing of Amazonian rainforest threatening the survival of plant and animal species. We readily recognise all of these as environmental problems, yet they are diverse in their physical or biological nature and geographic extent. We commonly distinguish them as *solid waste problems* or *local* or *global pollution problems* or *illegal harvesting problems* and so on, but what common elements do we recognise?

First, environmental problems involve nature—the natural world on Earth, including non-living physical features such as mountains, oceans and air, and all non-human non-domesticated living things. Some readers will immediately protest that evolutionary biology teaches us that humanity is a part of nature. True. In this book the view of 'nature' and of 'environmental problems' is anthropocentric. This is unavoidable. The problems that concern us are those that matter to people, and humans do have the technological and behavioural capacities to change the environment. The ethics of human–environment relationships are briefly examined in Section 1.5.

Second, when we use the word 'problem' to describe something happening in the environment, we generally think that human actions create or affect the event so described. Environmental problems involve human impacts on the natural environment. Thus harmful natural events beyond human control, such as hailstorms or droughts or cyclones or volcanic eruptions, are not viewed as environmental problems here. They are part of our ever-changing natural environment. However, the way people respond to such natural calamities can cause environmental problems, for example the adverse effects of drought-mitigating irrigation on downstream soil salinity.

Equally important is the distinction between human actions in the past, which cannot be undone, and current human impacts on the environment. What we call 'nature' today is, in part, a result of human activity in the past, including English fields and hedgerows, and Australian landscapes and wildlife shaped by Aboriginal firesticks and, recently, agriculture and rabbits. Tim Flannery's ecological history of the Australasian lands and people, *The Future Eaters*, provides a fascinating account of the shaping of Australian 'nature' by both Aborigines (Part Two) and Europeans (Part Three).[1] Flannery points out the absurdity of the view that Australia contains 'unmodified nature' or 'wilderness' after 60 000 years of human occupation. He also highlights the extent to which people's perceptions of 'nature' are determined by experience and culture, in particular the very different historical experiences and cultures of Aboriginal and other Australians.[2]

We cannot eliminate agriculture and rabbits; for better or worse, they are part of the current environment. Thus it is not sensible to see their existence in Australia as an environmental problem. On the other hand, the way Australians farm and the way they manage rabbits are things we can control now; features of the environment that we can do something about can be environmental problems.

Third, the notion that something is a 'problem' generally means that there is disagreement between individuals over appropriate actions. It is not sufficient to say that environmental problems occur when somebody damages the environment. What is damage, and is it a social problem? A farmer might apply fertiliser incorrectly, and thus reduce crop or pasture yield. A land developer may reduce eventual property sales returns by excessive subdivision. In each case, there is damage to the environment from the owner's perspective, but not necessarily from that of outsiders. If no one is hurt but the resource user, there cannot be a problem from a social point of view. On the other hand, if the farmer's fertiliser applications damage stream quality for swimmers and fishers, or the developer's tree clearing damages the views enjoyed by neighbours and passers-by, we do have a social problem, unless those harmed have agreed to bear the harm as a result of being compensated for damage suffered. Environmental problems involve a lack of consensus between the person harmed and the resource user.

Putting these ideas together, we can define environmental problems: *Environmental problems occur when some people are unhappy with other people's use of the natural environment, because it imposes harms on them to which they have not consented.*

The definition implies a distinction between harms, depending on whether or not there is prior consent to their imposition—in other words, whether or not they are legitimate. A red light legitimately 'harms' my progress at an intersection. As we will see in Chapter 2, legitimacy depends on prior community agreement about individuals' rights to scarce resources, such as road space in the intersection. Scarce resources mean that harms to others are unavoidable; the issue is not whether such harms occur but whether they are the subject of prior community agreement.

Our definition is not meant to rule out problems involving only one resource user and one person harmed. The plural 'people' is preferable because important environmental problems almost always involve many resource users and/or damage sufferers. As is explained at length below, this is what we would expect; it is usually relatively easy, and hence relatively cheap, to reach consensus about use of natural resources when only two or very few parties are involved.

More importantly, as defined, environmental problems do not necessarily have solutions; that is, consensus between the resource user and persons harmed cannot necessarily be achieved at acceptable cost. Suppose that I am particularly unhappy at the sight of highway litter, but that other travellers are less concerned. If I have to sacrifice too much time and money to locate and rebuke litterers, or to organise fellow travellers to lobby government to detect and punish litterers, my continued unhappiness may be the best outcome for the community as a whole.

Each of the environmental problems listed above involves a lack of social coordination or consensus between resource users and those harmed. Highway litterers are unresponsive to the scenic concerns of subsequent travellers. Pulp and paper mill executives commonly pay little attention to downstream pollution, which has no effect on profits. The Chinese middle class, proud of their new CFC-using refrigerators, may be unaware of the scientific consensus that CFC releases cause increased ultraviolet radiation in the higher latitudes. Bird smugglers pay attention to the policing and penalties that partially signal our concern about endangered birds, but the dollar signals from overseas collectors are stronger. Brazilian ranchers respond to the price of beef and government agricultural subsidies, not to the medicinal and genetic and carbon recycling values embodied in the tropical forest. In each case social coordination is lacking; resource users are either unaware of or disregard the desires of other people.

1.2 The reasons for coordination failures

Why do people fail to coordinate with others in their use of the natural environment?[3]

1 *People are self-interested.* People attend to their personal interests (which may include the welfare of others) ahead of the interests of others. Thus loggers will reveal more of what they know about the benefits of logging than about the damage it imposes on others, and environmentalists vice versa.
2 *Information about resource uses and values is dispersed and private.* Foresters know more than other people about the productivity and value of forests for timber; biologists know more about the capability of forests to maintain rare flora and fauna and ecosystems under different management regimes; environmentalists and other citizens know more about the true sacrifices they would be prepared to make to preserve old-growth forests from logging. This information is costly for decision makers to acquire, especially if some parties choose to conceal or distort the truth to serve their own interests.

Any society must deal with these barriers to coordination if it is to persist and thrive. In Chapter 2 we review the ways in which societies coordinate people's actions. Then we explore the reasons why coordination is not always possible, and the prospects for alleviating environmental problems by improved coordination between people. But first we need to understand human dependence on the environment.

1.3 Humans and Earth's environment

Environmental problems exist because human beings are both dependent on the natural environment and capable of consciously changing it. Our dependence on the natural environment is, crudely, twofold. First, the natural environment provides our life support system. We are sustained by a complex network of living organisms and non-living materials—the atmosphere, the oceans, minerals, organic matter and so on—which together maintain the flow of energy and the cycling of chemical elements necessary to life.[4] Second, the natural environment affects our quality of life, in both material and aesthetic terms. We feel better

Figure 1.1 Interactions between the climate and major chemical flows

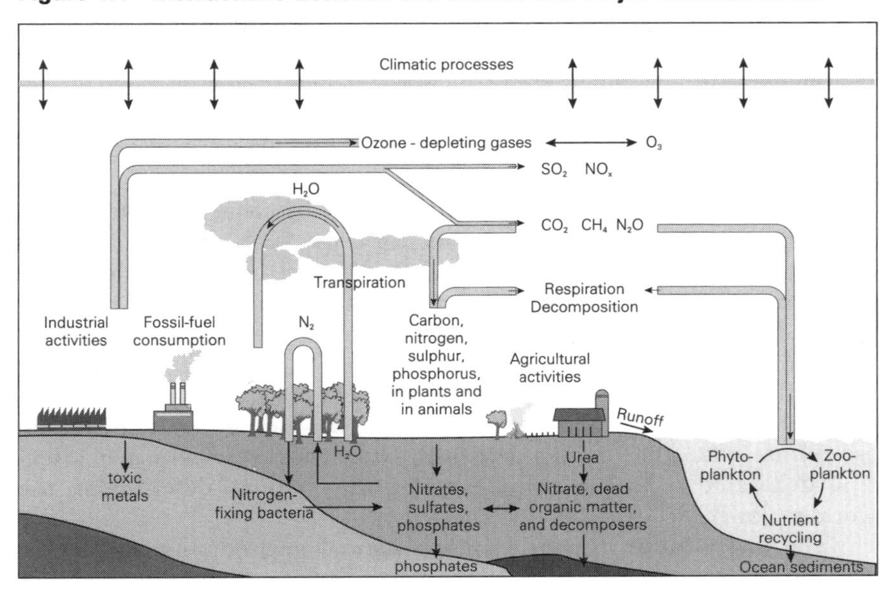

Source: W.C. Clark, 'Managing Planet Earth' *Scientific American*, Sept. 1989, p. 24.

off if timber and fish are abundant and therefore cheap, if the air smells fresh and if our view of the mountains is unobscured by smog.

Humanity's global life support system is composed of four interacting subsystems—climate, the cycling of the chemical elements necessary to life, the water cycle and living organisms—whose major interactions are illustrated in Figure 1.1. Changes in the natural environment, whether naturally occurring or due to human action, involve one or more of these subsystems.

The *climate subsystem* incorporates the atmospheric and oceanic processes that govern the global distribution of wind, rainfall and temperature. These processes are responsible, for example, for the major climatic changes associated with the El Niño effect in the southern Pacific and for any global temperature changes resulting from the accumulation of greenhouse gases.

The *cycling of chemical elements* involves the global circulation and processing of carbon, oxygen, nitrogen, phosphorus and sulphur. These elements are essential to life, and important in a wide variety of manufacturing processes. In compounds such as carbon dioxide, methane and nitrous oxide, they may have major effects on climate.

The *water cycle* involves evaporation, transpiration, precipitation and runoff. It circulates chemical elements between the land, the oceans, the atmosphere and living organisms. Water in the oceans plays a key role in determining the global climate.

Living organisms, and the ways they evolve, are dependent on the climate and the chemical and water cycles. At the same time, life other than humans is capable, usually over long time periods, of modifying the climate and circulation of chemical elements and water.[5] Photosynthetic plants emit oxygen and water vapour, animals emit carbon dioxide and methane, and animal grazing can alter evaporation and runoff.

To understand how humans can be harmed by changes in the natural environment, and how human actions can cause such changes, we must have some understanding of the functioning and interactions of these four systems. However, many of the physical and biological interactions within and between climate, chemical and water cycling and living organisms, and between these natural systems and human activities, are not well understood. For example, in the case of greenhouse gas emissions, the impact of emissions on global temperatures, and thence on agriculture and other human activities, is the subject of vigorous scientific debate, due to the complexity of the systems determining the world's climate.[6]

The major human impacts on the natural environment are due to agriculture, energy use, manufacturing, armed conflicts and, recently, recreational activities. The last century or so has seen unprecedented increases in the scale of human impacts. For example, since the mid-1800s, an area larger than Australia has been converted to permanent croplands, mainly from forests and woodlands.[7] Global energy use, mostly from fossil fuels, has grown about fifteen fold since 1900.[8]

The causes of these human impacts are twofold. First, there have been major technological changes in agriculture, forestry, manufacturing, transport, recreation and warfare. Second, there was a rapid increase in the human population, itself a result of technological changes, in particular improvements in medical technology. Activities such as diversion of large rivers like the Murray River in Australia and the Colorado River in the USA, the use of industrial chemicals in agriculture, fish harvesting using sonar and nylon nets, the burning of millions of tonnes of coal and oil to generate electricity and fuel cars and trucks, the use of chemical defoliants in warfare and the control of diseases such as malaria have no historical precedents. The environmental consequences of such recent human activities are frequently extensive in space (e.g. soil salinity, acid rain, stratospheric ozone depletion) and in time (e.g. radioactive and heavy metal contamination of sites, disappearance of species, deforestation, soil salinity, rise in levels of greenhouse gases in the atmosphere).

In environmental problems, nature mediates between those who use the environment and those who are harmed as a result. The environmental impacts and processes involved may be direct and simple, as in the case where one person neglects weeds, which spread to a neighbour's property. They may also be indirect and complex. Where environmental processes are complex, and environmental changes are extensive in space and time, it is often difficult to determine both human impacts on the environment and the consequent impacts of environmental change upon people. For example, pollution of the Great Barrier Reef by the fertilisers used by Queensland sugarcane growers has long-term impacts on reef ecology and the value of the reef to people; at present, these are uncertain. And, as mentioned above, there is scientific uncertainty about the impact of greenhouse gas emissions on global temperatures, and thus on agriculture and other human activities.

To understand environmental problems, we must understand what motivates people to use the natural environment as they do. In Chapter 2, we explain how economic systems—markets, government planning and traditions or customs—guide people's use of resources, including natural resources. But first, we must recognise two problems posed by the complexity and extent of human–environment interactions: how to decide the extent of the systems involved, and how to determine the ethics of human manipulation of the environment.

1.4 The problem of setting boundaries

If human–environment interactions are complex and extensive in time and space, how do we set boundaries to the investigation of particular environmental problems—how do we decide how far to search for impacts across people and the natural world?

Separate disciplines concerned with environmental changes have their own criteria for deciding how wide to cast the net: economics looks for significant impacts on human behaviour; ecology looks for significant impacts on the interactions of living and non-living components of the natural environment. Appropriate temporal and spatial boundaries are likely to differ across disciplines. For example, in the case of pollution of the Great Barrier Reef, the ecological impacts of farm fertilisers may be confined to the coastal plain, tidal and reef environments; however, because degradation of the reef affects interstate and overseas tourism, the economic impacts of pollution extend further in space than the ecological impacts.

Twentieth-century economists have generally ignored the effects of economic activity on the environment's capacity to assimilate human wastes and support human life. In effect, economists have assumed that the environment's life support services are freely available at levels that exceed any possible human demands for them.[9] In *The Future Eaters*, Tim Flannery argues that this was not always true of historical human societies, such as the Polynesian colonists of New Zealand and Easter Island.[10] According to Flannery, the early Polynesians were addicted to discovering new lands, overexploiting them briefly and moving on. In New Zealand and Easter Island, this resulted in ecosystem collapse and catastrophic resource shortages.

In contrast to economists in this century, classical economists such as Malthus and Ricardo, observing increasing scarcity of agricultural land relative to an increasing population, were pessimistic about the prospects for continued growth of existing human societies. However, succeeding economists recognised the importance of technical progress in providing substitutes for natural resources, and of prices in promoting technical change and economy in natural resource use, resulting in general optimism among economists about natural constraints on human progress and survival.[11] Not until the late 1960s and 1970s, following the publication of *Silent Spring*,[12] Rachel Carson's attack on the use of DDT and other artificial insecticides, and widespread fear of the consequences of nuclear war, did many economists begin to question the assumptions of costless environmental waste assimilation and life support services. More recently, events such as the Chernobyl explosion and the discovery of CFC-induced depletion of the stratospheric ozone layer have led to more widespread recognition that the systems subject to economic analysis should include the local or global environment, as well as the local or global economy.

Consistent with the change in views, a separate *environmental economics* subdiscipline, emphasising the importance of extending the boundaries of economic analysis to include major economy–environment interactions, has developed within the economics profession since about 1970. However, economists working in this field are not unanimous about where the boundaries to their studies should be drawn. The majority of environmental economists concentrate on methods of expanding private and public economic calculations to include environmental assets, treating environmental assets like commodities.[13] For example, it may be possible to create new property rights and markets for air pollutants (see Chapter 15) or fish stocks (Chapter 18), to price garbage disposal according to the cost of secure landfill sites (Chapter 13), to tax polluters according to discharge volumes (Chapter 15), and to provide public decision makers with estimates of the values of unpriced species threatened with destruction (Chapter 11). The aim is to achieve socially efficient use of resources (i.e. to obtain the maximum benefits for the community net of community sacrifices), taking into account *all* values gained and sacrificed, priced and unpriced, now and in the future.

Some environmental economists, while agreeing on the importance of efficient resource use, argue that achieving efficiency in the above sense will not guarantee that regional or national or global economic activity is consistent with ecological stability—with continued functioning of human life support systems. *Ecological economics* extends the boundaries of analysis of environmental problems to include both economics and ecology. In the words of some of its prominent proponents:

> ecological economics goes beyond our normal conceptions of scientific disciplines and tries to integrate and synthesise many different disciplinary perspectives . . . While the intellectual tools we use . . . are important,

> they are secondary to the goal of solving the critical problems of managing our use of the planet. We must transcend the focus on tools and techniques so that we avoid being 'a person with a hammer to whom everything looks like a nail' . . . Ecological economics sees the human economy as part of a larger whole. Its domain is the entire web of interactions between economic and ecological sectors.[14]

The macro goal of ecological economics is sustained functioning of the combined ecological–economic system. To that end, ecological economists are particularly concerned with the scale of use of natural resources relative to the dimensions of the ecosystems on which humans depend. (An *ecosystem* is a set of interacting species and their supporting non-biological environment, functioning together to sustain life by capturing the sun's energy and cycling water and the elements essential to life.) Sustainability can only be achieved if economic activity remains within the 'carrying capacity' of the ecosystems that support it; it follows that the state, which is responsible for the management of the macroeconomy, should step in to restrain economic activities that damage the functioning of ecosystems.[15] The complexity of interactions between the economy and the environment, and the resultant problems of defining and implementing sustainability, are discussed in Chapter 4.

1.5 The environment and ethics: Western and indigenous perspectives

Recall, from Section 1.3, that human beings are both dependent on the natural environment and capable of consciously changing it. Both attributes are products of our biological evolution; however, the latter is far more dependent on our cultural inheritance (i.e. on the deliberate transmission of information between people) than on our genetic inheritance. Dawkins, in *The Selfish Gene*, views human culture as a new kind of self-replicating entity.[16]

Human increase and exploitation of a finite natural environment is subject to two types of constraints. First, there are externally imposed physical and biological constraints, such as the amount of solar radiation received at Earth's surface and the ability of the plants within a particular ecosystem to capture that energy by photosynthesis. In the absence of culture, the future of humanity would be dependent on biological evolutionary change, as is the future of other species. Second, as self-conscious beings, humans can subject themselves to behavioural rules designed to avoid damage to our environmental life support systems. Pollution control regulations and laws to protect endangered species are familiar modern options. Flannery reports a much more venerable option, no doubt widely employed in human history: the Aboriginal people of the Cooktown area regard certain mountain summits as 'story places', not to be entered on any account. These sites are clearly defined areas of prime tree kangaroo habitat. Since the tree kangaroo was an important game animal for local Aborigines, who hunted it very efficiently with dogs, it would very likely have become extinct in the absence of the sanctuaries provided by the story places.[17]

Human cultural evolution involves the transmission of both instructions on the use of environmental assets and rules regarding permissible uses of those assets. Thus it is unsurprising that communities that have lived entirely separately for long periods of time, for example Aboriginal and European-descended Australians, have very different technologies and rules for using environmental assets. Our cultural inheritance substantially determines both our present ability to change the environment and our beliefs about the goodness and badness of such actions—our environmental ethics.

In one sense, environmental ethics is unavoidably anthropocentric. Among Earth's species, only humans have the technological and behavioural capacities to rapidly change the environment, and only humans can self-consciously determine the goodness or badness of their actions. For most people, these facts are sufficient to justify *anthropocentric environmental ethics*, in which the goodness or badness of environmental changes is judged according to the values that people attach to those changes, which include values attached to the continued existence of environmental assets, such as rare species and ecosytems.[18] Economics is anthropocentric in this sense. An extension of this view, still human-centred, sees present humans as stewards charged with managing nature on behalf of the current and future generations of humans plus (possibly) other higher species that are capable of sensing pleasure or pain.

Anthropocentric environmental ethics do not go unchallenged. Nash, reviewing the history of Western environmental ethics, points to a long-term trend in liberal societies, such as Britain, the USA and Australia, to expand the range of life forms accorded moral rights and interests, proceeding from the Magna Carta to votes for women to the US Endangered Species Act passed by the Congress in 1973.[19] In opposition to anthropocentrism, Aldo Leopold, drawing on the emerging discipline of ecology, proposed his 'land [meaning ecosystem] ethic', based on the idea that the individual is a member of a community of interdependent parts.[20] As Leopold put it:

> when we attempt to say that an animal is 'useful', 'ugly' or 'cruel', we are failing to see it as part of the land. We do not make the same error of calling a carburettor 'greedy'. We see it as part of a functioning motor.[21]

For Leopold, the land ethic:

> changes the role of *Homo Sapiens* from conqueror of the land-community to plain member and citizen of it . . . When we begin to see land as a community to which we belong, we may begin to use it with love and respect.[22]

Thus:

> there are obligations to land over and above those dictated by self-interest, obligations grounded on the recognition that humans and the other components of nature are ecological equals . . . [a land-use decision] is right when it tends to preserve the integrity, stability and beauty of the biotic community. It is wrong when it tends otherwise.[23]

How could Leopold's *ecocentric environmental ethics* be implemented? The lawyer Christopher Stone pointed out in his work *Should trees have standing?* that Western law includes provision for the interests of infants and other mental incompetents to be represented legally by human guardians; why not extend guardianship to trees and other environmental assets?[24]

Despite the changes in views of nature recorded by Nash, practically all environmental decision making, private and public, is anthropocentric, as is practically all economic analysis of such decisions.

Flannery reminds us that culture adapts to biological reality:

> one of its [culture's] key elements is the embodiment, in beliefs and customs, of actions which help people survive in their particular environment . . . cultures which we can call 'ecologically attuned', are the results of many thousands of years of experiencing and learning about a particular ecosystem.'[25]

Flannery regards Aboriginal culture, after about 60 000 years of experience of the resource-poor and climatically-variable Australian environment, as highly ecologically attuned.[26] Note that Flannery is *not* saying that Aborigines have always been ecologically attuned. On the contrary, he argues that, after their arrival, the Aborigines were responsible for environmental changes comparable to those that have occurred since 1788.[27] Conversely, Maori immigrants to New Zealand, and European and Asian immigrants to Australia, are culturally maladapted to the environments in which they now reside, as were the original Aboriginal immigrants 60 000 years ago.

Flannery argues that Aboriginal Australian culture and Australian ecosystems underwent a long period of *coadaptation* and *coevolution*, which eventually absorbed the Aborigines into a new integrated ecosystem. Thus, interpreting Flannery's argument in ethical terms, for traditional Aborigines, the distinction between anthropocentric and ecocentric ethics would be difficult to discern, if not absent. In the 1971 Northern Territory Land Rights case *Milirrpum v Nabalco Pty Ltd*, Mr Justice Blackburn commented:

> The evidence seems to me to show that the aboriginals have a more cogent feeling of obligation to the land than ownership of it . . . [I]t seems easier, on the evidence, to say that the clan belongs to the land than that the land belongs to the clan.[28]

In contrast, the value systems and resource-use practices of the Polynesians who entered New Zealand in 1000 to 1200 AD, and of European-descended Australians, put the interests of people above the integrity of existing ecosystems.[29] Maori overexploitation of New Zealand's unique bird species and marine resources produced a crisis in food availability within four hundred years of settlement. In Australia, one hundred years of irrigated agriculture have created major ecological problems in the Murray–Darling basin, but most present-day Australians still baulk at the prospect of cutting back irrigation water use to reduce future downstream salinity and to preserve riverine ecosytems.

Community values aside, for reasons explained in Chapter 4, when we allow for interactions between human society and the environment, it is unclear whether genuinely ecocentric environmental policies could be identified and implemented. In a world of imperfect scientific knowledge, plus constant change in the environment, people's preferences and technology, how do we know what is ecocentric?

1.6 A brief outline

This book explains environmental problems as results of failures of coordination between people in their use of natural resources. The remainder of Part I focuses on social coordination mechanisms and the relationship between the economy and the environment. Chapter 2 explains, in economist's terms, the need for social coordination in the use of limited resources, and the major systems of social coordination observed in human societies. Chapter 3 explains how coordination would work in the idealised market economy of the economics textbooks, and in an idealised planned economy, where government determines resource use. Chapter 4 moves to the real world, outlining the complex relationships both within and between the economy and the natural environment. The complexity is such that most environmental problems are best viewed as products of combined economic–environmental systems where, in principle at least, everything affects everything else. Chapter 4 emphasises the information collection and communication difficulties involved in coordinating people's use of the environment.

Part II is the analytical core of the book. It explains how costly information and/or individual self-interest can cause failures of coordination in use of the natural environment under both markets and government planning. Chapters 5, 6 and 7 deal with the limitations of market coordination, and Chapter 8 with the limitations of government coordination.

Part III explores the implications of costly information and self-interest for important tools used by economists. First, although Chapters 1 to 8 make only brief reference to the passage of time, the consequences of today's decisions about the use of natural resources often stretch far into the future. The problems of comparing benefits and costs arising at different points in time are discussed in Chapter 9. Chapter 10 reviews the most important of economists' tools for comparing alternative uses of environmental resources: cost–benefit analysis (also called benefit–cost analysis and project analysis). However, cost–benefit analysis, and other economics-based evaluation techniques, can only be useful to environmental decision makers if the techniques incorporate accurate information about the values people attach to different environmental outcomes. In Chapter 11 we review the various techniques for attaching dollar values to goods and services that are not priced in markets. We pay particular attention to the assumptions and techniques involved in these valuation tools, in order to assess whether their use is likely to improve the information available to decision makers.

Societies commonly measure their progress towards higher standards of living by the use of national accounting systems. It is widely understood that conventional national accounting measures, such as gross national product, fail to account for major changes in the natural environment that accompany modern economic growth. Chapter 12, the only chapter not concerned with the consequences of particular resource use decisions, discusses the problems of monitoring changes in the combined national economy and environment, and the advantages and disadvantages of modifying national accounting procedures to provide more information about changes in the environment.

The discussion of particular environmental problems begins in Part IV with problems that are more likely to be local in their impacts on people, and progresses in Part V to problems believed to affect people worldwide, and hence likely to involve the most complex economy–environment interactions. Chapters 13 to 15 deal with waste disposal, recycling and pollution control within a single country. Chapters 17 to 21 deal with environmental problems that frequently spill across national boundaries: global pollution, management of 'common pool' resources such as fish and wildlife, and conservation of biodiversity. Each type of problem involves particular types of coordination failure. In each case, an understanding of the reasons why people fail to coordinate their use of natural resources provides the basis for policies designed to improve coordination, so that environmental problems can be reduced or eliminated.

For the reasons discussed in Chapter 4, our lack of information about the possible future consequences of today's decisions is severe where the environment is concerned. The problem of uncertainty about future consequences of decisions is increased when the consequences of actions extend across national borders, but is most acute when the consequences of present actions stretch far into the future, since it is impossible to predict reliably future technologies, and the knowledge and preferences of future generations. Chapter 16, the first in Part V, addresses the problems of decision making in the face of this unavoidable uncertainty. The following chapters on global pollution, common pool resources and biodiversity illustrate particular problems arising due to long-delayed consequences of resource use.

Discussion questions

1. Aircraft noise imposes costs on airline employees at Sydney Airport, but this is not an environmental problem. The noise from aircraft approaching the airport imposes costs on local residents: this is an environmental problem. In each case, true, false or uncertain? Explain why.
2. Why do you think that it is likely to be more difficult and costly to reach consensus about the use of natural resources when more than a very few parties are involved?
3. In what important ways do the economy and the environment interact? List the major categories of services which human societies and individuals obtain from the natural environment. For each category of service you list, identify human activities which deplete and augment the stocks of the relevant environmental asset.
4. What would be some of the practical problems faced by a society which attempted to seriously implement Leopold's ecocentric environmental ethics, as sketched in Section 1.5?

2

Scarcity and systems of social coordination

Human societies face two basic coordination problems. First, where resources are scarce relative to people's wants, societies must define and enforce rules governing access to those resources. Second, societies must have means of coordinating individuals' production and consumption activities. This requires signalling of information about what other people want and can provide, and incentives for people to respond to the desires of others. Coordination is achieved by some combination of three alternative signalling and incentive systems: tradition, markets and central planning—the latter two dominate in existing societies. Thus, to understand environmental problems, we need to understand why markets and central planning may fail to reconcile people's interests in natural resources.

2.1 Scarcity and the need for social coordination

Social coordination is defined as individuals acting in a way consistent with the desires or needs of others in a recognised group, to obtain benefits for themselves or for the group. Within the animal kingdom, social coordination within a species occurs in procreation and the nurturing of young, in predation and defence, and in the defining and respecting of territory. The last of these would be superfluous if territory, and the resources which go with it, were abundant rather than scarce, relative to the collective needs of the species.

Fig. 2.1 The nature of scarcity

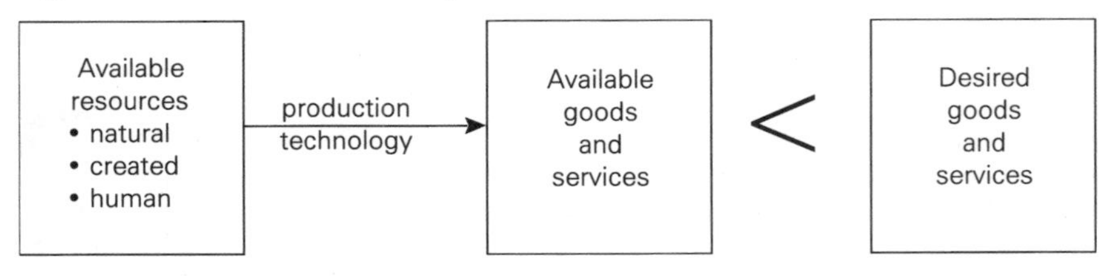

In the case of humans, desires for resources are based not only on physiological needs, but also on psychological desires, and are thus effectively unbounded. In affluent societies, the psychological usually dominates the physiological. Think of your dream house, sited where you would most like to live if money were no problem. How much of the value of the property would be attributable to your desire for shelter from the elements, as opposed to your desires for luxury, status, scenic views, proximity to work, recreation, etc.? Note that the judgment about unbounded desires applies to people in general, and not necessarily to every individual.

People's wants include privacy, clean air and water, preservation of flora, fauna and ecosystems and so on. How these wants are weighted relative to wants for powerful cars or hamburgers is determined by communities' ethical norms and the values which result from those norms, not by the economist, who takes people's likes and dislikes as given.

In all known societies, with desires unbounded, resources (natural, created and human) are inadequate to produce all the goods and services necessary to satisfy all human wants. This is the economist's definition of *scarcity*, illustrated in Figure 2.1. More effective production technologies based on advances in human knowledge, for example plants which are more efficient in converting nutrients to grain, or desktop publishing programs that reduce the time and labour and capital equipment required to produce manuscripts, alleviate scarcity but do not eliminate it.

Economic scarcity and economic analysis are anthropocentric; economists study how societies make choices that are unavoidable because human wants exceed .the resources available. There are other types of scarcity. For example, a biological scientist growing a bacterium in a petri dish may observe that the growth medium is scarce in relation to the physiological needs of the organism. However, this scarcity is of no concern to people unless human interests are involved, for example if the bacterium is being used to produce a cure for a human disease, or people care about its survival because of its rarity. Scarcity in terms of the bacterium's needs is not economic scarcity unless those needs happen to coincide with those of people. Again, imagine that biological control is successful in eliminating most of the cane toads from northern

Australia. Toads would then be rare, but there would be no scarcity of toads from the point of view of Australians, as distinct from the point of view of toads.

Scarcity of both resources (the means of producing goods) and goods (the things people desire) means that contests for access to and control over scarce assets (both goods and resources) are unavoidable. The most elemental contests involve outright conflict, where brute force is the deciding factor. This is what we often observe when young children first meet at preschool. It was frequently the case during European invasion and settlement of Australia and other parts of the New World.

The need for social coordination in the use of scarce assets arises because brute force or other unilateral action generally turns out to be too costly for the contending parties, in terms of life and limb and the resources used in offence and defence. Also, in conflict situations, where valuable resources may be suddenly lost, there is little incentive to produce or maintain those resources. A Bosnian Serb is unlikely to devote much effort to house repairs or maintenance if he or she may lose the house to Muslims or Croats at any time. Such costs of conflict and the resulting uncertainty create the desire for a 'peace pact' that establishes rules governing rights to use scarce assets and to transfer them between parties.[1] The 1989 Timor Gap Treaty between Australia and Indonesia, which divided the rights to seabed and undersea resources south of Timor between the two countries, facilitating oil exploration in the area, is an example of such a 'peace pact'.

Agreed rules about the use and transfer of assets provide the basis for social coordination in asset use. People now know what to expect when they deal with others over scarce assets. To illustrate the difference between the presence and absence of rules, consider the consequences of an absence of rules governing use of the scarce surface of the roads. Without rules governing right of way, turns and vehicle speed, we would be much less certain of the behaviour of other drivers. Progress would be slow and hazardous for most of us. Some would choose to rely on brute force, and travel in as large a vehicle as possible. The costs to road users, and their incentives to agree to establish a representative body to define and enforce driving rules, are obvious.

Rules governing asset use, such as driving rules, are useless if not generally observed. We often think of compliance as depending on the existence of monitoring and enforcement by other community members, or by public officials such as police, pollution inspectors and judges. However, this is not always the case; nations and people will obey rules without direct enforcement if they calculate that breaking the rules will cost them more than they will gain. Thus, for example, there is no international legal system with the power to enforce the provisions of the 1987 Montreal Protocol on Substances that Deplete the Ozone Layer, which commits signatory countries to costly reductions in their CFC production between 1986 and 1998. Nevertheless the signatories appear to be adhering to the Protocol. This is probably because they believe that if they defect, other signatories will retaliate by increasing their CFC production or by some other means, so that the defecting country will lose more than it gains.

Established communities (clans, tribes, nations) already have more-or-less representative bodies (elders, chiefs, parliaments) that are responsible for defining rights to scarce assets and rules of exchange. They also have administrative officials (reeves in the feudal manor, police, judges) who monitor and enforce the rules. Where these organisations exist, few people have much experience of the transition from conflict over scarce assets to an agreed set of rules governing access to and exchanges of assets. So we tend to take these rules, and the processes of social coordination that depend on them, for granted. This would not be the case if we lived in the former Yugoslavia or Somalia, where assets are often acquired by brute force in the absence of generally accepted representative bodies, rules and social coordination.

Our use of the natural environment is plagued by a lack of social coordination, either because there are no rules respecting scarce environmental resources, such as unpolluted air and rivers and rare flora and fauna, or because it is too costly to monitor and enforce such rules. The reasons for these deficiencies will be explored in Chapter 5. In the meantime, to get a feel for the issues, the whale-loving reader might consider the problems he or she would face as the 'owner' of a live blue whale, as opposed, say, to a horse or an emu. Can you fence and monitor your 'livestock' at an acceptable cost? You are used to the state (legislatures, police and courts) defining and enforcing your rights to enjoy your horse or emu; can the state do the same for a live whale, and if not why not?

2.2 Property rights and other rules

Humans devise rules (more formally, institutions) as constraints on behaviour that reduce conflict and uncertainty in people's dealings with others.[2] The rules with which we are concerned guide human interactions over scarce assets. Other rules cover other necessary forms of social cooperation, such as choosing a mate, raising children and procedures for bonding individuals to groups, such as tribes, firms and nations.

The rules governing human interactions over scarce assets include:

- *Property rights*—informal or formal legal rules that specify who can do what with particular assets, defined and enforced by the community group or the state.
- *Rules of inheritance* and *rules of exchange*—rules that specify the conditions required for socially sanctioned transfers of assets to heirs or trading partners.

- *Norms of good behaviour or politeness*—rules governing asset use that are informal and not usually actively enforced, but are nevertheless widely observed within a community (e.g. codes of behaviour on using your car horn, closing farm gates and leaving camp sites clean).

In Australia, we think of property rights and rules of inheritance and exchange as arising from law made by judges or parliaments, for example the common law of trespass and acts of parliament governing mining and environmental protection. However, especially in poor societies, these rules may also be the result of informal agreement between members of a community. For example, a group of commercial fishers may restrict fishing to certain times and places, with penalties privately imposed by members of the group. Recent judicial and legislative decisions concerning native title in Australia illustrate the distinction between these three sources of laws/rules. In the High Court of Australia's 1992 *Mabo* judgment, the Court decided that, in the absence of explicit action by the Crown to extinguish native customary title in Australia, such title continues to exist under the common law. Mr Justice Brennan's judgment contains a recognition of the significance of informal laws/rules:

> Native title has its origin in and is given its content by the traditional laws acknowledged by and the traditional customs observed by the indigenous inhabitants of a territory.[3]

Subsequently the Federal Parliament of Australia passed the *Native Title Act 1993*, which sets out the conditions for judicial determination of native title in particular cases.

There is no clear dividing line between property rights and norms of good behaviour. On the contrary, with changing attitudes to the environment and values of natural resources, what was formerly the subject of a norm (say, when your neighbour lit her backyard incinerator) may now be the subject of a defined and enforced property right (state or local government regulations governing fire lighting and rubbish disposal).

Road use illustrates the contributions of formal property rights and informal behavioural norms. Our major rights and obligations when using the roads are set out in the traffic laws. However, our way can be further smoothed by behavioural norms in such matters as use of horns and leaving the passing lane free for swifter traffic.

As explained above, property rights originate from the desire to avoid wasteful contests between would-be users when assets are scarce. An individual's property rights may include rights to use an asset in consumption, to use it to produce income, and to transfer it by sale or bequest or gift. Thus, for example, the private owner of a car may use it on holiday or to deliver pizzas for income, and may sell it to a new owner. A state public servant assigned a car for business use is likely to have just a limited right to use it for his or her consumption, say dropping the children at school or picking up the family dry-cleaning.

Property rights may be vested in individuals such as the private car owner, or in members of a defined group of people such as a group of public servants authorised to use a government car, members of a ski club authorised to use the club lodge, or members of a rural community authorised to collect timber or hunt in a community forest.

Property rights not only minimise conflicts over assets. They also affect incentives to protect or maintain or produce scarce assets. Thus the private owner of a car, who has the right to sell it and keep the proceeds, has a strong incentive to maintain the vehicle; the better the car's condition, the higher its sale price, and vice versa. This is less true of a public servant or company employee assigned a car; he or she does not receive the proceeds when the car is sold, and therefore bears only part of the benefits or costs of maintenance. The employer can, at cost to itself, enforce its property rights in the car by rewarding or penalising the employee according to the car's condition. However, note that the division of effective rights in the car between separate individuals creates conflicting incentives over its use since the employee cares less about mainenance than the employer, conflicts that are costly to resolve, and would not exist if a single individual possessed the sole right to usage, disposal and any proceeds from the car.

2.3 Property rights and incentives: how to preserve elephants by making them pay

The contrasting histories of the elephant herds in different parts of Africa during the 1980s illustrate the connection between well-defined property rights and incentives to protect and maintain scarce assets.[4] During the 1980s, hunting elephants was illegal in Kenya and Tanzania, and trade in ivory was banned. Yet elephant numbers in the two countries fell from about 380 000 in 1979 to 92 000 in 1989, due to poaching for the overseas ivory market. In contrast, in Zimbabwe, Botswana, Namibia and South Africa, where hunting is legal on state, private and communal lands, and trade in ivory and hides is legal, elephant populations increased from about 60 000 to 139 000 during the same period. These facts support the wildlife ecologist who stated: 'In emergent Africa you either use wildlife or lose it. If it pays its way, some of it will survive.'

Kenya and Tanzania do not allow elephants to pay their way, except through tourism, and few of the benefits of tourism go to agricultural villages or to underpaid park rangers. On the contrary, Kenyan and Tanzanian subsistence farmers suffer when elephants leave national parks and damage crops, and both farmers and rangers can supplement their incomes by assisting poachers. In effect, the people best placed to protect and maintain elephant herds have stronger property rights in (and hence greater incentives to produce) dead elephants rather than live elephants.

The situation is very different in the countries of southern Africa. In different areas of Zimbabwe, the state, wildlife concession owners and African subsistence communities all have property rights in both the live and dead elephants on their territory. They sell hunting and photographic opportunities to an international clientele. They sell the ivory and hides from elephants that die or are culled to deal with overpopulation and destruction of peasants' property. Farmers whose crops are damaged are compensated from these proceeds. Because elephants and other wildlife have a high commercial value, these owners have strong incentives to protect and maintain herds. Small wonder that poachers are shot on sight and over $300 per square kilometre is spent annually to protect wildlife.

What characteristics should property rights have, to minimise wasteful contests and to maximise social coordination in the use of scarce assets? They should be:

1 *clearly defined*, so that there is no ambiguity about the nature of the asset, the identity of right-holders, or the penalties for illegitimate use;
2 *strictly enforced*, either by physical exclusion ('fencing'), or by monitoring and penalising of illegitimate use. Then assets cannot be seized or used by others.

Social coordination will be further improved if assets are:

3 *exclusive*, so that all benefits and costs arising from a particular use or transfer of a defined asset accrue to an individual right-holder. This eliminates the conflicting incentives discussed above.
4 *transferable* to others, with the voluntary agreement of the right-holder ('owner'). This permits control of assets to move to those individuals who can use the asset in the most valuable way.

In practice, property rights never achieve this ideal. One reason is that communities believe that some assets should be shared. Another is that it is costly to write precise property laws, to monitor all property use, and to penalise infractions. In practice, individual and group rights are exclusive only insofar as the right-holders can prevent trespass and theft. If they cannot, the effective property rights are shared between the right-holders and trespassers and thieves. So communities put up with some conflict and waste in the use of virtually all assets, either in the interests of equity or because the costs of a more precise definition of rights or more careful policing of rights are thought to be too great. But some forms of property approximate the ideal more closely than others. For example, because cars are particularly valuable assets in modern societies, in rich countries rights to cars are quite precisely defined and energetically enforced. The same is true of rights to especially valuable assets, such as cattle or water sources, in traditional societies. On the other hand, your right to clean air, while valuable, is not so well defined and enforced, because the irregular nature and obscure origins of most air pollution make it too costly to do so. Thus, although cars are extensively used and numerous, driving is characterised by impressive social coordination and relatively few disputes. By contrast, city-dwellers commonly complain about the lack of social coordination in dealing with air pollution.

2.4 Alternative signalling and incentive systems

A society—that is, any group of people who cooperate on an organised basis over an extended period, such as a tribe or a religious order or a nation—is composed of large numbers of both consuming units (individuals, households) and producing units (individuals, households, firms). As previously explained, a society could not exist without agreed rules governing the use of scarce assets. A village comprised of farmer-irrigators cannot prosper; on the contrary, water and labour will be wasted and perhaps blows exchanged if much of each farmer's effort is devoted to diverting water flows away from others' fields towards his or her own. Beyond that, as explained in introductory economics texts, a society has to determine what to produce (e.g the amounts of grain versus milk versus clothing, for our farmer-villagers) and what methods of production to use (e.g. irrigated versus unirrigated crops, hand-hoes versus tractors, hand-sewing versus sewing machines). How can the activities of large numbers of separate producers and consumers be coordinated so that the resources required in production match those available and the goods produced match consumption requirements?

If individuals specialise in production (i.e. if some farmers concentrate on producing milk and some on producing crops), a society must have a *signalling and incentive system* to coordinate individuals' production and consumption activities. Such a system must:

1 *signal information* about availabilities and requirements of milk, crops, clothing and irrigation water to all producers and consumers;
2 *provide incentives* that motivate production and consumption of matching amounts of milk, crops and clothing, and direct scarce water to its most valued uses. This requires some system of rewards and penalties applied to producers and consumers.

We first experience such signals and incentives as toddlers when we begin to raid the kitchen table and counter for tasty morsels. Most of us learn quickly to read the signals of adult disapproval, backed by incentives ranging from smiles and frowns to a quick slap on the backside. As adults, our production and consumption activities are coordinated by various combinations of three alternative signalling and incentive mechanisms:

1 *tradition*, where recognition of signals and appropriate responses are instilled as part of the culture;
2 *central planning*, involving the collection of information by the planning authorities, its dissemination in the form of production and consumption plans, and a system of penalties and rewards related to adherence to plans; and
3 *markets and exclusive property rights for individuals*, involving market signalling of information and responses to price signals aimed at maximising the returns from one's property.

These three systems, in pure form, can be crudely represented using circular flow diagrams, which show information flows (including information affecting incentives) as well as material and money flows between producers, consumers and planners. These diagrams are shown in Figures 2.2, 2.3 and 2.4.

Virtually all societies employ elements of all three systems—what parent has not struck a market deal with his or her child, wearying of the task of establishing family traditions, or of imposing parental rules and sanctions, in order to get the child to complete some job or other? Thus the diagrams are not intended to represent any actual economic system; their usefulness for us is that they highlight contrasts in the way economic information is collected and used, and incentives operate, under alternative coordination systems. The absence of information flows between producers and consumers in Figure 2.2 does not imply that they fail to use incoming information, only that the individual's coordination with others is dictated by his or her own internalised values. Thus, regardless of the size of the harvest, the true believer sees it as right to give one-tenth to the church or the temple.

The three systems necessarily involve different distributions of property rights in assets. In a market system, individuals have exclusive rights to assets, or particular defined attributes of assets. For example, a tenant farmer has the exclusive right to farm the land and to receive part of the farm income, but the landowner retains the rights to hunt game on the land and to transfer it by sale. In a centrally planned system, the effective rights in most assets and attributes are divided between the planners and the on-the-spot users of the assets, since it is too costly for planners to specify and enforce every detail of asset use. This is the situation in the case of government and company cars, discussed earlier, and in the use of public parks, where littering, hunting and fire-lighting may occur despite explicit prohibitions by planners.

The division of property rights under central planning does not occur by design, but rather by default. To emphasise the point, think of the position of the slave in the slave-owning societies of the past. In such societies, the slave had *no* legal rights, to his or her labour or to any other asset. Yet, to the extent that the slave-owner found it too costly to supervise every single move on the job, the slave retained an effective property right to slack off or to consume the owner's produce.

Figure 2.2 Tradition

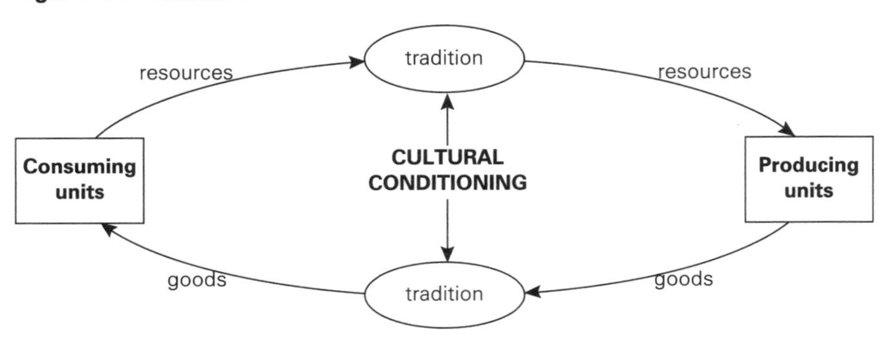

Figure 2.3 Central planning

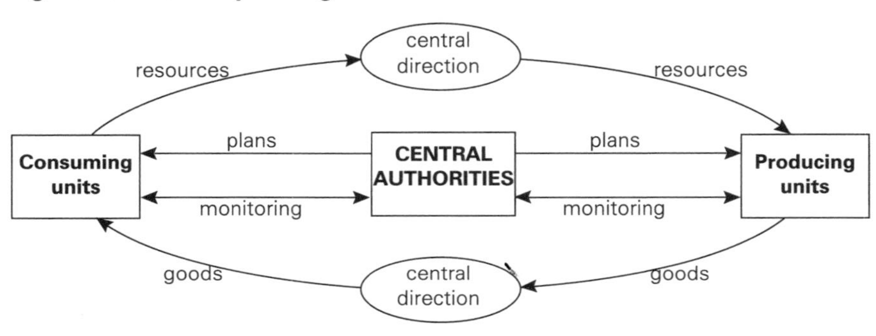

Figure 2.4 Markets

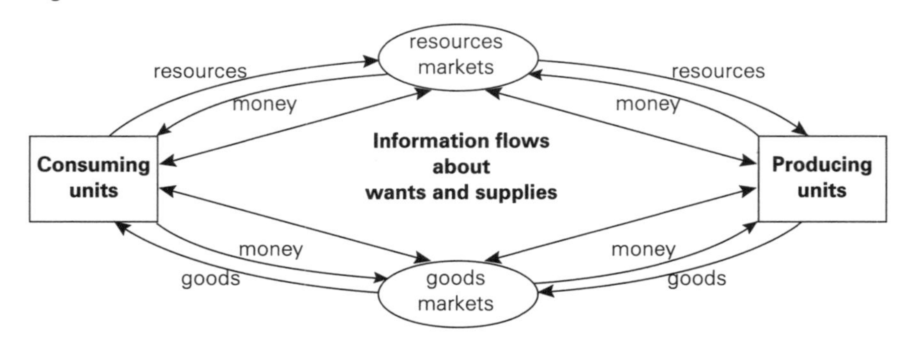

Under a traditional system, property rights in each asset could be exclusive to one or more individuals, depending on the traditions of the particular society. For example, in feudal Europe, rights to cultivated land were commonly exclusive to a particular household, while rights to grazing land were commonly exclusive to all households in a village.

It is common to have two or three signalling and incentive mechanisms involved in the allocation of the same asset. Consider the case of seating at football finals. Some seats are sold to the general public, and others are allocated to clubs and individuals by football planners in the league and in clubs. Some of the latter seats are reallocated in illegal or quasi-legal market deals to people who missed out earlier. Planners also may allocate certain seating areas to particular groups of patrons, such as those who do not wish to consume alcohol. Traditions also play a role in allocating seats within the stadium; for example, seating behind the goals is customarily the preserve of supporters of particular teams. At English soccer matches, ignoring such informal rules can be quite dangerous.

In the case of environmental assets, a mixture of signalling and incentive systems is more the rule than the exception. Historically, governments in Australia, and also in New Zealand, the USA and Canada, have not only controlled unclaimed mountains, deserts and tundra, but also highly valued forests, rangelands, rivers and inland and coastal waters. The use of these public lands, and associated natural resources, including timber, minerals, scenic attractions and flora and fauna, has commonly been determined by both bureaucratic plans and market prices. Tradition also has influenced the use of some public lands, notably areas of religious or cultural significance to indigenous peoples.

Privately owned assets are also frequently subject to a mixture of signals and incentives. Planning controls governing the design and siting of houses and factories, aimed at protecting public health and safety, were established in the 1800s. More recently, with the rise of environmental concern in the wider community, large segments of the public want a much wider say in the use of much private property. Politicians have responded by passing laws that have increased the impact of government-determined rules and penalties on the use of private property, and reduced that of market signals and incentives. Examples include laws on vegetation clearing, livestock and wildlife management, mine location and operation, and manufacturing plant and motor vehicle design and operation.

To understand environmental problems, we must understand why traditional, market and bureaucratic signalling and incentive systems can fail to fully coordinate people's use of scarce natural resources. In this book, most attention is concentrated on markets and central planning, which dominate Australian decisions about environmental assets. The limitations of market signalling and incentives are discussed in Chapters 5, 6 and 7, and of those bureaucratic signalling and incentives in Chapter 8. The value and limitations of tradition, for example as represented by the land-use ethics of Australian Aborigines or modern Landcare groups, are discussed in Chapters 7 and 18.

We can only appreciate the limitations of markets and central planning if we first understand how they are supposed to work under perfect conditions. In Chapter 3 we examine how markets and planning work in the idealised world of economics textbooks.

Discussion questions

1. Property rights make certain forms of competition for resources illegal, and people are better off as a result. True, false or uncertain? Explain why.
2. In a market-based economic system, defining and enforcing private property rights in natural resources—for example, mineral deposits, oyster beds, elephants—is likely to increase the benefits the society obtains from those resources. True, false or uncertain? Explain why.
3. Burning elephant tusks confiscated from ivory poachers may accelerate, rather than reduce, the destruction of Africa's dwindling elephant herds. True, false or uncertain? Explain why.
4. Freeway accidents typically produce traffic jams in the opposite-direction lanes, where the accident has created no obstruction at all, because motorists slow down to look. The resulting time delay to each motorist will usually exceed the benefits the motorist derives from slowing down for a look at the accident. Yet the great majority will slow down, delaying all those behind them. Why do signals and incentives fail to coordinate drivers' actions in this case?*
5. Consider the mainly free-market economy of South Korea and the mainly centrally-planned economy of North Korea, where individual consumer and producer decisions play a much smaller role in allocating commodities. In a 'command' economy, individuals are spared the costs of negotiating and enforcing contracts. Does it follow that commodity allocation in North Korea is achieved at a lower real cost to society? Explain.

* Adapted from a question in Paul Heyne, *The Economic Way of Thinking*, 5th edn, SRA, Chicago, 1987, p.12.

3

Social coordination in market and planned economies

Before we can understand why markets and central planning may fail to coordinate resource use, we need to know how each works in ideal circumstances. How do markets and central planning generate and signal information about the values people attach to natural resources and other goods and services? How does each provide incentives for decision makers to respond to the desires of other people? What assumptions about the information available to private and public decision makers guarantee full coordination between people in resource use? It turns out that coordination depends on the assumption that decision makers have precise information about resources and the values that other people attach to those resources, an assumption which often does not apply in real markets and real central planning. Thus, to understand the causes of environmental problems, we must investigate why decision makers have imperfect information.

3.1 Introduction

Markets and central planning are the dominant economic systems of our time. Environmental problems resulting from human use of the natural environment—such as driftnet fishing, the *Exxon Valdez* oil spill, Chernobyl and the shrinkage of the Aral Sea—occur under both systems. Commonly, although not always, this is because of the absence or distortion of signals about and incentives to respond to other people's concerns over the environment.

In fact, in an uncertain world, problems resulting from human use of environmental assets are unavoidable, unless we are prepared to retreat to the Stone Age and make minimal use of of the environment. Thus, so long as we use oil as a fuel, it must be transported. Even if oil companies and tanker captains have the correct signals and incentives regarding the effects of spills on others, some spills will occur; it will sometimes be less costly to compensate spill victims than to obtain accurate weather forecasts, to build hulls impervious to collisions (what happens when impervious hulls collide?) or to subject captains to weekly psychiatric testing.

In order to understand how signals and incentives can be deficient in actual markets and in actual planning, we need to understand the ideal operation of the market and ideal central planning. This chapter explains how each system generates values of resources and goods, how each provides incentives for decision makers, and the stringent assumptions required for each to achieve perfect social coordination.

The assumptions involved in an ideal market system are discussed in Sections 3.4 to 3.9. An ideal central planning system, discussed in Sections 3.10 to 3.12, is not so easy to define. Planning is inextricably entwined with political and administrative arrangements. Planners have to get their objectives from some individual or group, and they have to receive information from, convey instructions to and monitor all producers and consumers. Thus there are as many forms of central planning as there are distinct political and administrative arrangements. A Stalinist dictatorship with planners implementing the dictator's wishes is one; a democratic centrally planned economy, where planners' goals are dictated by the electorate via the legislature, is another.

We focus on democratic central planning because it corresponds to the situation of many environmental and recreation planners in Western economies today and our chief concern is environmental problems in modern democracies. Specifically, we assume that the objectives of government environmental planning are decided by elected politicians answerable to their constituents, and that detailed plans are produced and implemented by public service bureaucrats answerable to the legislature.

In order to explain clearly how markets and central planning operate under ideal conditions, this chapter avoids environmental problems. Ideal markets and ideal central planning are best explained using an everyday commodity divorced from major environmental issues. The commodity used here is the avocado, partly because it is familiar and partly because it is perishable, so that the issue of storage for future use does not arise. Thus we avoid the technical and behavioural complications (such as the difficulties of identifying the sources of air pollutants and of establishing the values that people attach to saving endangered species) involved in serious environmental problems. These issues are discussed in Chapters 5 to 8.

As explained in Chapter 2, markets and planning involve different effective distributions of property rights. In the ideal markets of the economics textbooks, property rights are exclusive to individuals. Under ideal central planning, the planners are assumed to have exclusive control of asset uses and the proceeds derived from assets.

3.2 Social coordination tasks

What is involved in coordination of production and consumption of a particular good, say avocados? Aside from the technical information that decision makers must possess in order to produce and use avocados, there are two requirements. First, the businesspeople or consumers or government planners who determine production and consumption of avocados must receive information about the benefits and costs of producing and consuming more avocados. In economics, cost or opportunity cost is the value of opportunities sacrificed by the person taking an action. The costs of more avocados are the values of the opportunities sacrificed by avocado producers and consumers—the values of the other crops that are not grown, other goods that cannot be consumed and so on. Thus the market or planning body must determine and communicate values—values of avocados, values of alternative household consumption goods, values of alternative crops, values of planting materials, labour, fertilisers etc. required to produce avocados and alternative crops—for decision makers to choose between avocados and other production and consumption alternatives.

Since both the opportunities sacrificed and the values of those opportunities vary across people (e.g. one avocado grower's most valued alternative use of time and funds may be merchant banking, and another's growing blueberries), so do costs. Where individual's valuations of alternatives are difficult to measure, so are the costs of sacrificing those alternatives.[1]

Second, the market or planning body must provide incentives for producers to supply an amount of avocados that matches the amount consumers wish to consume. This is the amount that balances the benefits and costs to people of having more or less avocados. When this happens, the society avoids situations where avocados are either so scarce that people have to make disproportionate sacrifices to obtain them, or so abundant that the sacrifices made to produce them outweigh their value to people (as when unwanted avocados are thrown away).

3.3 Economists' assumptions about behaviour

When we speak of incentives to produce more or less avocados, what sorts of incentives are involved? We assumed in Chapter 2, and both market and central planning coordination systems require, that people know what motivates their fellows. If you do not know what another person in your family or workplace or wider community desires or fears, your actions are unlikely to elicit the desired cooperation. Consider the social coordination problem faced by a hostage negotiator. It is when the negotiator becomes convinced that the hostage takers are unpredictable and impervious to promises or threats that the SWAT team will be called in.

What do economists assume motivates people? Economic thinking generally incorporates the assumptions that all choices are made by individuals, and that each individual, whatever his or her situation in life, acts in ways calculated to yield the greatest benefit net of the costs of the action taken.[2] For example, an individual who is both a committed conservationist and a football fan, faced with a choice between attending a forest demonstration and a weekend game, will decide according to the benefits and costs he or she anticipates from the alternative actions.

Note that the second assumption above does not imply that individuals are purely materialistic or selfish (they may get satisfaction from altruistic actions), or that they do not make mistakes (their calculations may prove wrong after the event), or that every individual acts this way all of the time.

What about actions that violate the societal rules discussed in Section 2.2? Economists commonly assume that property rights and other rules are observed, implying that the material or psychological costs of rule violations exceed their benefits.

The assumption that individual avocado producers and consumers compare benefits and costs of their alternatives and choose in their own best interests explains little on its own. To explain and to predict reliably people's behaviour, economists also have to identify benefits and costs that matter to the individual concerned—money gains versus money spent for the businessperson who grows avocados, dining satisfaction versus money spent to the avocado consumer, votes gained versus votes lost to the politician, wilderness saved versus time and money spent lobbying politicians to the ardent conservationist—and attach comparable values to the benefits and costs in each case.

Economists' benefits-versus-costs calculations often work quite well in predicting behaviour, for two reasons. One is that different businesspeople, consumers, politicians, conservationists and so on generally perceive their interests in similar ways, so that the actions of whole classes of decision makers can be predicted with reasonable accuracy. Second, in a market-oriented society such as Australia, the benefits derived from many things that people want or care about, and the values attached to many things that people sacrifice, can be at least partly measured in money terms. Goods and services are directly valued in money terms when they are traded in markets (e.g. the timber produced from a forest) and are indirectly valued in money terms when they are obtained or accessed by sacrificing other resources that are valued in markets (e.g. money donated to preserve forests, the earnings people sacrifice while participating in lobbying campaigns and the money spent on travel to visit forest parks).

Economists are sometimes criticised for an excessive interest in the money values of things. In fact, economists are simply using the least imperfect measuring rod, money, as a means of comparing, as far as possible, the benefits and costs of alternatives. Faced with a scarcity of resources, even decidedly non-commercial organisations (such as churches and conservation groups) use money values in comparing alternative courses of action. The issues of how far economists can and should go in attaching money values to environmental assets are examined in Chapter 11.

3.4 An ideal market

An ideal market for avocados, as found in economics textbooks, is based on stringent conditions. Imagine a market for avocados where it is possible to bargain and where there are plenty of avocado buyers and sellers, perhaps a large municipal market or a weekend produce market. Each buyer aims to maximise his or her net benefits from avocado purchases and each seller aims to do likewise for sales. Individual property rights in avocados and the money exchanged for avocados are exclusive, transferable and perfectly (costlessly) defined and enforced. All buyers and sellers know the rules governing market exchange, such as how to signal your buying and selling offers, and what sequence of messages accomplishes a sale. Suppose that all avocados on sale are identical and their quality is known to both buyers and sellers, so that all they are concerned about is price. All participants know immediately of the buying and selling offers made by others. Sales agreements and deliveries of avocados or cash must be simultaneous, eliminating worries about changes in the values of avocados or money before the deal is concluded. This unrealistic scenario approximates the textbook model of a market.

These conditions are asking a bit much for avocados, and indeed for any real commodity, as opposed to, say, shares in public companies. Avocados are not identical, and although we would like to know the flavour of each one we buy, and whether it has started to blacken inside, we certainly are not as well informed as the textbooks assume. The assumption that values do not change during the deal holds for avocados, but not for coal sales to Japan or prepaid holidays to Europe.

If ideal markets assume away all these complications, what do they assume that buyers and sellers don't know—what is left for the market to determine?

In a textbook market individual buyers and sellers of avocados don't know what prices (i.e. values) others are willing to accept (if potential sellers) or to pay (if potential buyers). The function of the market is to reveal information about others' willingness to pay for, or to accept monetary compensation for, avocados.

Figure 3.1 Individual willingness to pay and accept

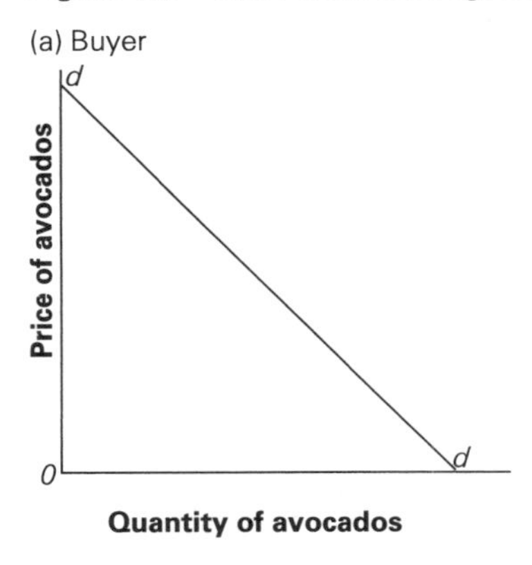

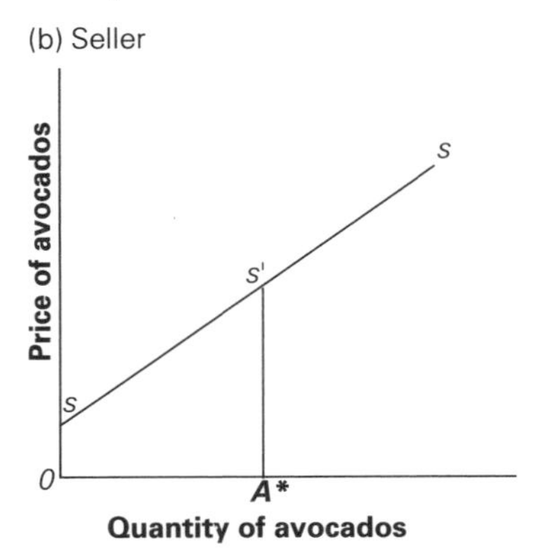

3.5 Market valuation

Before continuing, readers with limited knowledge of the economics concepts of demand and supply, and their basis in human behaviour, are advised to study Heyne, Chapters 2 and 3.[3]

How does such a market generate values for avocados? On the buying side, consider the behaviour of the individual who wishes to buy avocados for use during some time period. Consistent with the behavioural assumption in Section 3.3, the would-be buyer compares the expected benefit from each extra avocado enjoyed within the time period with its (opportunity) cost—the value of the alternatives, most likely other food items, sacrificed to obtain the avocado. The avocado price measures the value of the goods sacrificed in money terms. As a result of such calculations, each potential buyer of avocados privately knows the relationship between the amounts of avocados he or she would like to use during some period of time and the money sacrifices he or she would be prepared to make to obtain those amounts. Remember that the money is not valued for itself; it is simply the most convenient measure of the value of the other goods sacrificed in purchasing each additional avocado.

How will the individual's avocado purchases be related to the avocado price? As the price rises, greater amounts of other valued items have to be sacrificed for each avocado. Thus, as shown in Figure 3.1(a), the relationship between the number of avocados purchased in some time period (say, one week) and the avocado price will be negative; the number of avocados with expected benefits exceeding their cost falls as the avocado price rises. Figure 3.1(a) shows that if the price is high enough, the sacrifices of valued alternatives required to purchase a single avocado per week exceed its expected value, and no avocados will be purchased.

The curve *dd* in Figure 3.1(a) is the individual's *demand curve* or *demand schedule* for avocados for the given time period. Its height shows the maximum prices that the individual is (privately) prepared to pay (and hence bid) for successive avocados as the number of avocados he or she buys in that time period rises. Economists also think of the demand curve as the individual's (private) *marginal benefit curve* for avocados, measured in money; use of the term 'marginal' reminds the reader that the height of the curve measures the *additional* value attached to each successive avocado. Thus areas under *dd* in Figure 3.1(a) measure the maximum value, in money terms, of different numbers of avocados to the person concerned. For example, if avocados are free, the individual whose behaviour is depicted in Figure 3.1(a) will use O*d* avocados per week, which will cost that person nothing. The demand curve *dd* tells us how much that person would be willing to pay for each successive avocado; thus the individual's maximum valuation, in money terms, of O*d* avocados is represented by the area *dd*O (prices of avocados multiplied by numbers of avocados is a sum of money).

The story on the selling side is similar. The potential seller of avocados compares the expected benefit (money price received) from extra avocados per week with their opportunity cost, the value of other products not produced (say, blueberries) and/or resources used (say, chemical sprays) as the number of avocados provided per time period (again, say, per week) increases. As a result, each potential seller of avocados privately knows the relationship between the amounts of avocados the individual is prepared to provide to buyers per week and the amounts of money required to compensate for providing those amounts.

As the avocado price rises, the money reward for the sacrifice involved in providing additional avocados increases. Assume that each successive increase in the rate of avocado production requires the sacrifice of increasingly valuable alternatives. Thus, as shown in Figure 3.1(b), the higher the price paid for avocados, the more avocados will be provided per week. The curve *ss* in Figure 3.1(b) is the individual's *supply curve* or *supply schedule* for avocados for the given time period. Its height indicates the minimum prices that the individual supplier is prepared to accept (and hence offer to buyers) for successive avocados as the number of avocados provided rises. The supply curve is the individual's (private) *marginal cost curve* for avocados; its height measures the money values of the additional sacrifices involved as the number of avocados provided per week increases. Thus the minimum value, in money terms, of compensation required for the seller to part with any number of avocados (say A^* in Figure 3.1(b)) is given by the area under the seller's supply curve up to that number of avocados ($ss'A^*O$).

Our ideal avocado market is composed of many buyers and many sellers. The market demand and supply of avocados are obtained by adding up the quantities of avocados that buyers are willing to buy, and

Figure 3.2 Market demand and supply and the gains from efficient exchange

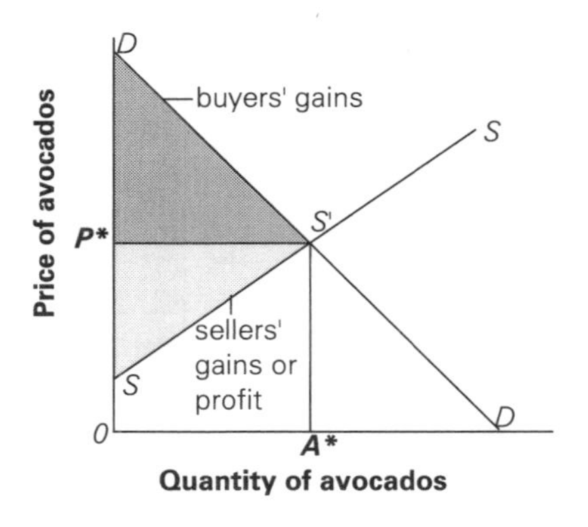

the quantities that sellers are willing to sell, at each avocado price. These curves are shown as *DD* and *SS*, respectively, in Figure 3.2. *DD* is the marginal benefit curve for avocado buyers as a group; its height at any point measures the maximum price that some buyer would be willing to pay for each additional avocado. *SS* is the marginal cost curve for avocado sellers as a group; its height at any point measures the money compensation required as some seller makes the sacrifices involved in providing each additional avocado. In an ideal market the price and quantity of avocados sold are determined by the intersection of the market demand and supply curves for avocados, where the price is *P** and the quantity *A**. This is the most efficient price and quantity of avocados from the point of view of market participants as a group, because at the demand supply intersection the marginal benefit someone obtains from the last avocado sold is just equal to the marginal cost someone else incurs in producing it. Every avocado whose marginal benefit exceeds its marginal cost is being produced and sold.[4]

3.6 Can the market generate true values?

Recall, from Chapter 1, that environmental problems occur when resource users are either unaware of or disregard the desires of other people. Are either of these outcomes possible in our ideal avocado market? If an ideal market cannot coordinate the actions of avocado buyers and sellers, the market social coordination system described in Chapter 2 could never overcome environmental problems. We consider the accuracy of market signalling of avocado values in this section, and the incentives of avocado buyers and sellers to respond to each others' desires in Section 3.8.

In our ideal avocado market, all buying and selling offers are known to all participants. Then it is possible to add up the quantities of avocados that buyers have offered to buy, and the quantities offered for sale, at each avocado price, to obtain the aggregate expressed willingness to pay of all buyers, and the aggregate expressed willingness to accept compensation of all sellers. But will these be the true market demand and supply curves? This seems most unlikely. Recall that the willingness to pay or accept of each buyer or seller is private information, known only to that individual. Recall also that we are considering a market where bargaining is the norm. The last time you were buying a wood carving in Bali, or a used car from a dealer, did you offer the maximum amount the carving or the car was worth to you? Did you think, at the time, that the Balinese stallholder or the car dealer was offering the item at the minimum price he or she would accept? Commonsense suggests that buyers will try to gain by offering prices below their maximum willingness to pay, and that sellers will try to gain by asking prices above the minimum prices they would accept. So it is not obvious that the market will elicit the private willingness to pay and to accept information required to determine the efficient price and quantity of avocados.

If the market is to function as explained in the textbooks, there must be some reason for buyers and sellers of avocados to reveal their true willingness to pay and to accept. Fortunately there is. Think of the likely bargaining process in the avocado market. Remember that avocados are identical, everyone knows all about them, there are many buyers and sellers and all buying and selling offers are immediately known. Buyers will begin by offering low prices, and sellers high prices. Buyers will assume that sellers are asking more than they are willing to accept, sellers will assume that buyers are offering less than they are willing to pay, and there will be few if any sales. In terms of the market demand and supply curves shown in Figure 3.2, buyers' bids will all initially fall below P^*, and sellers' offers will all be above P^*. No avocados will be exchanged.

Unsold avocados and unsatisfied desires for avocado salads and guacamole mean that no one is happy; after failing to get responses to their initial offers, buyers will raise their bids, and sellers will lower theirs. Competition among buyers and among sellers means that individual buyers and sellers have no choice in this; if buyers fail to bid higher, and if sellers fail to offer lower prices, they will get no takers. The range between buying and selling bids will narrow. This process pressures buyers who initially bid low prices (below P^*) for extra avocados to raise their bids and sellers who initially ask high prices (above P^*) for providing extra avocados to lower their bids. These buyers and sellers have to give up trying to buy and sell avocados at their initial offer prices; they must reveal prices closer to their true willingness to pay or to accept, and hope that this will bring forth a seller or buyer.

Buyers and sellers will observe the bid–offer spread narrowing, and eventually deals being done at or around P^*. In this way all participants will be informed that true demand and supply are being revealed around P^*, and will be willing to buy and sell at that price, because no buyer wants to pay higher than, and no seller wants to accept less than, the prices in sales already concluded. Thus an ideal market generates a precise value for avocados, because it forces buyers and sellers to reveal some (not all) of their private information about the values they attach to avocados.

What forces buyers and sellers to reveal their private values of avocados? The fact that our ideal market involves multiple buyers and multiple sellers. If there is only one seller of avocados, there is no competitive pressure to lower the initial high offer price. Since there is no other seller to undercut a monopolist, the single seller can choose to sit pat and wait for competition among buyers to raise their bids to whatever level the seller judges most advantageous.[5] Similarly, a single buyer of avocados can wait for competition among sellers to lower their bids. If there is just one seller and one buyer, neither faces competitive pressure to reduce the gap between the buyer's bid and the seller's offer. So there is no guarantee that true demand and supply information will be revealed, or that a deal will be done, even though the buyer's true willingness to pay exceeds the seller's true willingness to accept.

3.7 The gains from market exchange

The area under the market demand curve up to any number of avocados in Figure 3.2 measures the money value of buyers' total benefits from that number of avocados. The area under the market supply curve up to any number of avocados measures the money value of sellers' total sacrifices in providing that number of avocados. Thus net benefits, to buyers and sellers combined, of the exchange of any quantity of avocados are measured by the money value of the area under the demand curve and above the supply curve for that number of avocados. This area is maximised for the number of avocados where market demand and supply intersect, shown as the quantity A^* in Figure 3.2. Thus an ideal avocado market, as described in Section 3.4, maximises market participants' overall benefits from exchanges of avocados. Ideal markets are efficient in this sense.

An ideal avocado market maximises overall benefits from the production and use of avocados without anyone ever knowing enough private information about others' demand and supply schedules to identify the complete market demand and supply curves that determine net benefits from exchange. This happens because market bargaining forces buyers and sellers to reveal true information about demand and supply in the neighbourhood of P^*. Thus we know that an ideal market maximises net benefits from exchange of avocados without being able to estimate those benefits.

Economists commonly use areas under the demand curve and above the supply curve as measures of the net benefits to the *community*, not just market participants, from market exchanges of goods, in this case avocados. We will do so in subsequent chapters of this book. When we do this, we are making important assumptions, which must be kept in mind:

1 We are accepting a particular set of ethical judgments. The avocado demand and supply curves depicted above are based on a particular societal distribution of individual property rights, and the current distribution of incomes that results from those rights. For example, suppose that avocados are produced in a society where the ruling ethical standards allow legal slavery, and avocado plantations are major users of slave labour. The resulting effects on the distribution of community income and the costs of producing avocados will be reflected in the demand and supply curves for avocados.
2 We assume that each dollar of net benefits represents an equal amount of satisfaction or welfare for each individual avocado consumer or producer, rich or poor. Many people might argue that extra dollar benefits mean more to poor individuals than to rich individuals, and also that as the benefits to a particular individual increase, each extra dollar of benefits means less to that person.
3 We assume that *all* benefits and costs of avocado consumption and production, both measurable and subjectively recognised only by particular individuals, are reflected in the demand and supply curves (and in market bids). It is possible that benefits and costs of avocado consumption and production spill over to people who are not consumers or producers represented in the market. For example, suppose that individuals who eat avocados are less likely to be infected by a contagious disease; the resulting lower infection rates reduce disease risks to households who do not eat avocados. Since the latter do not buy any avocados, their benefits from avocado consumption are not signalled in the avocado market. Alternatively, suppose that avocado orchards are breeding grounds for a species of wasp that does not affect avocados, but can endanger the lives of people subject to allergies. The costs imposed on allergy sufferers by avocado production are not signalled in the avocado market.

The first two assumptions will not be examined further; they should be kept in mind when assessing market outcomes. The third assumption goes to the heart of this book; it is violated whenever there are environmental problems.

3.8 Market incentives

In the preceding section we equated market efficiency with maximising the net benefits from exchange. Can our ideal avocado market achieve it? In other words, can we be sure that avocado consumers and producers will be motivated to maximise net benefits, represented by the area between the market demand and supply curves? Based on the preceding argument about how the range of buying and selling bids for avocados will narrow and converge on P^*, it seems that this is the case, but what motivates buyers to buy as cheap as possible, and sellers to sell as dear as possible? The answer, already indicated in Chapter 2, is individual property rights. It is important to understand why this is so.

Suppose that you send your agent to the market to buy or sell avocados on your behalf. We assumed earlier that property rights in avocados and money are exclusive and perfectly defined and enforced. Therefore your agent has neither legal nor effective rights to your avocados and your money. If you cannot check that the agent follows your instructions to the letter, would you expect him to be as concerned as you to buy low or sell high? Obviously not, since the agent gets no direct benefits from those actions. Since bargaining is a hassle for most agents, we would expect buyer and seller agents to readily agree to prices other than P^* in Figure 3.2. Thus there are reasons for believing that the net benefits from exchange of avocados would be lower if the trade is conducted by agents.

The point is that it is only the individual property right holders, sellers before sales and buyers after, who benefit directly from the use and transfer of avocados and from use of income from avocados. What are the benefits that each group is attempting to maximise?

Prior to sales, avocado sellers have exclusive rights to avocados. They can deny anyone use of the avocados in the absence of compensation acceptable to them. Each seller aims to sell avocados at the highest possible prices. For each successive avocado whose sale price exceeds its marginal opportunity costs, indicated by the height of the market supply curve SS in Figure 3.2, the seller concerned gains a money surplus or profit. In Figure 3.2, the sale price is P^* and sellers' total sales revenue $OP^*S'A^*$; sellers' total costs of providing A^* avocados are $OSS'A^*$. Thus *sellers' gains from exchange* or *profit* is measured, in money terms, by the area above the supply curve and below the horizontal price line, the triangle $P^*S'S$. Property rights in avocados before sales give avocado providers exclusive rights to this profit.

Avocado buyers have exclusive rights to avocados after sales. Thus they can securely enjoy the avocados they have purchased. Each buyer aims to buy avocados at the lowest possible prices. For each successive avocado whose marginal benefit to the buyer (indicated by the height of the market demand curve DD in Figure 3.2), exceeds its purchase price P^*, the buyer concerned enjoys a net benefit or surplus because he or she is obtaining the avocado for less than the maximum price he or she would be willing to pay. Buyers' total expenditures on avocados equal sellers' total revenues, $OP^*S'A^*$; the total value they attach to OA^* avocados, in money terms, is $ODS'A^*$. Thus *buyers' gains from exchange* are measured, in money terms, by the area below the demand curve and above the price line, the triangle $DS'P^*$. Property rights in avocados after sales give buyers exclusive rights to these gains from exchange.

Now reconsider the bargaining process in the avocado market, this time from the point of view of the incentives of buyers and sellers. Buyers, aiming to maximise their gains from exchange, start by bidding low. Sellers, aiming to maximise their gains from exchange, start by bidding high. An absence of sales soon convinces each group to adjust their bids, since neither buyers nor sellers gain from exchange if no avocados are sold. In the ideal avocado market depicted in Figure 3.2, competition among buyers and among sellers, as described in Section 3.6, will force avocado prices to P^*. At this price, the net gain to buyers and sellers combined (the sum of buyers' and sellers' gains from exchange) is maximised. Note that our ideal avocado market achieves this socially satisfactory outcome despite the fact that neither buyers nor sellers aim to achieve it. Buyers would like to increase their gains by buying below P^*; sellers would prefer to increase their gains by selling above P^*.

3.9 Market valuation over time

Avocados are perishable; the buyer has only a short time after the purchase to enjoy them. Thus, for avocado buyers, parting with the money (the cost or sacrifice of alternatives, measured in terms of command over other goods) and savouring the guacamole (the benefit) occur at about the same time. This is not true for avocado growers or timber companies, who must wait years after planting trees before their first crop goes to market. It is not always true for consumers either; when you buy a car or a CD player or a dozen bottles of wine for future consumption, the costs and benefits resulting from the purchase are usually spread over years. We recognise the separation of costs and benefits in time when we say that a farmer or a consumer has 'money tied up in' an orchard, a car, a CD player or wine. The implication is that there is an additional cost involved in waiting for current sacrifices to yield future benefits. That cost is the interest that either must be paid (if production or consumption is financed by borrowing) or could have been earned (if our farmer or consumer 'ties up' saved funds in trees or a car or whatever).

The reason that interest is paid is that rights to goods or resources today are more valuable to people than rights to the identical goods or resources at some time in the future.[6] Put directly, even if there were no inflation to reduce the purchasing power of our money, all of us would prefer a $1000 lottery prize credited to our account today, rather than one year hence; in fact, we would be willing to pay the lottery organisers an amount of money to have next year's $1000 delivered now. That amount of money would represent interest.

Since future benefits and costs are valued less than identical present benefits and costs, they must be *discounted*, using the appropriate rate of interest, to arrive at their *present values*.[7] This permits direct comparisons of benefits and costs arising at different points in time, all valued in terms of present dollars.

Calculating present values of future benefits and costs is easy once the benefits and costs and the discount rate are known. In reality, avocado growers, timber company executives and car buyers do not know what their future benefits and costs will be. A host of factors, such as government logging restrictions, strikes, floods, disease outbreaks, changes in vehicle tariffs and petrol price rises and falls, combine to make future benefits and costs uncertain. Nor is the appropriate rate of discount certain; many interest rates coexist in the market at any time, reflecting allowances for the costs of negotiating loans, for inflation and for the riskiness of particular loans.[8] Thus market signalling of future values involves complications additional to those already discussed. These topics are addressed in Chapter 9.

3.10 Ideal planning

In markets, the people who make production and consumption decisions are the same people who provide value information and reap the rewards and penalties of those decisions. In central planning, the elected politicians and bureaucrats who make decisions are separate from the producers and consumers. This has two consequences. First, in planning, more information has to pass between more people to achieve the same result as a market. Second, planning outcomes may be affected by the incentives of a third group, the planners, who have no direct interest in particular production or consumption outcomes. This suggests that the conditions required for ideal planning will be still more stringent than those for an ideal market.

Economists commonly assume that individuals are the best judges of their own interests. The assumption in Section 3.1 that planner-politicians are answerable to voters, and planner-bureaucrats to the legislature, embodies this view, as opposed to the elitist or dictatorial view that the select few know what is best for the rest of us. Assume therefore, that the objective of ideal planning is to allocate resources so that the aggregate happiness of the community is maximised, with all individuals weighted equally. This can only be achieved under extraordinary conditions. In the case of avocados, planners must be fully informed about the characteristics of avocados and how they are produced, about individual avocado producers' production alternatives and capabilities and about individual consumers' desires for avocados and other goods. They must be able to monitor instantly, and have exclusive control over, all aspects of avocado production and consumption; in other words, they must have exclusive property rights to control the use of avocados and the proceeds derived from avocados. Finally, the planners' sole aim must be to maximise community benefits from avocado production and consumption.

If the above were true for all goods, and people were happy to avoid the hassles of making their own production and consumption decisions, political and administrative structures would be irrelevant. We could simply appoint someone to be society's planner for life, secure in the knowledge that he or she would do as well for us as we could possibly do for ourselves.

Note the similarities and differences between the benevolent, all-powerful, all-knowing planner and the ideal avocado market already discussed. Each assumes exclusive property rights that are perfectly defined and enforced. Each also assumes full information about avocados and their production and consumption alternatives, and immediate communication of information between the parties. This leaves two important differences. First, the planner is assumed to know producers' and consumers' willingness to accept compensation for, and to pay for, avocados, information that is private in the ideal market (note that the planner has to use an accounting unit, some form of money, to add up valuations across people and assets). Second, the planner is assumed benevolent, but not so market participants; rather, the market channels their self-interest so as to maximise net benefits for buyers and sellers combined.

Again recall, from Chapter 1, that environmental problems occur when resource users are either unaware of or disregard the desires of other people. In central planning, the planners are responsible for providing the necessary information and incentives, as opposed to the market generation of information and incentives explained in Sections 3.6 and 3.8. Benevolent, all-powerful and all-knowing planners can achieve perfect social coordination, thereby eliminating environmental problems—but what about real planners?

Real-world central planning falls short of ideal planning in both of the preceding respects; real planners do not know private values and are not necessarily benevolent. We need to examine valuation and incentives in real planning to understand how it is likely to fall short of ideal planning and ideal markets in coordinating avocado production and consumption.

3.11 Valuation in planning

How can planners obtain private information about willingness to pay for, and to accept compensation for, avocados? If collecting information was costless, they could mimic the market, by asking all potential consumers and producers to reveal their private information, promising to set the quantity and price of avocados at the intersection of the revealed market demand and supply curves, and to supply consumers and compensate producers at that price. If all consumers and producers believed the planners' promises, this would produce similar results to market bargaining, discussed previously. Consumers with low willingness to sacrifice for extra avocados, and producers with high costs of providing extra avocados, would be forced to reveal their true values, in order to maximise their respective gains from exchange.

In fact planners in democracies do not mimic market exchange in this way, at least partly because it is much less costly to permit individual property rights in avocados and market transfers of those rights, thus eliminating the planner from the exchange process. However, there are many goods in modern society that are not exchanged in markets, some for ethical reasons (e.g. orphaned children) and some because, as will be explained in Chapter 5, enforcing property rights and creating a market is impossible or very costly (e.g. clean air, scenic views). If mimicking market exchange is ruled out, by what other means can planners learn about values?

Consumer and producer values can be revealed in a variety of other ways: voting, opinion polling, lobbying by organised groups, creation of new political movements, direct action by individuals or informal groups (such as visiting or writing to politicians, writing letters to the editor, joining demonstrations and making political contributions), and indirect responses such as disobeying laws or moving to another state or country. All of these provide only limited information about the values of citizen-voters. To see why, suppose that the government creates an Avocado Production and Marketing Board to coordinate the production and consumption of avocados. The Board's task is to determine the number of avocados to be produced and made available to consumers, the price consumers are to be charged, and the compensation to be paid to avocado growers and those who transport avocados and distribute them to consumers. Would the forms of non-market signalling listed above provide the politicians and Board planners with people's true willingness to pay for avocados and true willingness to accept compensation for avocado production, transport and distribution?

Almost certainly not, for two reasons. First, the benefits and costs of non-market methods of signalling values will differ across avocado consumers and providers. As a result, some interested parties may find it worthwhile to signal values, and some may not bother, distorting the value information received by the planners. This may be a less serious problem with voting and opinion polls, where the costs of signalling are lower, but these are blunt instruments for the measurement of values across the community. A vote or opinion poll registers only the numbers of people for and against particular levels of avocado production, or particular farm or retail prices for avocados, not the values attached to support or opposition. Other methods of revealing values involve greater variations in anticipated benefits and costs. Busy householders are unlikely to lobby or write letters or demonstrate or make political contributions aimed at altering the price of avocados. On the other hand, the much smaller numbers of avocado growers, transporters and distributors have stronger incentives to signal their views to politicians and bureaucrats. They have a substantial financial stake in the price of avocados, and are likely to be organised in producer and trade groups, which lower their costs of signalling.

Second, will the values signalled to the politicians and Board planners be individuals' true willingness to pay for or to accept compensation for avocados? Recall from the discussion of market bargaining in Section 3.6, individuals are likely to distort the values they bid or ask if they believe that they can increase their net benefits from exchange as a result. The same applies to communication with planners. But unlike the market, under planning those who signal the benefits and costs of avocados are not subject to competitive pressure to raise buying bids and lower selling offers; the planners set the same prices for all. It is true that if the planners do not choose a price equivalent to P^* in Figure 3.2, where the quantities demanded and supplied are equal, the costs of misinformation will show up either as queues or rationing of avocados, or as unsold avocados. However, those distorting their true valuations of avocados do not necessarily bear these costs; such problems can be blamed on the planners, and are quite likely to be solved at government (taxpayer) expense. So planners cannot expect to obtain reliable valuations of avocados from the main interested parties.

The conclusion is that valuations in central planning will only be correct when signalling of values is costless and either:

1 consumers and producers tell the truth; or
2 planners mimic the market to determine the valuations and appropriate quantities of goods.

Planning can only be ideal when these conditions hold, and when the planners themselves have the appropriate incentives, to which we now turn.

3.12 Incentives under planning

Pure benevolence on the part of political and bureaucratic planners is not taken seriously in the era of 'Yes, Prime Minister'. If we drop the assumption that the planners are entirely benevolent, is it still possible that, given accurate information about the aggregate benefits of more or less avocados, they will act to maximise community benefits from avocados?

We assumed earlier that under ideal planning planners have exclusive control over all aspects of the production and consumption of avocados, i.e., they have exclusive property rights over avocados. Then it is only the planners' incentives that matter; cheating the planners will be impossible. Any avocado consumer who attempts to steal avocados will be immediately discovered and penalised, as will any avocado grower who attempts to supply diseased avocados.

Under democracy, the planners include both politicians and public service bureaucrats. The incentives of both groups affect planning outcomes, because effective property rights are shared between them. This is so because the politicians themselves have very limited capacities and abilities to collect and process information about people's desires for avocados and their costs of growing and distributing avocados, to monitor the implementation of plans, and to reward and penalise avocado consumers and providers. The planning bureaucracy performs these tasks for the politicians. However, since politicians are also unable to perfectly monitor and control Avocado Production and Marketing Board bureaucrats, the two groups share effective control of avocado production and distribution. We must consider the incentives of each.

Recall, from Section 3.3, economists' assumption that each individual acts in ways calculated to yield the individual the greatest benefit net of the costs of the action taken. In a democracy, we can reasonably suppose that for practising politicians, the most important rewards and penalties are votes for and against the individual and the party at the next election. The rewards and penalties that motivate tenured public service planners are less clear; we would expect them to consider the impact of their decisions on their career prospects within the service. If this is so, a politician deciding on avocado prices will consider not only the aggregate benefits and costs of higher or lower prices, but also the impact of the vote on the politician's and the party's chances of re-election. Bureaucrats involved in implementing avocado production and distribution plans are likely to consider not only their legislative instructions, but also how their actions will affect their career prospects.

The structure of democratic governments reflects limited faith in the benevolence of elected and bureaucratic planners. Politicians are subject to both formal and informal rules, including constitutions, common and statutory law, periodic elections and political party rules; bureaucratic actions are also subject to the law, and government departments must submit to regular audits. These are restrictions on the property rights of politicians and bureaucrats. They exist because politicians and bureaucrats are not assumed to be entirely benevolent.

Under what conditions will less-than-selfless planners act to maximise community benefits from avocados? If all citizen-voters were costlessly informed about the behaviour of politicians and bureaucrats, and could costlessly penalise them for not carrying out voters' wishes, planners' incentives would not be a problem. Then, if the Avocado Production and Marketing Board deliberately set avocado prices high to advantage growers at the expense of consumers, voters would signal the resulting costs to their political masters, and in due course Board members would be appropriately penalised. But we do not live in such a perfect world. This again illustrates our recurring theme, that costly information is the root of social coordination problems under both planning and markets.

Ideal markets and ideal planning are useful constructs because, once we understand the conditions necessary for perfect social coordination, the deficiencies of actual markets and actual planning, and the causes of environmental problems, are much plainer. Chapters 5 to 8 explain how and why real markets and real planners fall short of these ideals.

Discussion questions

1. If major new oil deposits are discovered on the North-West Shelf off Western Australia, the opportunity cost of Bass Strait oil used today will fall. True, false or uncertain? Explain why.
2. Market exchanges are driven by self-interest, not altruistic concern for others, therefore market exchanges rarely benefit the community as a whole. True, false or uncertain? Explain why.
3. Voluntary exchange in the market makes everyone better off. True, false or uncertain? Explain why.
4. Adherence to 'one vote–one value' in a parliamentary democracy ensures that the environmental policies adopted by government will be in the interest of the community as a whole. True, false or uncertain? Explain why.
5. Based on what you know about signalling and incentives in centrally-planned economies, do you think that they are likely to degrade the environment more or less than market-based economies? Explain why.

4

The economy and the environment

The economy and the environment are complex interdependent systems. Continued economic growth and even human survival are dependent on natural resources used in production and on the life support services of natural ecosystems, but our use of natural resources and discharges of wastes into the environment may threaten those ecosystems. Thus societies require feedback mechanisms to signal the health of their combined economic and environmental systems and to prompt timely corrective actions. Yet we do not know how the combined economy–environment works—how ecosystems function and the interactions between ecosystems and human behaviour. In the case of the environment, the imperfections of the information available to private and public decision makers are especially severe. So it is unclear whether the feedback mechanisms of market and centrally-planned economies can ensure that the growth- and life-supporting services of the environment will continue.

4.1 Economy–environment linkages

Figure 4.1 shows the relationships between the economy and the environment. The large box containing all the others represents the environment, which is self-contained except for the receipt of energy from the sun, and the occasional meteorite. Human society is contained

Figure 4.1 Economy–environment linkages

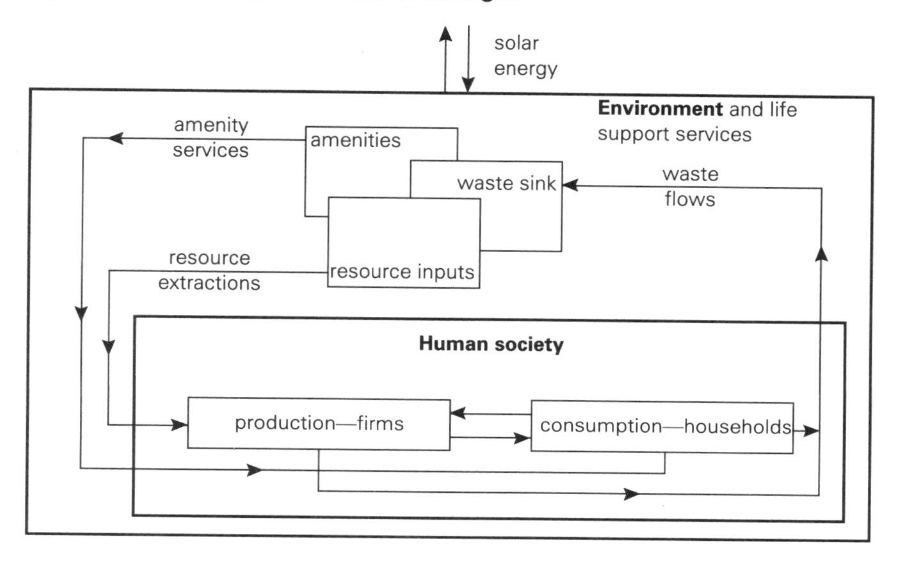

Source: Adapted from Michael Common, *Sustainability and Policy*, Cambridge University Press, Cambridge, 1995, p. 32.

in and dependent on the environment. Within the human society box, the economy is shown as a highly simplified circular flow of goods and resources. Producers supply goods for consuming households, who in turn supply producers with the labour and other resources required for production. The environment provides four types of services to people, three of which are represented by the three overlapping boxes in the upper part of the figure. It is a source of natural resource inputs into production, for example timber, minerals, fish and water for cooling in industrial plants. It serves as a receptacle or sink for our production and consumption wastes, for example vehicle exhaust emissions discharged into the air and domestic sewage pumped into rivers or seas. Third, the environment has a direct effect on our well-being or amenity—atmospheric oxygen is essential for our survival; we get pleasure from surfing or observing wildlife in a pristine rainforest; we are pained when we see salinity killing trees, or containers littering the highway.

The fourth service that the environment provides is human life support, the result of the combined functioning of the climate, chemical element cycling, water cycling and living organisms (described in Section 1.3). In Figure 4.1, life support is represented by the environment box itself—without it, humanity and the economy would not exist. *Natural resources* are those features of the environment that provide one or more of the four services.

The environment, resource input, waste sink and amenity boxes in Figure 4.1 overlap, to indicate that the different types of environmental services are not independent of one another. Commonly, limited quantities of environmental assets, and the effects of one use of a natural asset on its quantity and quality for other uses, mean that more of one environmental service is likely to mean less of others. Consider the services of the water in Australia's Murray River. The water supports human settlement and inland ecosystems that would otherwise not exist in the arid interior. The river's flow is used as an input in farm irrigation, as a sink for salt pumped from farming areas and for wastes from food processing plants, and as an amenity by swimmers, boaters and wildlife enthusiasts. When the rate of flow is high, or desires for irrigation, waste disposal and recreation are low, the river flow is not scarce, in either quantity or quality terms, and conflict between the various services is minimal. At other times, the use of more water for domestic supplies and irrigation leaves less available for waste disposal and for recreation and wildlife. Use of the river for disposal of more salt and more wastes forces costly treatment to supply domestic needs, and may reduce recreational enjoyment, wildlife populations and farm productivity downstream.

The interactions between the input, waste disposal, amenity and life support services of the environment can be complex, sometimes so complex that we do not fully understand them. Thus, in the case of the Murray, the relationships between irrigation water use and downstream salinity, farm productivity, wildlife populations and the condition of riverine ecosystems may be imperfectly understood, because we do not fully understand local soil conditions, or groundwater movements, or plant and animal responses to salinity or industrial chemicals.

The human society box in Figure 4.1 includes more than the economy. It also represents, although it does not show, the rules or institutions described in Section 2.2 that guide human interactions over scarce assets. Social rules, and the associated organisations that define and enforce them (such as families, clans, legislatures, police and courts) are essential parts of the signalling and incentive system of each society. This will be some mix of the three pure systems described in Section 2.4—tradition, central planning and markets. Such rules and organisations are essential underpinning for the economic system simplistically depicted in Figure 4.1.

In principle, the system described in Figure 4.1 might be an isolated subsistence village, an economically and environmentally isolated nation-state or the global economy. In practice, the exchange of resources and goods, and the mobility of people, resources and wastes, mean that no subglobal economic–environmental system is completely closed.

The laws of physics impose two constraints on economic and environmental activity within the world's economic–environmental system. The first law of thermodynamics states that neither matter nor energy can be created or destroyed. The waste flows in the system reflect the operation of this constraint. Temporary accumulations of commodities and equipment aside, whatever weight of materials goes into the economic system must come out, as goods or as wastes.

The diagram does not show the constraints on economic and environmental activity imposed by the second law of thermodynamics, which implies that some useful energy is always lost in conversions of matter and energy from one form to another. Thus, in the long run, when stocks of stored energy are gone, activity in the global economy and environment will be constrained by the availability of incoming solar energy.

Social coordination within the economic–environmental system is more demanding than social coordination within the economy alone. People now create costs and benefits for others indirectly, via their impacts on the environment, as well as directly, via their use of scarce assets in production and consumption. Coordinating the activities of people who are suppliers and customers for, or who compete for, timber, crayfish, Porsches or holidays in Bali is made easier by the fact that the interested parties have a readily recognisable interest in a particular resource or product. This common interest facilitates political and market deals. On the other hand, particular features of the environment incorporate so many valuable attributes, and the people affected by particular environmental changes are frequently so numerous, so diverse in their values and economic interests, and so dispersed over space and time, that communication and negotiation between all the interested parties will be very difficult or even impossible to arrange. Negotiations between Australian conservationists and commercial timber interests have proved very difficult. Negotiations between environmentalists in rich countries and Japanese who fish for whales and dolphins, or Chinese who discharge CFCs, or poor Latin Americans who clear rainforest for subsistence food production will be even more complicated and costly.

4.2 Types of natural resources

Natural resources come in a variety of forms—living and non-living, animal, vegetable and mineral, and so on. Our concern is with scarcity, and the resulting need for social coordination. Thus the categories most useful to the economist are those that distinguish resources according to their present and future availability relative to human demands, and thus suggest something about present and future scarcity. In this respect, economists distinguish two major types of resources, and two subtypes.

Exhaustible resources are those that exist in given stocks in the environment and cannot be replenished within time spans relevant to human planning. Thus coal and oil are exhaustible resources, even though new stocks can be created over geological time spans. In principle, it is possible for humanity to use up the entire stock of these resources. Thus it appears that the prime coordination task is to divide up the given stock between generations, in addition to determining current production and consumption uses. However, as the recent history of oil extraction and use makes plain, using the entire stock is highly unlikely; new discovery and extraction and reuse technologies, frequently the result of increasing resource prices and other signals of increasing scarcity, keep pushing back the evil day. In practice, then, the role of signals and incentives in expanding stocks of exhaustible resources is just as important as their role in dividing up given stocks.

Mineral stocks are the classic examples of exhaustible resources, but resources that are waste sinks and amenities may also be exhaustible. For example, a lake or inland sea has a finite and non-renewable capacity to absorb heavy metals before it becomes, and stays, biologically dead. Scenic features of the natural landscape can be permanently destroyed by mining or urban development. Genetic stocks created over millions of years of evolution, which may be valuable for production or waste assimilation or amenity purposes, are exhaustible in a similar fashion to minerals.

Renewable resources are those whose quantity is replenished by natural and/or human means. They may usefully be divided into two subcategories, *living* and *non-living* renewable resources. *Living renewable resources*, such as plant communities, soil micro-organisms and livestock and wildlife populations, are the result of evolutionary and (sometimes) human selection, and depend on solar energy, water flows and stocks of soil nutrients to sustain life. The rate of renewal of these biological resources depends on both natural circumstances and humans, who can alter both the genetic composition and environment of other living things.

Non-living renewable resources are those such as solar radiation, winds, tides, ocean currents and rainfall that are replenished by flows beyond deliberate human control. Such resources have to be utilised when available or captured and stored for later use, otherwise the value of the flow is lost. Capture and storage, for example in dams or by using solar collectors to charge batteries, converts uncontrolled flows to controllable stocks.

Both categories of renewable resources include production inputs (e.g. timber and water or wind or tidal power used to generate electricity) waste sinks (e.g. biological treatment of sewage and ocean currents used to disperse sewage) and amenities (e.g. atmospheric oxygen, dolphins and scenic vistas). The main tasks of managers of renewable resources are to manage populations of living resources and the capture and storage of non-living resources so that through time resource availabilities are responsive to changes in management costs and people's preferences.

Our classification of natural resources is not immune to changes in people's preferences and technology. A few centuries ago, few Europeans derived services from mountain wilderness; on the contrary, the high mountains were viewed as forbidding or threatening places.[1] When attitudes changed, the mountains became natural resources. The development of skiing technology also contributed to the change in the status of some alpine areas. The scientific discovery that taxol, a derivative of the bark of the Pacific yew tree, could be an effective treatment for ovarian cancers, changed the status of the yew from undesired to a highly valued resource.[2]

4.3 The sustainability issue: economic and ecological concepts[3]

The interactions between the economy and the environment prompt the question whether, over time, continued expansion of economic activity is consistent with ecological stability—with continued functioning of the ecosystems on which all human activity, and life itself, ultimately depend. A growing economy will use natural resource inputs and discharge wastes, progressively changing the environment on which it depends. The resulting reduction in the quantity and quality of natural inputs, waste sinks, amenities and life support services will endanger continued growth and gains in human welfare, perhaps even human survival, unless the system incorporates feedback mechanisms leading to timely corrective actions. Are the signalling and incentive systems of existing economies, incorporating various combinations of tradition, central planning and markets, capable of ensuring that the growth-supporting and life-supporting services of the environment will continue?

It is not hard to appreciate the sustainability issue: How do we achieve continued compatibility between economic decisions and environmental service flows? But there is no definition of sustainability that would permit measurement of national progress towards it or implementation of policies guaranteed to move an economy along a sustainable path. Economists think of sustainability in human-centred terms: *What is the maximum level of consumption (and hence well-being) that an economy can sustain forever*, assuming that it is possible to substitute renewable natural resources and human-made assets for exhaustible resources?[4] Ecologists think of sustainability in ecosystem-centred terms: sustainability is identified with *ecosystem resilience—the ability of the ecosystems to maintain their physical and biological functioning after disturbance*.[5] For example, an ecosystem is resilient, and therefore sustainable, if it can re-establish, with its biological functioning, if not all its constituent species, unchanged, after a cyclone or a volcanic eruption or an oil spill.

To clarify the significance of and relationship between the economists' and ecologists' definitions, consider a hypothetical small economy. Imagine a group of people, say an extended family or clan, marooned on a remote island (or planet, depending on the reader's taste) with no prospect of future escape or rescue.[6] Assume that the impacts of human activity on the island's natural resources and ecosystems are well understood. Long-term survival of the group is possible with careful management of the island's diverse but limited natural resources, some exhaustible (e.g. peat for fuel, ecosystems that maintain soil stability and fertility), some renewable (e.g. edible plants and animals, timber). The group produces the goods it needs to sustain itself—food, clothing, shelter etc.—using a combination of its own labour, capital it creates (such as tools, implements and improved production skills) and the island's exhaustible and renewable resources. It is possible, in fact essential for long-term survival, to substitute labour, created capital and renewable resources for any exhaustible resources on which the group initially depends; for example, members can plant fuelwood trees to replace the dwindling peat resource.

Some of the annual output of this small economy is consumed. Some is invested in created capital such as animals and plants retained for reproduction, tools and education of younger group members in production techniques, including protection of valued ecosystems. Over time, more and more of the exhaustible resources will be replaced by created capital and renewable natural resources.

Now, make two assumptions that greatly reduce the problems of attaining economic sustainability. First, having recognised the threat of population pressure on limited island resources, there is consensus on keeping group numbers stable. Second, all present and future group members have the same preferences; members consuming the same set of goods believe themselves equally well off.

What is a reasonable objective for the group's leader-planner? If he or she is far-sighted and benevolent, it is reasonable to assume equal concern about the well-being of present and future members of the community. Armed with information about the long-term impacts of human activity, including improvements in production technology, on the community's stocks of created and natural assets, and about the preferences of present and future group members, the planner will aim to maximise the level of individual consumption and well-being that the community can sustain indefinitely. He or she will try to equate per-person consumption across generations far into the future. This is the economist's idea of sustainability.

Unrealistic assumptions aside, our example highlights important features of economic thinking about sustainability. First, two types of recurring decisions are especially important for sustainability policy: how much of current output should be saved and invested, and how much of the stock of exhaustible resources should be used in the current time period. Second, to implement sustainability, it is crucial to know the extent to which exhaustible capital can be replaced by renewable resources, labour and created capital. If, due to an erroneous judgment about substitutability, an irreplaceable and essential environmental asset, say an ecosystem without which the island will gradually become inhospitable to human life, is eliminated, the objective of indefinite community survival cannot be achieved.

Turning to ecological sustainability, an ecologist would emphasise the importance of resilience of the island's ecosystems in the face of human activity. Asked what biological features indicate resilience, and thus what resource management policies the community should follow to preserve its natural life support system, an honest ecologist would confess that the answers are unknown.

Our example incorporates the main features common to the very wide array of definitions of sustainability and sustainable development:

1 the importance of economy–environment interactions;
2 concern about the impact of present activities on future options;
3 concern about the well-being of people now and in the future.[7]

Note that the last of these, an explicit concern with human well-being, is not central in ecologists' thinking. Humanity is just one species in, and deriving life support from, ecosystems. Ecosystem resilience does not require stability or even survival of all of the ecosystem's constituent species, including humans.

The dissonance between common (including economists') and ecologists' conceptions of sustainability reminds us that, for most people, sustainability is a human-centred, rather than a nature-centred, concept. The environment may change, but it should not change so much as to endanger human lives or living standards. According to most ecologists and evolutionary biologists, this sort of stability is not a natural property of environmental systems; rather they are dynamic and evolve over long periods of time.[8] Humans may be more comfortable with the notion of a stable environment, but in reality the processes of environmental change are chance-driven, with no inherent stability. And, since we live in a world governed by chance, we cannot calculate what nature will throw up next; sustainability policies that aim at desired future states of the world are not necessarily in harmony with nature.

4.4 System complexity and sustainability policy

Neither economists nor ecologists know enough about combined economic–environmental systems to understand all the future consequences of today's resource use. Economists pay little attention to the functioning of ecosystems; ecologists pay little attention to the behaviour of people whose actions may threaten ecosystems.

The complexity of economic–environmental systems can be appreciated by considering the physical and behavioural factors influencing the evolution of the system in Figure 4.1. On the economic side, complexity is due to the numbers of human agents and goods, the diversity of preferences, technologies, rules and organisations, and the abilities of people to learn from experience and to change their preferences, production methods, rules and organisations. In the case of the environment, complexity is a consequence of biological diversity and variations in the physical environment. There are so many possible interactions within and between the economy and the environment over space and time that it is simply impossible to know the future outcomes of today's use of the environment, as the following examples illustrate. Our ability to depict a combined economic–environmental system in schematic terms, as in Figure 4.1, does not imply that we know how to model it accurately or to manage it to benefit people. Consequently, even if we could define a future desired state of the world consistent with our idea of sustainability (say an increased area of tropical rainforests), we could not be sure of the policies required to achieve it, let alone whether our descendants would prefer it to other futures.

Examples of economic and environmental complexity

1 Bans on tropical timber exports. Some environmental groups have advocated that rich countries, such as Australia, ban the import of tropical timber, in order to protect tropical forests and forest-dwelling communities. The complex relationships between the economy and the environment in both exporting countries (e.g. Malaysia) and importing countries (e.g. Australia) make it very difficult to predict the consequences of such bans. How would the governments of countries such as Malaysia react to trade bans? How would local landowners, workers and businesses react to the change in their economic opportunities? Would the consequent illegal trade and quasi-legal trade via non-participating importers such as Korea or China be more or less environmentally harmful than trade involving well-known Japanese and Western companies? What would be the economic and environmental consequences of Western customers' switching to non-tropical timber and non-timber substitutes?

2 Sea otters and fish populations.[9] Sea otters, those furry stars of television nature shows, were hunted almost to extinction in the eighteenth and nineteenth centuries. There are presently two wild populations, in Alaska and on the California coast. Plans to increase their survival chances by establishing new colonies in the wild have led to conflict between fishers and conservationists, since otters feed on valuable commercial shellfish, including abalone and sea urchins. However, the role of sea otters in coastal ecosystems is more complex than this. Sea urchins attack and kill kelp, by eating the holdfasts that attach kelp to the sea floor; a study of otters in the Aleutian Islands found abundant kelp where otters were abundant, and little kelp where otters were absent. Many marine species, including commercially important fish, use kelp forests as breeding grounds or habitat during their life cycles. Thus fishers other than shellfishers may be economically affected by the number of otters, this time positively. In the absence of detailed studies of the interrelationships between otters, shellfish, kelp and other marine species, and of the economic values of the various components of the marine ecosystems involved, it is not possible to estimate the benefits and costs of a new otter colony. Even with such information, the long-term impacts of the colony on the environment and the long-term value of the various species involved will remain uncertain.

The island economy example of Section 4.3 indicates the futility of an economics-based national sustainability policy. Even the far-sighted and benevolent planner of our example would find it very difficult to establish the rates of substitutability between, and the relative values of, all the created and natural assets that benefit the island community. If we are realistic, and drop the assumptions that the impacts of human activity on the environment are understood, and that community members' preferences are stable over time, even our idealised planner has no hope of charting a sustainable path for the island economy.

Ecologists cannot help; they do not understand what makes ecosystems resilient in the face of disturbance. It seems that, until we know a lot more about the environment and its relationships to the economy, the best that communities can do is to address particular environmental problems as they arise. Having reviewed what is known about sustainability from both the economic and ecological perspectives, Common puts it this way:

> There is no blueprint for a sustainable society waiting to be discovered. The problem itself changes over time as the result of economy–environment linkages, and their repercussions in human societies. In so far as there is any solution to the problem, *it is successful adaptation to changing circumstances* [emphasis added].[10]

The discussion of economic–environmental systems in Section 4.1 emphasised that they evolve as a result of people's production and consumption choices. Those choices are guided by a society's signalling and incentive system. It follows that sustainability of an economic–environmental system is dependent on the rules and organisations used to inform and motivate producers and consumers. These include property rights, rules of exchange, moral codes, conventions, administrative rules and penalties, markets, elections, law and courts, legislatures and so on. If economic–environmental systems are so complex that it is impossible to know what actions will promote sustainability, the signalling and incentive system itself may be the best indicator of sustainability—of ability to adapt to changing economic and environmental circumstances.

Discussion questions

1. Once a natural resource, always a natural resource. True, false or uncertain? Explain why.
2. In a mixed economy, such as Australia's, what mechanisms operate to signal to private and public decision makers the impacts of economic activity on the natural environment? Consider both market and bureaucratic signalling mechanisms. In which areas of resource use do you think our mechanisms for signalling our impacts on the natural environment are most seriously deficient? Explain why.
3. What are some factors that might explain the finding that some living renewable resources (e.g. fish, forests) appear to be growing scarcer relative to minerals, which are an exhaustible resource?
4. What do you see as the major types of complexities of combined economic–environmental systems, that make it difficult to determine whether an economic–environmental system is evolving in a way that we might define as 'sustainable'?

II

REASONS FOR COORDINATION FAILURES

5

Limitations of market signalling and incentives: high costs of markets

My object all sublime,
I shall achieve in time,
To make the punishment fit the crime,
The punishment fit the crime.'

W.S. Gilbert, *The Mikado*

Markets cannot coordinate people's use of natural resources if there are no private property rights and markets for valuable assets. For example, users of clean air and live whales do not receive market feedback about all the consequences of their asset use for others, and therefore do not have the correct incentives to consider all the impacts of their actions on others. The main reason that no private property rights and markets exist is that precise property rights and market negotiations sometimes cost more than the anticipated gains from market deals. This is partly the result of limitations of current science and technology, which make it too costly to identify and measure resources, resource use and consequent impacts on others. Also, bilateral market deals may not be possible where many people benefit from or are harmed by the same action. However today's environmental problems may be resolved by market deals in the future, as scientific information, technology, resource costs and the values attached to environmental assets change.

5.1 Missing markets

In his science fiction novel *Childhood's End*, Arthur C. Clarke describes a future world where all the costs of decisions are signalled to the decision makers; all bullfight fans feel the bull's pain as the matador lunges for the kill. Clarke assumes that if individuals suffer all the consequential costs of their actions, they will have appropriate incentives to modify their behaviour in the interests of others. Such fantasies are not confined to believers in animal rights. Most of us have, at some time or another, imagined a world where the yokel who tosses cans out of the car window, and the rancher who bulldozes and burns rainforest, suffer punishments that 'fit the crime'.

An ideal market system, with precise individual property rights for every asset, would accomplish this. With ideal markets for every asset (not only avocados, haircuts and pick-up trucks, but also clean air and live whales) any producer or consumer who wanted to use these assets would have to compensate the owners of the rights to avocados or clean air or live whales via the market. Market bargaining would ensure that the true sacrifices involved in less clean air or less whales were signalled to the would-be beneficiaries and users of these valuable assets.

Unfortunately, as we saw in Chapter 3, the conditions required for ideal markets are stringent, too stringent to permit ideal avocado markets, let alone ideal markets for clean air or live whales. On the contrary, if you want cleaner air in your suburb and more whales, individual property rights and markets are no help; there are no such property rights and markets. The costs you experience when sulphur dioxide from a power-generating station fills your lungs, or when you learn that the whale population has fallen, have to be signalled to the power company, or to Project Jonah and Japanese whalers, by means other than markets.

Much of this chapter and Chapter 6 are devoted to examining the characteristics of valuable assets (such as clean air and live whales) that prevent the creation of property rights and markets in such assets. But first we must understand the consequences of the absence of property rights and markets.

Assume for the moment, that coordination of resource use is based solely on market signalling and incentives. If there are no individual property rights and markets for assets such as clean air and live whales, people who use these assets do not experience all the consequences of their asset use. Assuming that people respond to the benefits and costs they themselves bear, the result is that users do not have the correct incentives to consider all of the impacts of their actions on others.

Our daily lives are full of activities with consequences, both good and ill, that are not borne, directly or indirectly, by the asset user. When we spray mosquitos and tend scenic gardens or farms, the resulting absence of mosquitos and attractive vistas benefit our neighbours and passers-by, with no likelihood of compensation. People who throw cans along the roadside or allow their dogs and cats to roam and kill native

Figure 5.1 Consumption and production externalities

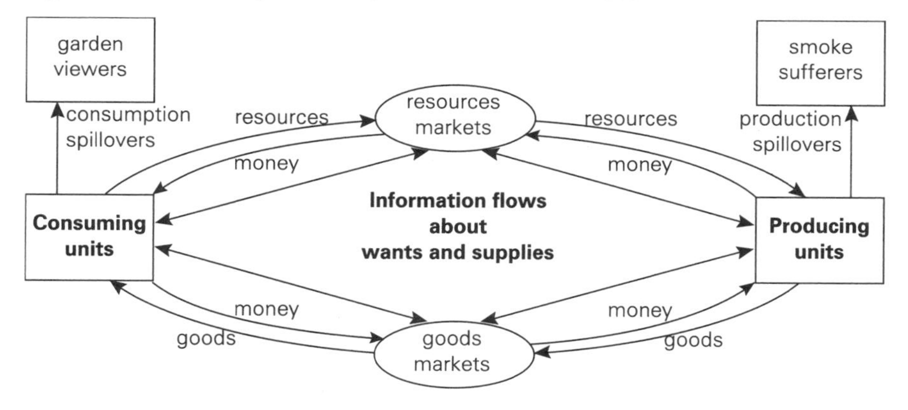

animals, businesses that discharge industrial effluent into rivers, and fishers whose driftnets kill dolphins and seabirds commonly escape many of the costs that they impose on other people. They will almost certainly escape market penalties. We need to understand precisely why there are no individual property rights and markets to sheet home all the consequences of actions, both good and ill, to the decision makers. Why don't markets reward us when our gardens brighten the days of passers-by, or penalise us when smoke from our factory aggravates people's asthma?

5.2 Lack of market feedback: externalities

Figure 5.1 illustrates the above problem using the modified circular flow diagram introduced in Chapter 2. It is assumed that market signalling and incentives work for all commodities except attractive gardens (produced by households) and smoke (produced by factories). In the absence of markets for garden viewing and clean air, and assuming, as the diagram does, that there is no other communication between gardeners and garden viewers, or between factories and those who suffer from the smoke, gardeners are not influenced by the desires of garden lovers, and factories by the desires of smoke sufferers. In other words, there is no feedback and control mechanism to inform and motivate individual gardeners and factories to take account of all of the effects of their actions on others.

As described in Figure 5.1, garden viewing and suffering caused by factory smoke are what economists call *externalities* or *spillovers*. These are *consequences (benefits or costs) of actions (consumption, production or exchange) that are not borne by the decision maker, and hence do not influence his or her actions*. Negative externalities that are by-products of people's use of the natural environment are commonly termed *pollution*.

The existence of externalities means that individual property rights must be imperfectly defined or enforced or non-transferable. Recall, from Section 2.3, that social coordination in the use of scarce assets is maximised when property rights satisfy four conditions: when they are clearly defined, strictly enforced, exclusive and transferable. Suppose that this were the case for garden viewing and clean air—that individual rights to garden viewing and clean air were costlessly defined and enforced and transferable, as was assumed for avocados in Chapter 3. Then it would be possible for rights to garden viewing and use of clean air to be bought and sold. Gardeners could sell viewing rights to passers-by, and to continue in operation, a factory would have to purchase clean air for a dump site in the same way that it might purchase vacant land to bury solid wastes.

With property rights and markets for garden viewing and clean air, market signalling of the benefits and costs attached to garden views and alternative uses of air would proceed in the same fashion as for avocados in Chapter 3. The would-be buyers of garden viewing and clean air (garden lovers and factories producing smoke) would state their willingness to pay. The would-be sellers (gardeners and clean air owners) would state their willingness to accept compensation. Assuming many buyers and sellers, all would be forced to reveal accurate values. Exclusive property rights to garden viewing and to clean air would give gardeners and clean air owners the opportunity to gain by selling those rights, and garden lovers and factory owners to gain by purchasing those rights, in the same way that avocado producers and consumers gain by trading exclusive rights to avocados. The property rights would create the incentive for market deals, the interdependence between the parties would be recognised, market valuations of garden viewing and clean air would provide gardeners and factories with feedback about the benefits and costs of their asset use for others and they would be motivated to adjust their outputs of gardens and smoke accordingly.

Where externalities exist, market demand and supply curves no longer reflect all the benefits and costs of consumption and production, as was assumed for avocados in Chapter 3. Consider the distortions involved when there is no market for clean air, and industrial decision makers are therefore able to ignore the costs due to smoke damage. Suppose that we have a steel-making plant located in an airshed like the NSW South Coast or the Hunter Valley. Assume that the steel maker is a price taker; that is, in a competitive steel market, the price P_{st} it gets for steel is the same regardless of the amount of steel sold. Thus the demand curve or marginal benefit curve for steel from the plant is a horizontal line starting at P_{st}, as shown in Figure 5.2. The private marginal cost curve for steel from the plant is shown as MC_p; it is based on the value of the alternative opportunities plant management sacrifices by using labour services, raw materials and the funds tied up in buildings and equipment to produce steel, rather than other products. Considering only those costs signalled to it in the marketplace, plant management

Figure 5.2 Private determination of the output of smoke

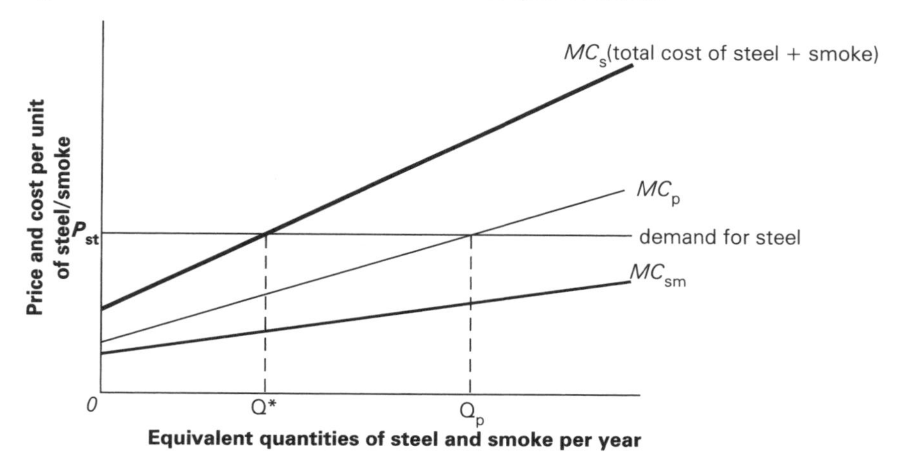

will maximise profits by producing Q_P tonnes of steel per year, equating the price it expects for its steel with the private opportunity cost of the last tonne it produces. Looking at the decision another way, in deciding whether or not to produce extra tonnes of steel, the managers are guided by the extra profit they can make, which is given by the vertical distance between the demand curve and the private marginal cost curve. That extra profit falls to zero when the annual steel output reaches Q_P.

Assume that the plant produces steel and smoke in fixed proportions, that is, every tonne of steel is accompanied by the emission of (say) 100 kilograms of smoke particles. Then the horizontal axis in Figure 5.2 measures equivalent quantities of steel and smoke. In practice, plant managers will have ways of varying the proportions of steel and smoke, for example using smokestack scrubbers that remove particulate matter from gas emissions. These options are discussed in Chapter 15.

Aggregating across all the people in the airshed who bear costs as a result of the smoke (costs such as breathing difficulties, medical expenses, dirtier homes, cleaning bills and obscured views), the marginal costs of smoke from the plant are equal to MC_{sm}. This curve measures, in money terms, the additional costs imposed on all people in the airshed as smoke discharges per year increase.

The community as a whole values both steel, which generates community income, and cleaner air. Thus, from the community point of view, the total sacrifice, or total social cost, involved in steel production includes not just the values of the resources used up, but also the costs of the resulting smoke emissions. The total social marginal costs of steel and smoke produced by the steel plant are MC_s in Figure 5.2, the vertical sum of the marginal costs of both steel and smoke pollution. From the community point of view, the appropriate comparison of marginal benefits and costs is then between the price of extra steel P_{st} and MC_s, not the price of extra steel and MC_p, which picks up only those costs signalled in markets. The annual net benefits from steel and smoke combined are a maximum when steel and smoke output is Q^*, where the value of the last tonne of steel equals the sum of the production and pollution costs it creates. If the rate of steel output from the plant exceeds Q^*, the price received for the extra steel is inadequate compensation for the combined sacrifices involved in additional steel inputs and smoke damage. However, in the absence of property rights and markets for clean air, the costs of smoke pollution will not be signalled to plant management, and steel and smoke output will be Q_p, not Q^*.

5.3 Non-market feedback mechanisms

If there are no markets to coordinate people's use of scarce assets, societies must rely on the alternative signalling and incentive systems described in Section 2.4, tradition and central planning. Externalities such as those depicted in Figure 5.1 may be overcome by creating non-market mechanisms to provide the necessary feedback to decision makers. As shown in Figure 5.3, the feedback is provided by non-market institutions, meaning either general rules or rules operating in formal organisations such as clubs or business firms. The mechanism may be as straightforward as warm smiles and enhanced community prestige to reward creative gardeners. It may be as complex as an array of rules governing factory siting, manufacturing technologies, operating times and procedures, and limits on airborne emissions, monitored and enforced by a pollution control authority and the courts.

Non-market feedback is distinguished from market feedback by the absence of bargaining based on direct offers of willingness to pay versus willingness to accept compensation, described in Chapter 3. Non-market feedback to overcome externalities can take many forms, including:

1 *customs* or *conventions* (e.g. conventions concerning when it is acceptable to stand at concerts and sporting events);
2 *politeness* or *neighbourliness* as behavioural norms (e.g. because you and your residential neighbours anticipate a long-term relationship, you are careful to give prior warning or consult on matters such as large parties, lighting fires out of doors and removing trees);
3 *formation of recreational or social clubs* that share a site or facility (e.g. a swimming pool or a ski lodge; club rules are designed to prevent externalities, such as admission of complete strangers or watching of television, that harm club members but cannot be prevented in the outside community);
4 *formation of firms that encompass the externality* (e.g. if a company owns the homes near its steel plant, the rent the company can expect is affected by the amount of smoke produced by its plant;

Figure 5.3 Non-market signalling

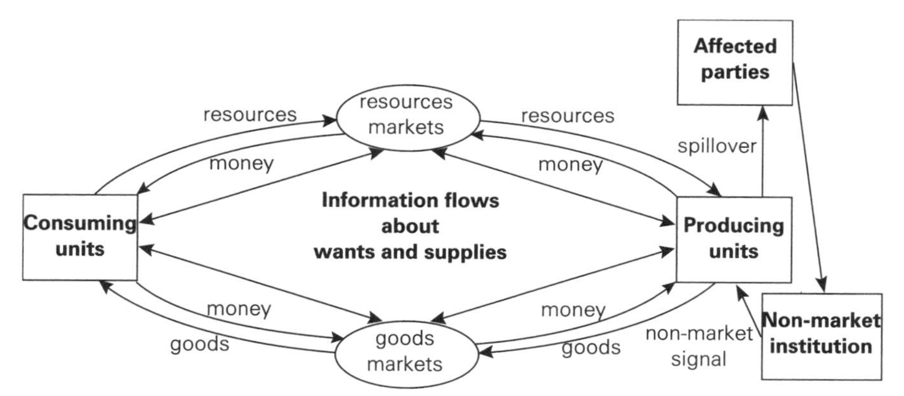

5 *common law and statutory entitlements to sue for damages* (e.g. UK court rulings granting damages to the owners of fishing rights damaged by upstream polluters, and legislation allowing property owners to sue builders to protect access to solar radiation);
6 *government-created signals and incentives*, which include direct government regulation (e.g. motor vehicle emissions controls that restrict the actions of both vehicle manufacturers and vehicle users), taxes on polluters and subsidies to reduce pollution and increase beneficial externalities (e.g. money payments or other assistance to encourage the owners of outstanding gardens to open them to the public).

Recall the classification of signalling and incentive systems in Section 2.4. Customs, conventions and behavioural norms are examples of coordination by tradition. Government regulations, taxes and subsidies are examples of central planning. Individuals cooperating in clubs and firms in effect agree voluntarily to subject themselves to the central planning rules determined by those mini-societies. Entitlements to court-determined damages involve non-voluntary exchanges of rights via a third party, the court, rather than voluntary market exchange.

5.4 Why are there no markets in cases of externalities?

Assume, for the moment, that society does not prohibit markets in garden viewing or clean air on ethical grounds. In a predominantly market economy, such as Australia, what other reasons might explain the absence of these markets? The reader may suggest that one of the forms of non-market allocation just listed may be a less costly means of signalling the benefits or costs of garden viewing or smoke. This is true, and it is important to grasp the full significance of the comment. There are really two related points. First, outstanding home gardeners may be very economically rewarded and encouraged by non-market mechanisms such as warm smiles and community prestige. Second, and more important, market coordination of production and consumption itself involves the sacrifice of resources that have alternative uses, a fact ignored in the ideal market model outlined in Chapter 3.

Once we recognise that market allocation is costly, an obvious reason for the absence of markets for valuable assets such as garden viewing and clean air is that the expected costs of market deals exceed their expected benefits. The benefits from market exchange were identified in Chapter 3. Buyers gain the difference between what they are willing to pay and what they actually pay; sellers gain the difference between their costs and the price they receive. If all prospective buyers and sellers of garden viewings and clean air judge that the costs of market exchange will be greater than these gains, there will be no market.

The idea that market exchange is costly in terms of time, effort and money is no news to anyone who has recently sold a used car or purchased a house or a package tour to Europe. Organising classified advertisements, taking telephone calls from potential buyers, haggling with used car dealers, inspecting houses when you could be watching a weekend football game, paying for title searches, calling airlines and travel agents and poring over travel brochures are all costs of market exchange (except where people enjoy some of these activities). Economists commonly term such costs *transactions costs*. Since a variety of costly activities are involved, it will help if we identify what they are. Our aim is to understand which particular costs of market exchange prevent the existence of certain markets.

5.5 The costs of markets

What are the essential activities involved in any market exchange? Consider a familiar form of market communication, newspaper classified advertisements, for example an advertisement for a used car. To generate responses, the advertisement must provide the following information:

1 the identity of the would-be seller;
2 measures of the attributes of the vehicle, such as make, age, details of previous owner(s), kilometres driven and so on;
3 a price that the seller would accept.

The advertisement conveys information from the seller to potential buyers. Before a sale can occur, potential buyers must also identify themselves to the seller and reveal their willingness to pay for the car.

The advertisement need not specify the seller's precise property rights in the car, but it must be consistent with people's understanding of current property rights. This will not be the case if someone places an advertisement offering to sell rights to view their garden or to clean air in their suburb. Readers will not believe that such rights can be defined (how does one identify and measure a view or a parcel of air?) and enforced (how can we measure use and identify users?). Implicit in the car advertisement, or any other selling offer, is the understanding that the seller has specific legal rights in the vehicle, and that those rights are transferable and can be enforced on behalf of a new owner. All of this would be untrue if the car was stolen. The buyer will want to confirm these understandings prior to completion of the purchase. After all, what is really being transferred, and what really motivates market exchange, is a bundle of property rights in the car.

Specification of the property rights being exchanged is an essential part of any market transaction. The specification of rights includes the degree of enforcement of those rights; for example, fishers purchasing quotas to harvest a certain number of abalone will be extremely concerned about enforcement of regulations banning poaching of abalone.

All of the above matters must be attended to in any market exchange. Thus the costs of market exchange include the costs of:

- potential buyers' costs of identifying would-be sellers and sellers' costs of identifying would-be buyers;
- measurement of the quantity and quality of the asset being transferred;
- revealing potential buyers' willingness to pay and potential sellers' willingness to accept;
- specification of property rights and transfer of rights.

As we shall see in Sections 5.8 and 5.9, if these costs of market exchange are judged to exceed the benefits from exchange, no market will exist.

The high costs of market exchange explain the existence of bureaucratic organisations, in particular the business firm, in predominantly market economies. Consider the case of a car assembly plant or a privately owned sporting team. There are clear gains from specialisation and team work in each case, but it would be prohibitively costly for all the individuals involved to coordinate their efforts by market contracts between individuals. How could they possibly all agree on their respective contributions, day by day or match by match? It is far less costly for all to sign long-term contracts with a single firm manager, who is then responsible for task allocation and monitoring, rewarding and penalising each worker. In the terms of Section 2.4, the firm is a centrally planned society where workers voluntarily agree to submit to central planning by the manager, who determines what and how to produce.

5.6 The costs of specifying property rights

What are the costs of specifying property rights? As explained in Section 2.3, for property rights to be well specified, they must be clearly defined and strictly enforced, with enforcement including physical exclusion measures, monitoring of use and penalising of unauthorised users. Suppose that you were considering the purchase of an unusual property, say a salmon farm in a bay in southern Tasmania. To identify the rights being transferred, you would want to identify the current owner(s), and to identify and measure the assets involved (land, water in the bay where the growing pens are located, spawning tanks and buildings onshore, fish stock etc.) and the precise rights held in each. To assess enforcement of the rights on offer, you would want to know about the measures (fences, guard dogs etc.) and costs involved in physical exclusion of possible poachers, the possibilities of identifying poachers and other human and animal intruders and measuring consequent losses (extent of private and public surveillance etc.), and the possibilities that poachers can be penalised and the likely penalties. Thus the main costs of specifying property rights are costs of identification (of assets, right holders and asset users), of measurement (of assets and asset use), of physical exclusion and of penalising illegitimate use.

If specification of property rights is costly, it will sometimes be more efficient to buy and sell vaguely specified rights rather than precisely specified rights. This is because buyers and sellers will gain more from the reduction in costs of defining and enforcing rights than they will lose from failure to specify rights more precisely. For example, a weekly lift pass at a ski resort is a much less precise method of allocating scarce lift capacity than per-ride tickets whose prices can be altered during the week, but the costs of selling and checking single-ride tickets are far greater.

Note that precise specification of rights to the salmon farm does not require both the physical exclusion and the identification and penalising of unauthorised users. If would-be poachers can be prevented from gaining access at low cost, there is no need for their identification and subsequent prosecution; if poachers could be unerringly identified and penalised at low cost, there would be no need for 'fencing'. In practice, right holders will commonly use some combination of these techniques for protecting their property, depending on relative costs. Scrambling transmissions and providing decoders to paying customers is feasible for pay television operators broadcasting by satellite; broadcasting free and tracing and penalising users who do not pay is impractical. Therefore pay television operators will adopt the exclusion technology. With the latest transponder technology, identification and charging of individual motorists using a tollway promises to be far less costly than exclusion of motorists who refuse to pay at tollbooths.

Not all of the costs of specifying property rights are directly borne by market participants, and so affect the existence of particular markets. Taxypayers bear much of the costs of enforcement through government funding of police, prosecutors, courts and prisons. Also, to the extent that the government provides participants with information about the definition, ownership and enforcement of property rights, some identification and measurement costs are shifted to taxpayers in general. In

Table 5.1 Costly activities required for social coordination

Activities	Coordination mechanisms: Property rights	Coordination mechanisms: Market exchange
Identification/ monitoring	Of: • assets • right holders • asset users	Of: • items transferred • right holders • trading partners
Measurement	Of: • assets • asset use	Of: • items transferred
Exclusion	Of: • non-right-holders by some sort of 'fencing'	
Penalising/rewarding	Of: • illegitimate users, by social pressure, courts etc.	
Preference revelation		By: • potential buyers and sellers, via offers to pay and accept

Australia, state funding of the Torrens system of registration of land titles and motor vehicle registration reduces the private costs of transferring land and cars, probably increasing the volume of trade in land and motor vehicles.

Exclusive property rights are necessary for markets, but may exist in the absence of markets. For example, if individual possession and use of marijuana is legal, individuals can have exclusive rights to marijuana grown for personal use, even if sale of marijuana is prohibited. Nevertheless, since both markets and property rights require identification of interested people and measurement of the asset and/or asset use, there is a close connection between the presence and absence of well-specified property rights and the presence or absence of markets. If costs of identification and/or measurement are high for one, they will be high for the other. Thus it is not very expensive to identify cars and the owners and users of cars, and to monitor the use of cars. Consequently property rights in cars are quite well specified, would-be buyers and sellers of cars judge that their gains from exchange will exceed their costs of identifying trading partners, measuring car quality, haggling over the price and confirming property rights, and a market for cars will exist. On the other hand, it is prohibitively expensive to identify particular specimens of clean air and the owners and users of that air, and to monitor the use of that air. Consequently there are no exclusive property rights in clean air, and no markets where clean air can be bought and sold.

Table 5.1 lists the costly activities required for market coordination of asset use, involving both specification of property rights and market negotiations. Note the similarities in and the differences between the informational and technical requirements for the two processes.

5.7 A market solution for acid rain?[1]

Acid rain is precipitation containing sulphuric and nitric acids. These acids form in the atmosphere as a result of emissions of sulphur dioxide and oxides of nitrogen, mainly from power stations and industrial plants. Wind patterns suggest that emissions from America's Midwest and Ontario are the major causes of acid rain in the eastern USA and Canada, and that emissions from Britain and Western Europe are primarily responsible for acid rain in Scandinavia. Areas receiving large amounts of acid precipitation have experienced extensive forest dieback and declines in aquatic life in acidified lakes.

In the unrealistic world of ideal and therefore costless markets, the acid rain problem would be solved in the following manner. First, the respective property rights in clean air of tree and lake owners and power companies and other industries would need to be specified. Second, the owners of forests and lakes would have to confirm that the damage to their property is, as strongly suspected, due to emissions of sulphur dioxide and nitrogen oxides. Third, the sources of the emissions would have to be identified. Finally, the parties would negotiate to determine a quantity and price of clean air that would be allocated to industries for waste disposal. If the forest and lake owners initially possessed the rights to clean air, industries would have to pay them to purchase clean air for waste disposal. If the industries already possessed the rights, the pollution sufferers would have to purchase clean air rights to reduce pollution. At the end of the negotiation process, the environmental problem of acid rain would be resolved, because no one would be experiencing costs to which they had not consented.

Is such a market solution to the problem of acid rain possible, or just an economist's flight of fancy? For an answer, we must consider the likely costs of the various activities listed in Table 5.1, as they apply to clean air and the emissions believed to cause acid rain. We do this in the following sections. We first consider the costs of market coordination when just two parties are involved, and then the more realistic case, when there are many emitters and many recipients of acid precipitation.

5.8 The costs of a market solution for acid rain: two parties

Imagine a single power station located in an isolated valley and surrounded by a huge area of forest all owned by the same company. The station emits sulphur dioxide, all of which ends up as acid precipitation on the forest; no other individual or organisation experiences the resulting pollution. Now consider the costs of the activities listed in Table 5.1, first those required for property rights specification and then those required for market negotiations.

Costs of specifying property rights As previously discussed, there are in fact no well-defined exclusive rights to use clean air, due to past prohibitive costs of identifying air parcels, their owners and users, measuring air use, and either excluding non-owners or enforcing penalties for the illegitimate use of clean air. However, in our simple two-party case, the relevant property rights are not so difficult to specify. The power station is responsible for all of the emissions; the forestry company holds rights to all trees that may be damaged by those emissions. Assume that clean air is not valued for itself, but simply serves as the transport medium. Then a court determining property rights simply decides who has the right to determine the amount of emissions released into the air—the power company or the forestry company. The court's decision determines which party bears the costs of any forest damage *due to emissions*. If the forestry company has the right to control air use, the power company can only continue to emit pollutants by negotiating to compensate the forestry company; if the court awards the right to the power company, the forestry company can only reduce pollution damage by paying the power company to reduce emissions. The consequences of such alternative assignments of property rights are discussed at length in Chapter 14.

If the court goes further, and awards damages to the forest owner based on emissions damage, it faces the difficult and costly tasks of identification and measurement of emissions damage. How does it identify and measure the extent of forest damage caused by emissions, as opposed to naturally occurring damage? These information problems are the same as those faced by market negotiators, discussed below.

Costs of identifying potential trading partners Identification requires scientific determination of the causes of damage to the forest. It is not enough to strongly suspect that emissions of sulphur dioxide are the cause. The link has to be established scientifically because neither the power company nor the forestry company will be willing to pay to prevent damage that has no proven connection to their use of their property.[2] If there is no connection, there is, in reality, no pollution and no coordination problem.

If the connection between emissions and forest damage is established, identifying trading partners is no problem; all the emissions come from one source and there is only one sufferer.

Costs of measuring the item transferred The party who has to pay up will insist on measurement of the impacts of its actions. If the power company has to pay for fouling clean air, it will require reliable measures of how changes in its emissions affect the forest; if the forest owner has to pay the power company to reduce emissions, it will require accurate measurement of emissions reductions. In either case, costly measurement procedures will be required to assure parties that they are getting what they pay for.

Costs of revealing willingness to pay and to accept If property rights and the connection between emissions and forest damage are established, and impacts of parties' actions are measured, the power company and the timber company are in a position to negotiate a price and quantity of emissions released into the air. In this simple neighbour-to-neighbour pollution scenario, there is just one impediment to a speedy market deal that satisfies both parties; neither can be sure that the other is revealing its true willingness to pay or to accept.

Recall that in a market with many buyers and sellers of identical avocados, competition from other buyers or sellers forces individuals to reveal their true willingness to pay or to accept. Here there is only one buyer of rights to release sulphur dioxide into the air, and only one seller. Whichever is the potential buyer, power company or forestry company, will likely have some scope to understate its gains from, and hence willingness to pay for, more or less emissions of sulphur dioxide, since the potential seller does not know the buyer's production alternatives or finances. Conversely, the potential seller of emissions rights will likely have some scope to overstate the value it attaches to those rights. If each doubts the other's offers, market negotiations may be protracted, and correspondingly costly. (Problems of two-party negotiations over pollution levels are discussed at length in Section 14.4.)

Summarising, in the two-party case, the major impediments to a market solution of the acid rain problem are partly technical and partly behavioural. Scientific identification of a causal relationship between emissions and forest (or other) damage, and measurement of that relationship, is costly. And with only one buyer and one seller, it is difficult and costly for each of the parties to establish whether the other has revealed its true willingness to pay for, or to accept payment for, more or less emissions. If the power company or the forestry company perceives that the resulting costs of market exchange exceed the likely gains from trading emissions rights, there will be no market deal, and no market coordination to reduce the acid rain problem. However, with just two parties, a market deal may be possible.

5.9 The costs of a market solution for acid rain: many parties

With many emitters and many recipients of acid precipitation, the costs of market exchanges to resolve the acid rain problem are far greater. The technical and behavioural reasons for the higher costs will be fully explained in Chapters 6 and 7. Here we simply describe the differences from the two-party case.

To begin to get an idea of the costs of market coordination in the many-party case, suppose that sulphur and nitrogen oxides emissions are released from one hundred power stations and steel mills in America's Midwest, and that all of the resulting damage is suffered by one hundred forest owners in New York and New England. If each forest holding receives all of the emissions from a single source, and only from that source, market coordination over acid rain might be achieved by specifying separate property rights in the emissions passing between each of the emitter–recipient pairs, and negotiation of one hundred two-party deals along the lines described in the preceding section. This would increase costs, due to the large number of separate sets of property rights and market negotiations, but market deals might still achieve a reduction in the acid rain problem.

The reality of atmospheric pollution is not so straightforward. Air movements mix as well as transport emissions, so that emissions from each midwestern power plant or steel mill are deposited on many, if not all, forest holdings. On the recipient side, each forest owner receives acid precipitation from many, if not all, emissions sources. What does this imply for the costs of market coordination?

Costs of specifying property rights Recall, from Section 5.6 and Table 5.1, that for property rights to be meaningful, it must be possible to identify the owner and users of an asset, to measure asset use, and to enforce penalties on illegitimate users. For airborne emissions, this is possible in the two-party case, where all emissions have the same source and affect the same recipient. The total quantity of emissions is all that matters. This is not the case with multiple emitters and recipients. Given current monitoring and tracing technologies and resource costs, uncontrolled air movements and atmospheric mixing of emissions make the necessary identification and measurement activities prohibitively expensive. Consider the problem of a forest owner with notional property rights to control dumping of emissions into clean air. The forest owner cannot exclude airborne emissions. With atmospheric mixing of emissions, it will be very costly, if not impossible, to identify all the upwind sources of damage to the forest, and to measure the separate contributions of each source. Thus the forest owner has no effective property rights over incoming emissions.

With no well-specified individual property rights, social coordination between emitters and recipients requires multi-party, not two-party, deals.

In the North American situation discussed by Dolan (see note 1), clean air is the subject of conflicting rights claims, not conclusively adjudicated by US courts. On the one hand, acid rain sufferers and environmentalists claim that residents of the eastern USA and eastern Canada have property rights to unpolluted air, and consequently have rights to compensation from sources that pollute 'their' clean air. On the other hand, midwestern industries claim that, since they have made large investments in plants that met all emission standards at the time of construction, they have property rights to use clean air to dispose of their emissions up to the level permitted by those standards. If they are required to reduce emissions below the previously permitted standards, they have rights to be compensated for the loss of 'their' clean air, which they have been using as a cheap means of disposing of their wastes.

In the absence of agreed property rights, when large numbers of parties are involved, political lobbying and litigation aimed at obtaining favourable definitions of rights will take precedence over market negotiations.

Costs of identifying potential trading partners Once the scientific connection between emissions and forest damage is established, connections between particular emission sources and sufferers must be identified. As just explained, the complexities of atmospheric mixing make precise estimates of connections impossible at any reasonable cost; contrast the difficulties and likely costs of identifying the sources of acid rain damage with those of identifying the source of damage when a tree falls across your boundary fence, or a herbicide spray drifts onto your garden.

Over time, new markets come into existence because technological breakthroughs reduce the costs of markets. For example, it may soon be possible to tag airborne emissions from major sources with distinctive isotope tracers, greatly reducing identification costs for such sources.[3] This is one illustration of the important point that the costs of market exchange are susceptible to advances in technology.

Costs of measuring the item transferred Along with the problem of connecting sources and sufferers, there is the problem of measuring the quantities of acids deposited at particular sites that result from emissions originating at particular sources. As before, the complexities of air movements and mixing make accurate measurements impossibly expensive in the absence of advances in technology. Even then, continuous monitoring of emissions and acid deposition at large numbers of sites will be very costly.

Costs of revealing willingness to pay and to accept Truthful revelation of willingness to pay and to accept is a far greater problem with multiple emitters and recipients of airborne pollution. There are dozens of electric power stations in the US Midwest, and thousands of industrial emitters. The downwind recipients of acid rain number in the millions. Each emitter may affect millions downwind, and each sufferer may receive acid precipitation from thousands of emitters. Consider the incentives of acid rain sufferers to reveal their willingness to pay when emitters have the rights to use clean air for waste disposal. If an individual sufferer purchases rights to clean air from one or more emitters, will he or she get an appreciable reduction in acid deposition? No, because this person is only one of millions downwind, and appreciable reductions depend on similar purchases by many others. So, acting in isolation, this sufferer is likely to save his or her money, and the values he or she attaches to reductions in acid rain will not be revealed in the market.

What if the sufferers initially have the rights to be free of acid rain? This will give any individual downwind the right to enjoin upwind emissions. So even if all sufferers but one have sold their rights to clean air to emitters, the remaining holdout can prevent all emissions, and consequently demand a very large share of power stations' and steel mills' gains from being able to use clean air for waste disposal. If most sufferers realise this, most will hold out for very high market prices, far higher than the true values they attach to clean air, and few market deals will occur.

The technical and behavioural impediments to, and the resulting costs of, market coordination over the acid rain problem are far greater in the multi-party case. Science and technology are now required to determine and measure not only the aggregate relationship between emissions and forest (or other) damage, but also all the huge number of pairwise relationships between each individual power station, steel mill etc. and each individual downwind sufferer. And even if this technological feat could be accomplished at acceptable cost, individual emitters and sufferers, acting in isolation, are unlikely to reveal their true willingness to pay or to accept compensation for more or less emissions. Thus there is no present possibility of a market solution to the acid rain problem when many parties are involved. Whether a market solution might be possible in the future depends on future changes in technology and resource values. The impacts of such changes on the costs of market coordination will be discussed in Chapter 6.

5.10 The scope and limitations of markets

The case of acid rain illustrates a number of points about the scope and limitations of market signalling and incentives in solving environmental problems. First, obscure cause-and-effect relationships (which prevent identification of causes of damage, of asset users and of possible trading partners) and difficulties in measuring asset use and its effects, are major impediments to both the specification of property rights and market negotiations. Second, scientific and technical advances lower the costs of specifying property rights and of market exchange by improving our knowledge of environmental causes and effects and our ability to measure transfers between people. Thus scientific and technological advances extend the ability of markets to solve environmental problems. Finally, market solutions to environmental problems are likely to be most costly, and therefore most improbable, when the number of parties involved is large, and the problem cannot be handled by splitting it into a large number of separate two-party deals between resource users and sufferers.

Difficulties in identifying causes and effects in environmental problems, and difficulties in measuring transfers, can occur whether the number of parties involved is two or a million. They pose no greater intrinsic problems for markets than for non-market signalling and incentive mechanisms; both have to somehow identify parties and measure transfers. On the other hand, large numbers of parties whose fortunes are inseparable pose special problems for markets. We saw in Chapter 3 that market signalling and incentives work by defining individual property rights and facilitating separate two-party deals between large numbers of individual consumers and producers. If two-party deals are ineffective because large numbers of people suffer the same collective fate regardless of their individual behaviour, as in the case of downwind sufferers from acid rain, the advantage of the market in forcing individuals to reveal their true valuations is lost. For a full understanding of the deficiencies of markets, we need to understand precisely why some benefits or costs of people's activities, such as acid rain, are consumed or borne collectively, rather than individually. Then we need to understand the consequences of collective consumption for the signalling of environmental costs, and for incentives to respond to those costs. This is the subject of the following chapter.

Discussion questions

1. Negative externalities such as air or noise pollution can only be overcome by defining property rights and creating a market for the pollutant concerned. True, false or uncertain? Explain why.
2. Explain as precisely as you can why no market exists to reconcile the interests of:
   - myself and my downhill neighbour who builds a house which blocks the view from my living room;
   - hikers who want rural vegetation preserved and farmers who want to clear vegetation for agricultural use.

   Hint: Can you identify well-specified private property rights in each case? Why not?
3. It is sometimes argued that government should subsidise energy-saving insulation of homes in order to save energy. Is this a case of an externality, justifying government intervention—in other words, is this a case where the market fails to fully account for people's wishes? Explain why or why not.
4. Car exhaust pollution, produced by hundreds of thousands of cars and commercial vehicles, harms human health and affects visibility in metropolitan Melbourne and Sydney. Explain precisely why private market deals to achieve the efficient amount of exhaust pollution would be prohibitively costly.

6

Limitations of market signalling and incentives: non-excludable goods

The prospect of being excluded from the enjoyment of a good is what motivates people to reveal true values and to pay up. However, not all goods and services are like oranges, where one person's consumption of a unit automatically excludes others. Where many people benefit from the same action, for example saving live whales, individuals have less incentive to reveal their values and pay up. Such goods and services have two crucial characteristics. First, unlike oranges, one person's enjoyment does not exclude others; the same live whales give pleasure to all. Second, it is too costly, relative to the values people attach to live whales, to exclude non-payers. It is the combination of these characteristics which leads to the need for multilateral (as opposed to bilateral) bargaining, and to breakdowns in market coordination of resource use.

6.1 Avocados and wild cranes

Recall the discussion of incentives in the avocado market, in Section 3.8. With perfectly defined and enforced property rights, the seller has exclusive rights to avocados prior to the sale, and the buyer exclusive rights after the sale. Others are excluded from using avocados; the only way they can legally enjoy avocados is to compensate a current owner for the transfer of his or her rights. The prospect of being excluded from the enjoyment of the avocados is what motivates people to reveal the value they attach to avocados and to pay up. (Theft—that is, the appropriation of avocados in violation of clearly defined property rights—is an obvious alternative. Returning to the subject of Section 2.1, theft threatens a society's stability by undermining the rules that are the basis for social coordination in the use of scarce assets. Our concern here is with the exclusion of those who do observe society's rules.)

Not all goods are like avocados. Suppose that I get a warm inner glow from knowledge of the successful captive breeding and release into wildlife refuges of an endangered species of crane. A representative of the organisation responsible for the program comes to my door selling sponsorships to fund the crane preservation project. Unlike the avocado producer, this producer of wild cranes does not hold property rights in wild cranes that I can purchase, thereby compensating it for some of its costs of providing cranes. If the organisation controlled access to refuges, my sponsorship could buy me crane viewing rights. But if my satisfaction comes from the mere knowledge that the wild crane population is rising, it is impossible for the organisation to exclude me from enjoying the results of their activities, whether I buy a sponsorship or not. And I cannot honestly believe that my $50 will make any difference to the number of cranes they release. So, if the representative catches me in a mean or frugal state of mind, I can readily justify a refusal to contribute on the basis that my sponsorship will make no difference, and I will enjoy the benefits of crane preservation regardless. Unlike the avocado case, I have no incentive to reveal that I value wild cranes or to pay up. This is what economists term *free riding*: *obtaining benefits without paying a corresponding share of the costs of obtaining those benefits*.

6.2 Non-rivalry and non-excludability

There are two important differences between avocados and crane preservation. The first concerns the relationships between consumers. Avocados are a *rival good*: one person's enjoyment of avocados reduces the amount available to others. Crane preservation is a *non-rival* or *collective good*: one person's happiness at knowing that the cranes are increasing in no way reduces the happiness of other crane lovers. Similarly, one person's enjoyment of additional crane viewing opportunities does not reduce viewing opportunities for others until the viewing sites begin to be congested.

Note that, unlike the case of avocados, as the number of people who value crane preservation increases their desires can be satisfied by the existing crane release program, since the same cranes (but not the same avocados) give pleasure to all. Crane preservation, and other non-rival goods, display what we might christen 'the magic pudding effect'; it is not necessary to sacrifice extra resources to satisfy the desires of extra consumers.

There are non-rival 'bads' as well. Acid rain and other forms of dispersed pollution harm many people, and the harm I suffer does not reduce the harm to you. In this case, production measures that reduce pollution produce a corresponding non-rival good, pollution control.

Non-rivalry is a fundamental physical characteristic of particular forms of production. It exists when many people can enjoy the benefits of the same units of production, because the effects are naturally dispersed in space and across people. It results from the spatial mobility of something—people, information, electromagnetic radiation, air, water or fauna. For example, the benefits of crane preservation are dispersed across people by the passage of information. The benefits of radio broadcasts are dispersed by electromagnetic radiation, those of air pollution control by air movements, those of plague locust control by insect movements, those of scenery preservation by the movements of people and so on.

For a particular form of production, the extent of dispersion in space and time affects the number of people involved in the non-rivalry. If smoke from a factory chimney is confined to a particular valley and lasts a few days, during those days all valley residents experience the same reductions in clean air and the same beautiful sunsets. If CFC emissions from a particular refrigeration plant quickly equilibrate across the whole stratosphere and continue to break down ozone for one hundred years, all residents of the higher latitudes during the next one hundred years experience the consequent reduction in protection from ultraviolet radiation.

The second important difference between avocados and crane preservation concerns the availability of economic means of excluding would-be beneficiaries who do not pay. Relatively few people want to produce or protect assets if the costs of excluding non-payers and non-contributors exceed the benefit to the producer or protector. Avocados are an *excludable good*: there are technical means (e.g. signs, locks, fences and surveillance cameras), backed by enforcement organisations such as police and courts, to exclude non-payers at costs less than the gains to the avocado owner from sale of the fruit. Crane preservation is a *non-excludable good* (except to the extent that people can be denied access to important feeding and nesting sites). There is no present technical means of excluding non-payers from information about the crane population, or from viewing cranes within most of their natural range, at costs comparable to the values people attach to cranes.

Recall, from Section 5.6, that property right holders can enforce their rights to control the use of their property in either or both of two ways: by physical exclusion and/or identification and penalising of unauthorised users. Either or both can achieve *exclusion of non-payers*, which is what is meant by excludability in this book. If these activities cost more than the relevant asset is worth to prospective owners, effective property rights are unlikely to be enforced, and the asset is non-excludable.

Non-excludability is economically determined. The costs of exclusion of non-payers, enforced by property right holders and the state, vary according to the costs of physical exclusion, monitoring of asset use and of enforcement organisations. Thus the costs of exclusion can be reduced by scientific and technological advances. Lower exclusion costs permit better definition and enforcement of property rights, and the consequent extension of market exchange. For example, the invention of barbed wire greatly reduced the costs of excluding other people's livestock from grazing land in Australia and the American West.[1] The introduction of devices that can scramble and decode television signals has created a market for pay television programming. On the other hand, declines in the value of assets can make exclusion uneconomic. For example, the ancestors of today's wild camels in central Australia were valuable private property; their value declined with the rise of alternative transport technologies.

There are a great many non-rival goods that are non-excludable because 'fencing out' beneficiaries is currently too costly, relative to the value of the good. We are familiar with appeals for funds from non-commercial private charities, conservation organisations and radio stations; in each case many beneficiaries of these activities choose not to pay—to free ride. Other familiar examples of non-rival and non-excludable goods include national defence, police protection and suburban parks.

Most important negative externalities (e.g. acid rain and the effects on nature lovers of commercial whaling and rainforest clearing) are also non-rival and non-excludable 'bads'. Production measures that reduce these 'bads', such as installation of emission control equipment, reductions in whaling catches and selective logging, are non-rival and non-excludable goods. In these cases, unlike crane preservation, private production of the non-rival and non-excludable good involves direct trade-offs between two or more products, some of which have commercial markets (e.g. electricity, steel, whale meat, whale oil, wood pulp and timber), and some of which have no market or only rudimentary markets (e.g. cleaner air, live whales, preserved flora and fauna and ecosystems).

Economists' standard term for goods that are both non-rival and non-excludable, such as crane preservation, is 'public goods'. This term is misleading, because non-rival and non-excludable goods are not necessarily provided by government. *Non-rival and non-excludable goods* is admittedly a bit of a mouthful, but we will continue to use it here because it identifies the characteristics of goods that lead to failures of market coordination.

6.3 Is non-excludability dependent on non-rivalry?

Non-excludability rules out exclusive property rights. Without such rights there can be no markets, and no market coordination of production and consumption of non-excludable goods, as in the case of crane preservation. Is this always ultimately due to the natural characteristics of non-rival goods? Do rivalry and excludability, and non-rivalry and non-excludability, always go together, as is the case for avocados and crane preservation?

The short answer is no to both questions, but it turns out that all important environmental problems do involve some form of non-rivalry leading to non-excludability and the absence of markets. The reasons for this are explained below and in Chapter 7.

We begin with examples of non-rival goods that are excludable. These are non-rival goods that are sufficiently valuable to justify their 'fencing off' and/or monitoring of use and penalising of unauthorised users. Familiar examples include performances such as rock concerts and sporting events, where fans can be excluded from a facility at relatively low cost, private wildlife parks, and pay-per-viewing television. Exclusive property rights and markets in these goods exist, giving buyers and sellers incentives to reveal information about the values attached to the rights.

In the reverse case, a good may be rival but not valuable enough for exclusion to be worthwhile. For example, passers-by may pick fruit from my trees along the street, but I judge that the costs of erecting a fence or taking them to court to enforce my legal rights would exceed my benefits from the fruit saved. However, such cases are rare and trivial, because few people will incur the costs of producing goods to which they have no effective property rights, because of the costs of enforcing exclusion.

Cases where rival but non-excludable goods are valuable natural resources, not produced by anyone, are much more important. Many mobile natural resources, such as fish in large bodies of water, birds, groundwater and oil, are rival because one person's use of them reduces the amount available to other consumers or producers, but are non-excludable because it is too costly to fence them off or to monitor use and penalise unauthorised users. In the absence of effective exclusion, there are no effective property rights to such resources until someone captures the fish or the bird or pumps the water or the oil. Sometimes no one claims exclusive rights to such resources, as is the case for resources in most of the world's oceans. Often, as with ocean resources within the 320 kilometre exclusive economic zone asserted by most countries, legal claims are made but cannot be effectively enforced. Such resources are frequently described as *open access* or *common pool resources*; open access because of the absence of exclusive rights, and common pool because of shared access to the system containing the resources—a lake or the air or an underground reservoir.

Here we use the term open access to describe a wider range of circumstances than the term common pool resources. Open access accurately describes any goods or natural resources, rival or non-rival, mobile or immobile, that are not the subject of exclusive rights because exclusion is too costly relative to asset value. This is the case for resources of little value, such as streetside fruit, and for resources situated in hostile, and therefore high-cost, environments. For example, with current technology the net returns from mining coal deposits in Antarctica are so low (negative) that it is not worth incurring the costs of defining and enforcing exclusive property rights.

Common pool resources are a category of open access resources, which due to the mobility of resources (e.g. fish, liquids, gases) or resource components (e.g. plant seeds) exhibit *both rivalry and non-rivalry* among resource users. Each common pool resource is part of a *resource system* that includes a stock of individual units of the resource plus their supporting environment.[2] Examples include the stock of tuna plus the marine ecosystems on which they depend, a stock of prawns plus the coastal waters and seabed, a species of bird plus the forest in which they forage, nest and breed, and a groundwater pool plus the geological strata that allow recharge and enclose the pool.

How is use of common pool resources both rival and non-rival? First, individual units of common pool resources are rival and excludable in use (i.e. when they are withdrawn from the pool). Second, because common pool resources or resource components are mobile, their capture or augmentation affects productivity throughout the resource system. Actions at any point within the system have non-rival and non-excludable impacts throughout, and hence impact on all other users of the resource. For example, the size of one fisher's catch reduces the number of fish in the pool, and hence the degree of success of other fishers. Conversely, if one fisher undertakes measures to increase fishery productivity—for example, if a scallop fisher releases scallop larvae—he or she provides non-rival benefits for all fishers, assuming that the larvae or the scallops move freely throughout the common pool. Similarly, the amount of oil pumped by the owner of a single well tapping a common oil reservoir affects subsurface oil pressure throughout the oil pool, and hence the productivity of all wells tapping the same reservoir. Thus the management of unharvested common pool resources, as distinct from the ultimate use of resources from a common pool, involves both non-rivalry and non-excludability equivalent to the case of crane preservation. The nature of common pool resources, and the deficiencies of market signalling and incentives in the production of common pool resources, are discussed at length in Chapters 7 and 18.

6.4 Why are few non-rival and non-excludable goods provided privately?

Private producers of non-rival and non-excludable goods must be compensated for the sacrifices involved, unless they do so for the love of it or out of a sense of civic duty. We saw in the case of crane preservation that non-excludability allows individual beneficiaries to avoid paying—to free ride. Consider the consequences of free riding for private market production of such goods. Suppose that a lover of classical music wishes to provide radio broadcasts free of advertising, but requires revenue to finance a transmitter, studio and staff. Many classical music lovers will benefit from such broadcasts, but many may choose to free ride when asked to subscribe. If subscriptions fail to cover the musical entrepreneur's costs, either the station will fail to get off the ground or transmission time will be severely limited. As a result, there may be little or no market provision of advertisement-free classical music broadcasts.

Non-excludability is a necessary condition for free riding, but it is not the sole requirement. Recall the isolated forest owner of Section 5.8, suffering the costs of acid rain. Control of the sulphur dioxide emissions of the adjacent power station is a non-rival and non-excludable good, but since the forest owner is the sole recipient of the resulting acid rain, he or she recognises that no one else will act to reduce emissions. *It is impossible to free ride when you are the sole beneficiary of a non-excludable good.*

Returning to crane preservation, let us examine the free rider's economic reasoning more closely. As implied in Section 6.1, the individual recognises that he or she is one of a large number of crane lovers. In deciding whether or not to contribute, the individual presumably compares anticipated marginal benefit from purchasing a $50 sponsorship with the marginal cost of the sacrifice. A formal version of the reasoning given earlier would go:

1 Regardless of the number of crane lovers who contribute, my $50 will make no difference to the number of cranes released. So the (marginal) benefit I expect from my contribution is zero.
2 I can think of many valuable uses for my $50. Thus the (marginal) cost of my contribution is positive.
3 Since the cost of contributing exceeds the benefit I expect, it is best not to contribute.

If all crane lovers reason similarly, there will be no contributions, and hence no preservation, despite the fact that large numbers of people do value cranes in the wild.

The situation is not always as bleak as this, for two reasons. First, sometimes a single individual or organisation has a very high willingness to pay for a particular non-rival and non-excludable good. The single supplier may then provide enough of the good to substantially benefit non-payers. For example, very wealthy individuals and companies sometimes fund major non-commercial wildlife preservation programs.

Second, some people may preserve cranes, save whales, broadcast classical music or control their pollutant emissions because they like to or because they firmly believe that it is the right thing to do, whether financially compensated or not. Some people who benefit from such activities are likely to contribute voluntarily for the same reasons. Also,

Figure 6.1 The demand for crane releases

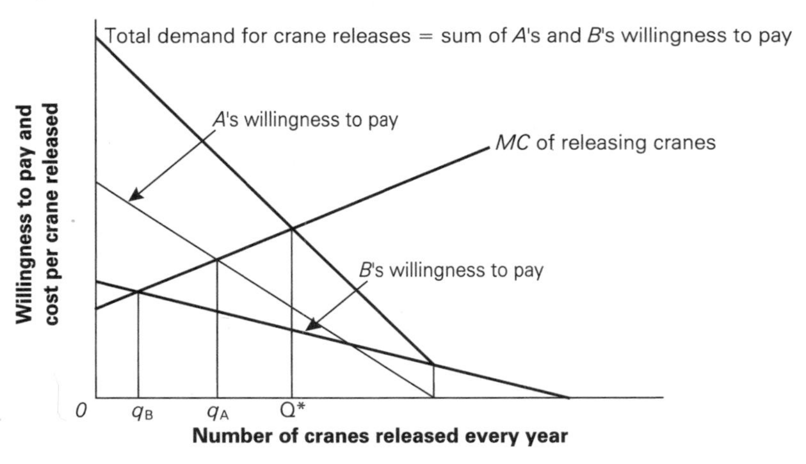

if people observe others contributing to a non-excludable activity, their sense of fairness may prompt them to contribute more. Thus we do observe organisations whose main mission is to produce non-rival and non-excludable goods surviving on subscriptions and donations. Examples include charities, international relief organisations, international conservation organisations such as Greenpeace and The Nature Conservancy, and local volunteer organisations such as community radio stations and fire brigades.

Notwithstanding such voluntary efforts of both producers and consumers of non-rival and non-excludable goods, in normal circumstances we do not expect to get as much crane preservation or broadcast hours or clean air as we would if the beneficiaries could be excluded and therefore forced to pay for these goods. Consider the incentives of the beneficiaries and the prospects for revealing their true willingness to pay to preserve cranes in the highly simplified situation depicted in Figure 6.1. Assume that cranes are raised and released by a private organisation that truthfully reveals its costs of doing so, and hence the compensation required. Its marginal costs of releasing additional cranes are represented by the *MC* curve. There are only two beneficiaries of the releases, *A* and *B*; their respective demand or marginal benefit curves are also shown. These curves show the maximum amounts that *A* and *B* are willing to pay for each additional crane released. Since crane preservation is non-rival, *A*'s enjoyment from the knowledge that additional cranes have been released this year in no way detracts from *B*'s enjoyment of the same preservation activity, and vice versa. Thus the total demand or willingness to pay for each additional crane released is the *vertical* sum of *A*'s and *B*'s separate individual demands for the *same crane*.

If all parties reveal their true valuations market exchange would reveal the *MC* curve and the individual and total demand curves depicted in Figure 6.1, and the net benefits from preserving cranes would be maximised by releasing Q* cranes per year, where the total value *A* and *B* attach to the last crane released is just equal to its cost to the organisation releasing it. But will *A* and *B* contribute this much? They might, if they knew one another, in which case they could negotiate a cost-sharing deal that recognises that each is dependent on the other. Also, if *A* observes *B* contributing, *A* may also contribute, or contribute more, out of a sense of fairness. But if *A* and *B* do not know one another, and would find it far too costly to identify one another and negotiate a cost-sharing deal, as is probably the case for most of the world's crane lovers, they will make their decisions in isolation. Note that, if *A* were the only person benefiting, *A* would pay for, at most, q_A cranes. If *B* were the only person benefiting, *B* would pay for, at most, q_B cranes. But in the absence of exclusion, each automatically gets the benefit of whatever preservation is paid for by the other. Thus, when each realises that another (unknown) beneficiary exists, a likely reaction, operating in isolation, is to reduce contributions. In fact, if *B* suspects that *A*, in isolation, would pay for q_A, which is more than *B* would pay for in isolation, *B* may assert that he or she is uninterested in cranes and refuse to pay anything. Thus, negotiation about cost sharing aside, and assuming that *A* and *B* do not get pleasure from altruistic contributions, the number of cranes released will be at most q_A, and very likely less.

Even if *A* and *B* do know one another, negotiation over sharing the costs of crane releases will be difficult if each acts selfishly. Neither knows the other's true willingness to pay, and neither can stop the other benefiting from her or his contributions, as is the case for excludable goods. (If you want my avocado, my eating the avocado will definitely exclude you, and you have no choice but to reveal some willingness to pay for the avocado.) Negotiation over cost sharing will be much more difficult for ten people, let alone ten thousand or ten million.

6.5 Decentralised production of non-rival and non-excludable goods

In the cases of crane preservation and classical music broadcasts, the non-rival and non-excludable good is the deliberate product of a single organisation. It is not difficult to identify the producer and measure output, only to arrange compensation and divide the costs among the beneficiaries. However, many non-rival and non-excludable goods and 'bads' represent the accumulation of the output of many separate individuals and/or organisations, each producing a product that is indistinguishable to beneficiaries or sufferers. Examples include individual car owners and factory owners who install emission controls, individual irrigators who reduce water use and therefore reduce downstream salinity levels, and homeowners whose renovations improve streetscapes. Then we have the additional problem of identifying the producers who produce goods or act to reduce 'bads', and measuring their contributions, before individual producers can be compensated.

Sometimes it is possible to identify the separate producers and to measure their contributions at an acceptable cost. For example, it may be economic to monitor each factory's emissions of the sulphur dioxide and nitrogen oxides contributing to acid rain; aerial photography may permit measurement of the contributions of particular land holders to forest damage and rehabilitation. Sometimes identifying and measuring contributions will be prohibitively costly. For example, the fertiliser applications and detergent use of farmers and households in Victoria's Yarra Valley all contribute to algal pollution of the Yarra River, but identification of their separate contributions to this 'non-point pollution' problem may be a scientific as well as an economic impossibility. Another case where identification of separate contributions is prohibitively costly with current technology is motor vehicle exhaust pollution in metropolitan areas.

When the separate contributions of the producers of a non-rival and non-excludable good cannot be identified, high identification and measurement costs, as well as free riding by consumers, prevent market provision of the good. This is the case for acid rain, discussed in Chapter 5. Suppose that there is no free riding because there is only one beneficiary of the non-rival and non-excludable good. Market compensation of producers of the good may still be too costly due to the high costs of identifying the contributions of separate producers.

Imprecise measurement of individual contributions may promote free riding on the part of the producers of a non-rival and non-excludable good. For example, suppose that a single forest owner who receives acid rain from neighbouring industrial plants offers to compensate the plants for making the air cleaner without being able to measure accurately their separate contributions. Each emitter will claim compensation, but rely as far as possible on other emitters to reduce total emissions. In behaving thus, the producers of non-rival and non-excludable goods are in effect acting as free riders. They are trying to obtain compensation benefits while avoiding the additional pollution control costs that reduce acid rain and thus lead to compensation payments.

The notion of a single pollution sufferer facing many producers of a non-rival and non-excludable 'bad', such as acid rain, and being willing to pay compensation for pollution control without accurate measurement of contributions does seem far-fetched. However, what if the single sufferer is a government, representing individual people, and under strong political pressure to provide a quick solution? Then the possibility of some form of compensation to politically powerful industries, without adequate monitoring, does not seem so far-fetched. Such issues will be discussed in Chapter 8.

6.6 Why it is rational to pollute oneself?

In pollution problems, the same people are often both sufferers from and producers of the same non-rival and non-excludable 'bad', and hence both beneficiaries and producers of the corresponding good. Consider your likely behaviour as owner of an old cheap car. Motor vehicle emissions of carbon monoxide and other gases are major contributors to air pollution in metropolitan areas. Mandatory installation of emissions control equipment on all new vehicles reduces these emissions, but not for older or poorly maintained vehicles, which are the most important individual contributors to the problem. If you installed emissions control equipment in your old car, you would be providing a non-rival and non-excludable good to yourself and all other residents of your metropolitan area. Presuming that you will not do so out of sheer altruism or a sense of civic duty, and that you do not expect your actions to influence others, your reasoning might go:

1. My car is one of half a million vehicles in the city (or, if friends remind me that older vehicles like mine produce most of the emissions, one of 150 000). Therefore my efforts to control my car's emissions will make no discernible difference to my breathing or to visibility where I live or work. Also, since my car's contribution to total air pollution levels is not identified, I will not suffer any penalty for not acting or obtain any reward for acting. So the benefit I expect, either as a sufferer from pollution or as a controller of pollution, is zero.
2. The cost to me of modifications to the car is hundreds of dollars, far more than zero.
3. I am better off doing nothing, at least nothing more than keeping the car properly tuned, which benefits me in other ways, such as savings in fuel costs.

The prospect is that many owners of older vehicles will reason similarly, and elderly vehicles will continue to be major contributors to metropolitan air pollution, despite the high costs of the pollution to residents. In this case, from the consumer perspective, you choose to free ride by not making a contribution (in kind) to pollution control. From the producer perspective, you choose not to produce a non-rival and non-excludable good, clean(er) air, because there is no market or other reward for its production.

Costly as motor vehicle pollution in cities may be, our individual willingness to produce the same pollutants we suffer from can be put in economic perspective by recalling the prohibitive costs of exclusive individual rights in clean air. Unfortunately we do not each carry around our own exclusive volume of air; if we did, most of us would quickly adjust our car emissions into *our* air to balance car costs against pollution costs. As discussed earlier, with present technology, the costs of defining, enforcing and exchanging individual rights to clean air far exceed the costs of the pollution caused by individuals.

The situation can be quite different for large polluters. In March 1993, the US Environmental Protection Agency sponsored the first auction of permits to emit sulphur dioxide, after allocating 5.7 million permits to the 110 worst polluting industries. Where a government agency defines and enforces property rights to discharge wastes into the air and the volumes traded are large, the gains from market exchange of rights to cleaner air may exceed the costs of market exchange.

Discussion questions

1. Once a non-excludable good, always a non-excludable good. True, false or uncertain? Explain why.
2. Many visitors to public parks fail to clean up their litter when they leave. This indicates that they would vote to abolish the litter-control regulations which apply in the parks. True, false or uncertain? Explain why.
3. Which is more important for the existence of markets—non-rivalry or non-excludability? Explain why.
4. Asia's remaining wild tigers are disappearing due to clearing of habitat and poaching to supply the black market in tiger parts, used in traditional Asian medicines. How do the concepts of non-excludability and free riding help to explain why the market economic system fails to preserve the dwindling stock of wild tigers, despite their value to tourists and lovers of wildlife the world over?

7

Limitations of market signalling and incentives: common pool resources

Valuable natural resources which are accessible to anyone with the equipment to harvest them are typically subject to 'finders keepers'. Where exclusion of unauthorised users is too costly, as in the case of ocean fish or groundwater in large underground basins, there is a rush to acquire rights to the resource by harvesting it, and usually neglect of the overall productivity of the resource system—the fish stock or underground basin. With no rights to the unharvested resource, users are under the constant threat of 'use it now or lose it'. This 'tragedy of the commons' can only be overcome by either fencing off individual parcels or by the imposition of rules for sharing the harvest.

7.1 Milkshakes as common pool resources

Common pool resources have been likened to a large milkshake shared between several children, each drinking from his or her own straw. The consequences for the rate of consumption of the milkshake are obvious; it will disappear very quickly, so quickly that each child is unlikely to enjoy it nearly as much as if he or she had been given the same amount in a separate glass, and was able to drink at his or her preferred pace. In similar fashion, if loggers, fishers, groundwater users and oil producers cannot be excluded from a common pool, they all have incentives to harvest resources as quickly as possible, before others do. However, with natural resources the consequences of rapid harvesting are typically more serious. A competitive rush to harvest involves greater input costs than a leisurely harvest. Precipitate harvesting of living resources will change the age distribution and may reduce reproduction rates; if harvesting is cheap enough, it may even lead to extinction of the resource, as with America's passenger pigeon. Rapid pumping from underground reservoirs, without careful attention to subsurface pressure changes, can seriously reduce groundwater recharge and oil extraction possibilities.

7.2 Common pool property rights and incentives

What motivates the rush to harvest and the neglect of the overall productivity of the common pool? Consider the effective property rights of the children and of the fishers and oil producers. Until they capture the milkshake or the fish or the oil, they have no property rights in it; any left by one harvester is available to others. Once captured, the resource is their exclusive property. Consequently the individual fisher or oil producer has strong incentives to maximise his or her net gains from the immediate capture of fish or oil. This means fishing or pumping up to the point where the cost of the extra fishing or pumping just equals the value of the extra fish caught or oil pumped.

On the other hand, acting alone, the fisher or oil producer expects little or no gain from actions to maintain or augment the stock remaining in the common pool (e.g. throwing back young fish or pumping slowly to maintain subsurface pressure). This is because he or she has no rights to any part of the stock so preserved. (Compare the situations of a farmer and a fisher asking their banker for a loan, the farmer to purchase young breeding animals, and the fisher to release young fish into the common pool. Which proposal offers the banker the more secure basis for loan repayment?) Yet the individual's fishing or pumping behaviour affects the fishing success or pumping costs of others harvesting the common pool. Thus individual users create externalities; they impose costs on other users, costs that they themselves do not bear in the absence of market or other feedback about the impacts of fishing and pumping on productivity, and that do not influence their harvesting behaviour.

In Chapter 5 we explained that externalities result from an absence of property rights and markets. In common pool problems, the markets that are missing are markets for the *unharvested* resource. Compare fish in the ocean or oil in a large oil reservoir with cattle on farms or coal underground. The owner of cattle or coal is not under constant threat of 'use it now or lose it', because exclusive property rights allow him or her to choose between harvesting now or saving the resource for another day when the cattle will have multiplied or product prices are expected to be higher or harvesting costs lower. Suppose that another person disagrees with the owner's choice of how many cattle or how much coal to harvest now versus in the future, because that person expects the resource to be much more valuable in the future. The second person can bid in the market to take control of the resource, in order to preserve it for the future. This signals the value information to the resource owner. There are no such markets, and no such direct signals of future values, for common pool resources.

If an individual is the sole owner of a resource pool, with power to exclude, the externality disappears. Suppose that you are the sole owner of a lake stocked with trout, able to exclude anglers by fences and by patrolling the shoreline. Because the number of anglers you admit will affect the success of each, now and in the future, it will affect the fees that you can charge today and in the future. Therefore you will take account of the impacts of additional anglers upon the successes of other anglers when deciding the numbers that you admit.

A milkshake is excludable at trivial cost; we merely have to divide it among several containers and police consumption to give each child exclusive property rights. Common pool resources are non-excludable because, with current technology and input costs, it is too costly, relative to the expected benefits from exclusive use and exchange, to define and enforce exclusive rights to resources in separate sections of the pool. For example, despite the value of Victoria's scallops, the costs of legally defining private 'scallop farms', of fencing them or patrolling 'farm boundaries', and of identifying and prosecuting trespassers, are apparently higher than the value of the additional scallop harvests that might result. But this is not the case for New South Wales oysters, or Tasmanian salmon. Property rights in these kinds of fish farms are well specified.

It is possible that scallop fishers, a relatively small and easily identified group, may informally agree to define and enforce 'property boundaries' within the common pool, so that each fisher does bear some of the consequences of his or her fishing practice. Such community-based arrangements are discussed in Section 7.6 and Chapter 18. It is also possible that changes in technology and in the values of scallops and other resources may make private farms feasible in the future.

7.3 The existence and scale of common pool problems

As with non-excludability in general, what is a common pool resource now may become private property in the future or vice versa, depending on changes in natural resource values, exclusion technology and the costs of inputs used in fencing and monitoring. The bison herds of the American Great Plains were common pool resources before barbed wire, but today's herds can be fenced in. Northern Australia's crocodiles used to be a hunted common pool resource, but protection of wild crocodiles, and increasing values of the skins and meat, have made crocodile farming profitable in recent years.

As explained in Section 6.3, naturally-occurring common pool resource systems include a stock of individual units of the resource plus their supporting environment. The geographic extent of such systems ranges from global to local. The stratospheric ozone layer is a global common pool resource. Ocean fisheries such as the southern bluefin tuna fishery are subglobal but very extensive. At the other end of the scale, valuable non-excludable resources may be non-migratory or confined to a small area, for example abalone, scallops in Port Philip Bay, local oil and groundwater pools, and local pastures and forests which remain in common use by community consent.

What determines the existence and geographic extent of a common pool resource system? Partly nature, partly technology and economics, and partly social attitudes. The spatial mobility of mobile resources (such as atmospheric gases, fish, birds and groundwater) is in the first instance naturally determined. Common access to a resource system is also partly determined by the technologies and costs of fencing out would-be users or monitoring and penalising illegitimate use. Finally, common access may persist even if exclusion is feasible. Communities may choose to maintain common access for community members on grounds of morality, equity or economics. This was the case in the medieval commons, and is in Swiss alpine commons still. For example, it may be technically and economically possible to give a single organisation the exclusive right to harvest scallops from Port Philip Bay. This should alleviate the problems of overharvesting of scallops and associated damage to bay ecology, but it may be socially unacceptable to grant a sole right to this natural resource. Alternatively, common access for community members will be maintained if the costs of splitting the resource system into individual holdings and excluding individual community members (including any output sacrificed because community members no longer work together in grazing, hunting, fishing and so on) are high relative to the costs of excluding or monitoring and penalising outsiders.

Many valuable natural resources remain subject to open access for social and political reasons, despite exclusion being economically feasible. Australia's ocean beaches could be fenced by private owners, as are some beaches in California. As in the case of Kenya's and Tanzania's elephants discussed in Section 2.3, government decisions can make natural resources effectively open access by lowering local people's benefits from resource use and/or increasing the costs of fencing out or monitoring and penalising intruders. Many forests in Third World countries are effectively open access resources for these reasons. The forests are legally the property of national governments, but excluding or monitoring forest access and use by local people or squatters or illegal loggers and hunters costs far more than governments can afford. Local inhabitants, who would have the lowest costs of monitoring forest use, have few property rights in, and hence little incentive to maintain, forest resources. The result is continuing deforestation or forest degradation, as forest users extract what they can in the short run, ignoring the long-run productivity of forests.[1]

7.4 Market signals and overharvesting of common pool resources

How does harvesting of a common pool resource, where markets fail to signal the productivity impacts of harvesting, differ from harvesting of the same resources by a single owner? A simple example involves individual herders' decisions about the number of cows grazed on a common village pasture to which all herders have free access. Suppose that the village contains one hundred or so herders, none of whom could ever afford to graze more than a few cows, so that each individual's herd is tiny compared to the total village herd. Assume that all cows are identical and cost the same amount (the costs of raising or purchasing the cow, plus herding costs), that all herders have identical management skills and that the output of the total herd is measured in terms of milk, which sells at a constant price. Now consider the effects of altering the long-term intensity of grazing, measured by the number of cows continuously grazing the fixed area of pasture.

Figure 7.1(a) shows the relationship between the number of cows grazing and the total returns from milk production and the total costs of the herd. As the number of cows increases from zero, total milk production first increases at a diminishing rate, as cows begin to compete for pasture. Production reaches a maximum as grazing becomes more intense and subsequently declines as cows are increasingly stressed and milk production per cow falls enough to outweigh the increasing numbers of cows grazed. (Beyond some number of cows, cows will begin to die and measuring the number of cows continuously grazing becomes problematic, but we are not concerned with these grazing intensities.) Since all milk brings the same price, the curve describing the relationship between total milk production and the number of cows grazed will look just the same as the curve that shows how total return from the herd changes as the number of cows grazed changes.

Because the cost of adding an additional cow is the same for all herders and cows, the total cost of a herd of any size is depicted as an upward-sloping straight line beginning at the origin in Figure 7.1(a).

If both cows and pasture were owned by a single milk producer, like the trout owner of Section 7.2, and that person's aim was to maximise net returns from the production and sale of milk, the number of cows continuously grazed would be Q*. With Q* cows grazing, Figure 7.1(a) shows total returns from the herd exceeding total costs by the maximum amount possible.

Now shift to Figure 7.1(b). The horizontal axis still measures the number of cows grazing the common pasture; the vertical axis now measures returns and cost *per cow*. If we divide the total milk returns shown by the curve in Figure 7.1(a) by the number of cows producing those returns, we get the average return per cow shown in Figure 7.1(b). For convenience, the total return curve is drawn so that the average return curve is a straight line. The additional or marginal return from each additional cow added to the herd continuously grazing the pasture,

Figure 7.1 Determination of use of an open access resource

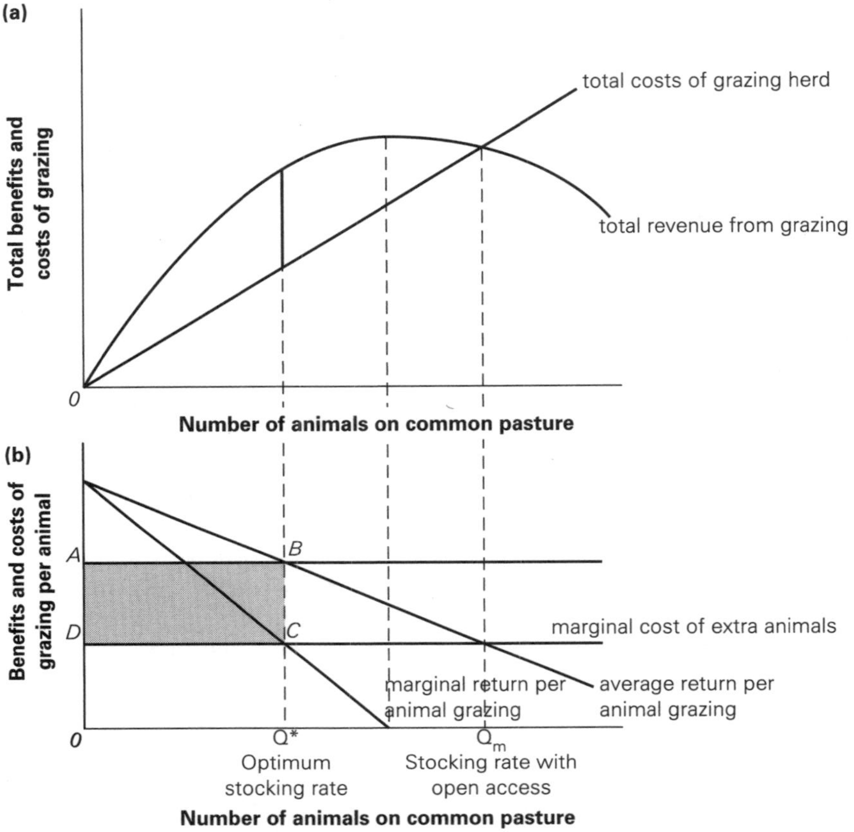

(a)

Total benefits and costs of grazing

total costs of grazing herd

total revenue from grazing

0

Number of animals on common pasture

(b)

Benefits and costs of grazing per animal

A

B

D

C

marginal cost of extra animals

marginal return per animal grazing

average return per animal grazing

0

Q*

Optimum stocking rate

Q_m

Stocking rate with open access

Number of animals on common pasture

equivalent to the slope of the total returns curve in Figure 7.1(a), is also a straight line.

Since all cows cost the same, the average cost per cow grazed, and the marginal cost due to each additional cow, are identical and shown by the same horizontal line in Figure 7.1(b).

In the per-cow terms of Figure 7.1(b), our sole owner will add cows to the pasture up to the stocking rate Q*, where the marginal return from the last additional cow grazed is just equal to the marginal costs of adding that cow, at point C in Figure 7.1(b). The owner's net return from grazing Q* cows is shown in Figure 7.1(b) in two ways: as the area below the marginal revenue curve and above the marginal cost curve to the left of their intersection at C or alternatively as the average profit per cow, *AD* ,multiplied by the number of cows, Q*—the area *ABCD*.

Now consider the situation when individual herders act independently, and the pasture is a common pool open to all. In deciding whether to add an additional cow, the individual herder will note the condition of the pasture as it is currently stocked, and compare the benefits and costs *to him or her* of adding the cow. What are these benefits and costs as seen by the individual herder? We take them in turn.

1 Assuming that the herder judges that his or her cow is as good as other cows, the herder will estimate that the (marginal) return from the additional cow will be the same as the *average return observed from the cows already in the herd.*
2 The herder's (marginal) cost of the additional cow is the common marginal cost for all.

The herder *does not* consider another cost of adding the extra cow. With pasture production fixed, the herder's cow will take some pasture from each of the hundreds of cows already in the herd, slightly reducing each cow's milk output. The herder may take this into account if he or she has one or two or three cows in the existing herd, increasing the marginal cost estimate slightly to reflect the impact of the additional cow on their production. But since the herder owns only about one-hundredth of the total herd, this adjustment will be small. What the herder does not take into account, because the other herders have no ability or right to exclude the extra cow, and therefore cannot signal their costs to the herder, is the reduction in their milk returns caused by his extra cow.

If all herders reason similarly, and all base their calculations on the same estimates of average returns and marginal costs, additional cows will be added to the pasture as long as the expected average return per cow exceeds the marginal cost of a cow. For example, suppose that the common pasture is grazed by Q^* cows, the number that would be chosen by a single owner. A herder considering the addition of another cow will anticipate a return from that cow equal to the current herd average, measured by the vertical distance Q^*B in Figure 7.1(b). The herder's corresponding cost will be Q^*C. Therefore the cow will be added. The same logic applies for each additional cow added as the stocking rate increases above Q^*. The process will stop at Q_m, where pasture availability per cow has fallen to the point that *every* cow in the herd, including the last cow added, produces just enough milk to cover its costs. So *the result of unrestricted entry to the common pasture is zero profits for all herders*.

At the stocking rate Q^*, the true additional return from the extra cow, after taking into account the slight reduction in average returns from all other cows on the pasture, is given by the marginal curve, not the average curve. Thus the true return to an extra cow at Q^* is Q^*C in Figure 7.1(b). The individual herder, acting alone, ignores the costs imposed on other pasture users, measured by the distance BC in figure 7.1(b). BC is an externality that is not signalled due to the absence of a market for the unharvested pasture.

7.5 Overcrowding of beaches as an overharvesting problem

Our analysis of overharvesting also throws light on the problem of frequent overcrowding or congestion of beaches, roads, parks and other public sites and facilities. Such sites and facilities are commonly open to all because it is either illegal or too costly to exclude people. Take the case of people coming to a popular beach in summer. In that case the horizontal axis in Figure 7.1(b) measures the number of people spending the afternoon on the beach, and the vertical axis measures the resulting benefits and costs per person. The average return curve is now the average benefit per person on the beach (measured in terms of willingness to pay), and the marginal return curve measures the effect of an additional person on the total satisfaction of all beachgoers. The curves assume that all people have identical preferences and all dislike crowding.

Following the same reasoning as for the common pasture, when a person considers whether or not to join those already on the beach, he or she expects to experience the level of crowding he or she observes. Thus this person compares the average benefit from the observed situation with the cost of joining in. He or she ignores the slight loss of satisfaction that his or her presence will impose on every other beach user because of greater crowding. If all beachgoers reason similarly, people will keep coming until the average benefit to everyone is driven down to equality with their marginal cost of joining the throng (which may be near zero). The result of such behaviour, if all beachgoers have similar preferences and dislike crowding, will be a very crowded beach where everybody gets zero net benefit from their beach experience. In other words, everybody would just as soon be somewhere else. The absence of a market for beach space results in overcrowding.

A private beach owner with the power to exclude people would never permit a situation where everybody would just as soon be somewhere else, because then no one would be willing to pay anything to use the beach. By charging beachgoers an entry fee that discouraged some people and reduced crowding, the owner would increase the average benefits of those who were willing to pay the entry fee. The owner would get revenue, and those beachgoers who pay to enter would be no worse off, otherwise they would not be willing to pay for access. If the owner was a good judge of people's willingness to pay to avoid crowding, he or she would eliminate beach congestion externalities by providing people with accurate signals about the costs of their presence for other beachgoers.

Common pool problems, as defined in Section 7.3, do not require direct interaction between people. My fishing affects your catch because we draw on the same resource system, not because our nets tangle or because I scare away the fish that you are trying to catch. Congestion problems involve direct psychological or physical interactions. For example, my arrival at the beach reduces your enjoyment of the view and disturbs your peace and quiet; my entry to the crowded freeway further slows your journey to work. And since what is true for you is also true for other beachgoers and other drivers, my use of the site or facility imposes non-rival harm on all other users.

7.6 Local solutions: common property arrangements

Our previous example of herders sharing a common pasture was simplistic in many ways, none more than its assumption that the villagers were incapable of getting together and devising rules to reduce the overgrazing problem. After all, groups of neighbours have been sharing common resource pools—hunted animals, grazing land, lake and coastal fisheries, irrigation water—throughout human history. The same groups have been banding together to produce non-rival and non-excludable goods, for example mutual defence, drainage, dams and flood barriers.

The fire protection provided by local volunteer fire brigades in rural Australia is an example of a non-rival and non-excludable good produced by individual contributions, which are policed (more or less) by cultural norms and pressure from within relatively close-knit communities. The traditional rule is that each farmer should have a firefighting unit available to fight fires in the locality, and be on call when a fire starts.

The success of such community arrangements depends on characteristics of both the asset and the community. The local community concerned must get all or most of the benefits of actions to augment the common pool, or of the non-rival and non-excludable good produced. This means that non-rivalry must be restricted to a small area that the community can police, and the community must be able to exclude outsiders at low cost. Thus such arrangements may work for small lakes, and for marine species that do not move about much, such as crayfish, but not for migratory marine species such as whales and tuna.

Where communities are able to define and police exclusive rights of access for community members to common pool resources, for example a local fishery, the resources are appropriately termed *common property resources*. Unfortunately the term 'common property resources' is also used to refer to common pool resources where no exclusive rights exist. Readers should be alert for this misleading terminology.

What social circumstances are likely to favour the creation of common property rights? The community must be small enough so that all members can participate in joint decision making. People's preferences and production alternatives must be sufficiently similar so that they broadly agree on the use of common resources, or on the non-rival and non-excludable good to be produced. If people want to graze different types of animals, or use different methods of irrigation or fishing techniques, it will be more difficult to get agreed rules. The community must be stable enough so that all have a long-term stake in adherence to the rules. Since community members police one another, monitoring and enforcement of rules depends on close contact between community members at work and at home, so that rule violators will be easily observed and community sanctions can be applied at low cost.

We are familiar with the use of common property arrangements to resolve local environmental problems in private recreational clubs, such as golf clubs and ski clubs. Developers of exclusive housing estates in Melbourne and Sydney, and resort estates in Queensland, often provide common property resources, such as golf courses and recreational centres, for the use of property owners. To protect the common property, and thus increase home prices, home buyers must agree to observe a set of rules devised by the developer.

Can common property arrangements assist in reducing more important environmental problems in modern democracies? Possible uses of common property arrangements to resolve common pool problems are discussed in Chapter 18.

7.7 Non-rival and non-excludable goods, common pool resources and environmental problems

Major environmental problems that resist market solutions all involve both non-rivalry and non-excludability. Non-rivalry because major environmental problems are those where the consequences of the same production or consumption actions are dispersed across many people—wildlife lovers, air pollution sufferers, fishers, and so on. Because many are affected by the same action, non-rivalry requires multilateral coordination between producers and consumers, unless exclusion costs are low enough to enable producers to exclude non-payers. If a non-rival good is non-excludable, ruling out exclusive property rights and market deals, market coordination of production and consumption of that good is practically impossible.

As explained in Chapter 3, markets achieve social coordination by enabling bilateral exchanges of exclusively held property. Only in these conditions do buyers and sellers have incentives to reveal their identity and true willingness to pay or to accept. Multilateral market bargaining over a non-rival and non-excludable good is likely to lead to distortions of preferences and production. Figure 6.1 demonstrates that this is true even in the simplest three-party case, let alone in the multilateral bargaining situations that apply to problems like acid rain and preservation of whales and old-growth forests.

If the chief failure of markets to coordinate people's use of natural resources involves non-rival and non-excludable goods, where do common pool problems such as overfishing and too rapid pumping of oil fit in? Recall, from Sections 6.3 and 7.2, that the management of *unharvested* common pool resources involves non-rivalry and non-excludability.

Actions taken to deplete resources in the common pool are non-rival and non-excludable 'bads'. Conversely, actions taken to augment resources in the common pool, including simple restraint in harvesting, are non-rival and non-excludable goods. The herder who does not graze an additional cow, and the fisher who restricts fishing days, are slightly increasing the pasture available, and the fishing success, of all other herders and fishers. Unfortunately, unlike the owners of farms or fishing lakes who save pasture or fish to augment *their* future harvests, the herder and the fisher have no property rights in the greater stocks that they create, cannot exclude non-payers, and therefore cannot be rewarded in the market.

The parallels between common pool resources and non-rival and non-excludable goods can also be explored from the opposite perspective. The organisation that breeds and releases wild cranes into nature's 'common pool' is increasing the 'harvesting success' of all of us who care about wild crane populations—in a very tangible way if we happen to be birdwatchers.

7.8 Searching for market solutions

The route to market solutions to environmental problems caused by non-rivalry and non-excludability is to find ways of lowering cost, technological and organisational barriers to exclusion. Once exclusion of non-payers is economically feasible, exclusive property rights can be defined and enforced, either by formal legal organisations or by informal community action. Then market exchanges can provide the incentives and signals necessary to coordinate production and consumption of non-rival goods. This has been going on throughout history; goods become and sometimes cease to be excludable, depending on current costs, technology and values.

Economists studying behaviour of people in affluent countries observe that people's demands for environmental protection rise rapidly with income once annual income per head passes about US$5000 (in early 1990s dollars).[2] So the amounts that people are able and willing to pay to physically exclude or to monitor and penalise unauthorised users of valued environmental assets increase with economic growth. Thus, where technological and organisational solutions to the exclusion problem are presently conceivable but too costly, exclusion and market deals between parties may become economic in the future.

Where technological and organisational changes leading to market solutions to the exclusion problem appear impossible, as in the cases of knowledge of wild cranes and biodiversity loss, those who bear the costly consequences of the absence of markets will turn to the alternative social coordination mechanism, government planning. However, in doing so, we must be mindful of the fact that major environmental problems also occur where elected and bureaucratic planners determine the uses of natural resources. The question is, can central planning do better at reconciling people's concerns about non-rival and non-excludable goods than market signalling and incentives? To begin to answer this question, we must study the limitations of central planning in a democracy, to which we turn in Chapter 8.

Discussion questions

1. The invention of barbed wire resulted in large increases in the productivity of grazing lands of Australia and the American West. True, false or uncertain? Explain why.
2. It is sometimes efficient for a common pasture to be grazed at a stocking level exceeding the point of maximum sustained pasture yield. True, false or uncertain? Explain why.
3. Writing in the *Age* in 1991 about the shark fishermen of Bass Strait, Graeme O'Neill explained that fishermen had been taking younger and smaller fish for the past decade, and that many fishermen refused to believe that the fishery was over-exploited, despite scientific evidence to the contrary. Do you think that shark fishermen act the way they do because they are ignorant or illogical? If not, suggest an alternative explanation for their behaviour.
4. The Industry Commission estimated in 1993 that urban road traffic congestion costs about $4 billion a year in Melbourne and Sydney. In the case of road congestion, what is the common pool resource and what is the precise nature of the externality which the individual road user creates for other road users?

8

Limitations of government signalling and incentives

Where environmental problems occur because private property rights and markets are too costly, can government planning do better? Government has one great advantage in social coordination; it can coerce contributions from people who would not contribute voluntarily. For example, it can tax whale lovers and fine air polluters. But the same coercive power is also a great weakness; interest groups have incentives to distort the information supplied to planners. And where planners' rights and responsiblities are not precisely defined and enforced, planners commonly neglect costs which they impose on others. In central planning, as in markets, accurate feedback to decision makers about all the benefits and costs of their decisions is often too costly at present, so environmental problems occur under central planning. The challenge is to find the most satisfactory mix of two costly, and therefore imperfect, signalling and incentive systems.

8.1 The advantage of government's coercive power

The rules that government defines, and the organisations that monitor and enforce them, are non-rival and non-excludable goods, protecting all members of the community against no-holds-barred contests over scarce resources. Recall from Chapter 2 that the existence of exclusive individual property rights and markets is the result of an original and continuing social compact whereby individuals agree to restrain their demands for scarce assets, in return for community recognition and enforcement of their agreed rights. In this sense government decision making complements, rather than competes with, private decision making.

Not all members of society play by its rules; some still wish to acquire assets by cheating or by brute force. Therefore enforcement of society's rules sometimes requires coercion; that is, securing individuals' cooperation by taking away some of their rights. For example, a business may be fined for illegal dumping of hazardous wastes, or an individual imprisoned for illegal capture and smuggling of endangered species. Government, acting through legislators, police and the courts, is the one organisation in society with the generally accepted legal right to coerce adults.

Market exchange relies on voluntary cooperation, induced by offering people exclusive property rights that they value more than those they presently have. An avocado seller values the exclusive right to the money more than his or her present exclusive right to use the avocados; the reverse is true for an avocado buyer. But where there are no property rights and no markets because goods are non-rival and non-excludable, it is rarely possible to get people to cooperate voluntarily to produce the desired amount of the good. Many lovers of wild cranes will contribute little or nothing to preservation, preferring to free ride on the contributions of others. Only a foolish fisher will return fish to the common pool to breed, if he or she believes that others will get all the resulting benefits. Here government's coercive power can make a difference. If government can enforce payment by all beneficiaries of crane preservation, many more wild cranes will be released. Government may not be able to enforce precise limits on the fish caught, but it probably can achieve increases in fish stocks by enforcing limits on the total catch or fishing days or the type of nets used. Government coercion achieves the production of goods that would be undersupplied, or not supplied at all, if we relied on people's voluntary cooperation.

It is government's coercive power that makes the difference; planners have no other major advantage over market exchange in coordinating production and consumption. If there is no technology that permits cranes or fish to be 'fenced in' on identified holdings, or that permits those who enjoy or use the cranes or fish to be traced, both private owners and planners will have difficulty enforcing exclusive rights. If it is technically difficult and costly for private conservation organisations to measure the population of wild cranes or of fish, or to identify all the wild crane lovers or fishers, it will be similarly difficult and costly for government planners. In practice, due to limited budgets for scientific research and for monitoring and policing of resource use, planners often rely on information supplied by interested parties, such as conservation organisations, foresters, fishers, industries that emit pollutants and members of the public.

8.2 The danger of government's coercive power

Government's coercive power is its great advantage in social coordination; it is also government's great weakness. Interest groups can distort values and technical information supplied to planners in attempts to get property rights and other rules changed in their favour. When government changes such rules, people gain or lose property rights, as they do in market exchanges, but with the difference that gainers do not directly pay, and losers are not directly compensated, for rights thus acquired or lost. Consider disputes over logging versus preservation of Tasmanian forests. The timber industry can assert that regional economies will be devastated by the cessation of logging and that ecological systems will quickly re-establish after logging. Greens can assert that large numbers of Australians put high values on recreation and preservation of flora, fauna and wilderness in old-growth forests, and that the ecology will be permanently altered by logging. Neither side is required to raise funds to bid for defined property rights in the forests, which would force supporters of both sides to consider biological and value trade-offs between timber production and ecological system maintenance. Putting it another way, when property rights are exchanged in markets, people who value those rights have to 'put their money where their mouth is', but this is not so when rights are acquired through government rule changes. (Economists use the term 'rent seeking' to describe activities aimed at acquiring rights through government rule changes.)

8.3 Free riding by citizens

Planners suffer from more than just deliberate distortions of information. Distortions also occur because some people affected by planning decisions do not bother to communicate their knowledge and values to planners. The reason? They do not believe that their input will make any difference. Important planning decisions—those that affect large numbers of people, such as decisions to allow logging in Queensland rainforests or to prohibit mining at Coronation Hill or to discharge sewage into the sea off Sydney Heads—are non-rival and non-excludable goods (or 'bads', depending on how they affect you). If you believe that your input will have no effect on the planners' decision, you probably will choose not to signal your views by lobbying, writing to politicians, demonstrating or whatever. You choose to free ride on the signalling efforts of others.

Voting on important issues suffers from the same problem. If we do not believe that our vote will make any difference, we are unlikely to research the issues before voting, and may not vote if there is no penalty for not voting. So the information communicated in our vote on, say, banning sand mining along the New South Wales coast is as likely to signal our ignorance and prejudices as considered values. Another problem with voting as a means of signalling to planners is that people rarely get to vote on a single issue; voters almost always have to choose between individuals or parties that bundle many issues together in a policy platform. An individual's vote at a general election tells the planner nothing conclusive about the voter's views on banning mineral exploration along the New South Wales coast if the political parties also differ on a goods and services tax and retaining the monarchy, and the voter feels strongly about all three issues.

8.4 Imperfect political feedback: externalities under planning[1]

Ideally, if you are a politician, you should lose votes and political influence if your planning decisions harm the community in aggregate more than they help it. If you are a bureaucrat, your salary or seniority or career prospects should be adversely affected. If this were the case, your decisions would reflect overall community desires, because what would be best for the community would also be best for you. In practice, because it is costly for citizens to monitor and reward or penalise your every move, you will have some freedom to pursue personal goals inconsistent with citizens' desires.

In economists' terms, citizen-voters in a democracy have a principal–agent problem. In this case, citizens are the principals, and politicians and bureaucrats their agents, with some discretion to act in ways contrary to citizens' interests.

Exploitation of your planner's privileges may be as trivial as using your government telephone for personal calls. On the other hand, it could involve favouring lobby groups who have promised to support your re-election campaign, or promised you employment when you leave public service. Because you do not personally bear all the consequences of your actions, you are responsible for *externalities*—costs or benefits imposed on citizens that they cannot sheet home to you. Lack of political feedback to planners causes externalities equivalent to those that exist when there is no market to signal benefits and costs to market decision makers.

We are assuming that both citizens and planners know the connection between planning decisions and the benefits and costs that the citizens experience. Often neither citizens nor planners will be fully informed about such matters. For example, in the case of acid rain (discussed in Chapter 5), no one may be certain of the effects of planning decisions on industrial workers or downwind sufferers from acid rain.

Unequal political feedback may result also in externalities under planning. For example, the victims of the steel industry pollution described in Section 5.2 could be mostly householders who find it very costly to monitor and reward or penalise the politicians and bureaucrats responsible for pollution control policies. Steel producers and users and steel industry unions, all of whom benefit from cheap steel (and thus cheap smoke disposal), may be well organised and able to monitor, reward and penalise planners at relatively low costs. As a result, planners place disproportionate weight on the benefits to industry and its employees.

Externalities are unavoidable in both market and government signalling and incentive systems, and for the same reason: perfect feedback about the benefits and costs of decisions is too costly. In the case of markets, it is too costly to create exclusive property rights and markets. In the case of planning, a system where every citizen knew all the consequences for him or her of each political and bureaucratic decision, and could appropriately reward or penalise the planner, would be impossible at any cost.

Where exclusive property rights and markets are feasible, externalities are likely to be a far greater problem in planning than in market decision making. It helps to understand why if we return to our avocado example of Chapter 3. Compare the monitoring problems faced by a consumer who buys avocados in the market and the same consumer when buying avocados from the Avocado Production and Marketing Board, which is responsible to elected officials. In each case, the consumer needs to measure the quality of the avocados and identify the responsible party to reward or punish, depending on quality. This will be no problem for the consumer when buying avocados in the market; if the avocados are inferior the consumer can punish the supplier directly by buying less, offering less, or, most drastic, switching to an alternative supplier.

Now consider the same consumer receiving inferior avocados from the Avocado Production and Marketing Board. The Board planners, and the elected planners who relay signals from citizens, are third parties relatively remote from avocado producers and consumers. The individual consumer's costs of monitoring and rewarding or penalising planners for bad avocados or long queues or delayed delivery are generally high. It is costly for the consumer to identify the politician or bureaucrat responsible, by tracing the problem up the public service hierarchy to the decision maker. Unlike face-to-face dealing with a market supplier, who has full legal responsibility for the product, the consumer may find it hard to pin down the blame. It is also costly for the consumer to penalise the individual responsible, by contacting his or her member of parliament or by registering a formal complaint.

The consumer's enthusiasm for identifying and penalising the planner responsible may be undermined by the knowledge that he or she is one of many consumers getting bad avocados. The consumer may well decide that individual efforts will make little or no difference. If others think the same, there will be little or no effort to trace and penalise the planner responsible. Monitoring and penalising of planners are non-rival and non-excludable goods, with the usual incentives for individual citizens to free ride on the contributions of others. Finally, if the Avocado Production and Marketing Board is the sole supplier of avocados, as is common with centrally planned coordination of production and consumption, the consumer can only switch business to a competing supplier by making a costly move to another state or country.

The individual consumer faces similar problems of remoteness from responsible decision makers, and similar incentives to free ride, when dealing with large bureaucratic private suppliers such as banks, insurance companies or airlines. However, a monopoly government bureaucracy is subject to less direct feedback than a bureaucratic private supplier because of the consumer's higher costs of shifting to an alternative supplier. Another difference is the greater significance of profitability in sheeting home rewards and penalties to private sector managers. Thus externalities are likely to be more significant with monopoly government bureaucracies than with private bureaucracies.

When government supplies non-rival and non-excludable goods, where no private markets exist, there is usually no alternative supply, as there is for avocados. But the difficulties in monitoring, rewarding and penalising planners remain. The high cost of ascertaining the facts and penalising planners in alleged cases of police misconduct is a good example.

8.5 The costs of planning

We saw in Chapter 3 that ideal planning, which would maximise community benefits from the production and use of (say) avocados, is impossible because the extraordinary conditions required are not met in practice. Ideal planning is impossible for the same reason that ideal markets are impossible and some markets do not exist: planning is a costly business. And the types of costs are mostly the same; like market participants, planning agencies have to bear the costs of:

- identifying interested parties;
- measuring whatever is produced, consumed and transferred;
- identifying valuations;
- defining and enforcing planners' property rights.

In addition, because planners are third parties with their own incentives, citizens bear the costs of monitoring and rewarding or penalising the planners.

Suppose that an environmental planning agency is trying to resolve the problem of acid rain, discussed in Chapter 5. The agency must incur costs in attempting to identify emitters and sufferers, and to

Table 8.1 Costly activities required for social coordination

Activities	Coordination mechanisms: Property rights	Market exchange	Political exchange
Identification/ monitoring	Of: • assets • right-holders • asset users	Of: • items transferred • right-holders • trading partners	Of: • items transferred • resource users • those affected by use • responsible planners
Measurement	Of: • assets • asset use	Of: • items transferred	Of: • items transferred • planners' performance
Exclusion	Of: • non-right-holders by some sort of 'fencing'		
Penalising/ rewarding	Of: • illegitimate users, by social pressure, courts etc.		Of: • planners, by voting, political contributions etc.
Preference revelation		By: • potential buyers and sellers, via offers to pay and accept	By: • asset users and those affected by use, via voting, hearings, lobbying etc.

measure the amounts of emissions and acid deposition. To control acid rain, the agency must enforce its exclusive rights over the use of 'the community's' air for disposal of sulphur dioxide and nitrogen oxides. This involves the costs of identifying and penalising emitters who are 'trespassing' by illegally using that air for waste disposal. The agency and the government must also incur costs in attempting to discover what emitters and sufferers are truly willing to pay or accept for more or less emissions, by holding referendums, elections and public hearings, by undertaking surveys, by listening to lobby groups and so on. Finally, citizens must incur costs in identifying and monitoring and either rewarding or penalising the agency and political planners responsible for acid rain control, according to the costs and benefits they create for others. In all cases, perfect information is unattainable at acceptable cost so the agency will never achieve complete coordination of the desires of acid rain emitters and sufferers.

Table 8.1, an expanded version of Table 5.1, lists the costly activities required for political coordination of asset use, and compares them with the activities required for specification of property rights and for market exchange.

8.6 How do we assess democratic planning?

It is not enough to list the advantages and disadvantages of democratic planning in the abstract. In order to understand why particular environmental problems exist where planners are responsible for social coordination, we need some framework, a checklist of planning requirements or a formal model of the planning process, to be able to systematically assess the planning process in each case. The formal economic models in Figures 5.2, 6.1 and 7.1 provide such assessments for particular environmental problems that occur under market social coordination. When all the assumptions of the models hold, they measure the ways in which the absence of exclusive property rights and markets distort market signals and incentives, and the money value of the resulting net losses to the community. Of course, the textbook assumptions almost never hold, and the actual values of the demand and cost curves upon which calculations would be based are very difficult to identify, because there are no markets to force people to reveal willingness to pay or to accept. Nevertheless, the economic models provide a framework for assessing the effects of missing markets on production and consumption, and a starting point for estimates of the costs of missing markets.

We do not have equivalent formal models of the planning process applied to environmental problems. Even if we had formal models that incorporated 'political' benefit and cost curves, the political process does not yield a lot of reliable observable data about voters' and politicians' and bureaucrats' values on which we could base calculations. Also, political decisions frequently are based on motivations that do not yield obvious net benefits to decision makers: concern that planning processes be seen as 'fair' in the sense of open to all community members; moral motivations (probably an important reason for the strong public support for environmental organisations such as the Australian Conservation Foundation and Greenpeace in recent years); and a desire to participate (another likely explanation of the support for environmental organisations).

This said, there is so far no better basis for explaining the decisions of political actors—voters, politicians, lobby groups and public service bureaucrats—than the economist's standard assumption, stated in Sections 3.3 and 3.12, that people act so as to maximise the net benefit to themselves from the action chosen. This assumption does not imply that political actors are always selfish. Self-interest does not rule out satisfaction derived from actions which benefit others, such as donations to famine relief. Rather, in explaining political actions, we believe that the most reliable guide is to first try to identify an actor's dominant objective and then to assume that the person acts to achieve that objective. So we must begin by asking what a political actor is most likely to be trying to maximise; depending on the actor, it might plausibly be money income, his or her vote at the next election, the agency's budget, preservation of indigenous flora and fauna, and so on. We then assume that the individual acts to achieve that objective. Thus, in the absence of contrary evidence, we expect to find students voting for the party that promises lower university fees, the Greens insisting on logging and mining bans, the plastics industry making larger campaign contributions to a party that opposes compulsory recycling of containers, and primary industry department bureaucrats being more sympathetic to resource development than are environment department bureaucrats.[2]

The reader may protest that the preceding assumptions about the objectives of particular political actors are simplistic, and unfair to many individuals. True, but as usual in economics, we are using a very simple assumption about individual behaviour to suggest explanations for the behaviour of typical members of a large group, not particular people. It is most unlikely that any other simple behavioural assumption (e.g. that people rank others' welfare equally with their own) would do as well in predicting the behaviour of groups of political actors.

In the absence of formal models of planning, we assess democratic planning of natural resource use by examining, first, the particular environmental problem and, second, the processes that generate information about the problem and create incentives for people to respond. We do this by asking four questions about the particular problem and planning process. We assume throughout that individuals act in what we believe to be their own best interests. The four questions are:

1 *How important are non-rival and non-excludable goods where government coercive power is an advantage in overcoming free riding?* (See previous discussion in Sections 6.4 and 8.1.)
2 *What are planners' sources of technical and value information, and what are the incentives of these sources to tell the truth?* (See Sections 3.11, 8.2 and 8.3.)
3 *What are the incentives of the elected planners to act in the interests of all their constituents?* (See Sections 3.12, 8.4 and 8.5.)
4 *What are the incentives of the public service planners to act in the public interest as defined by elected representatives, and to implement plans at the least cost to the community?* (See Sections 3.12, 8.4 and 8.5.)

In the remainder of this chapter we illustrate how each of these steps in the assessment of government environmental planning might proceed, by examining a Commonwealth government foray into environmental planning, the issue of mining at Coronation Hill.

8.7 The dispute over Coronation Hill[2]

The dispute about whether or not to permit mining at Coronation Hill in the Top End of the Northern Territory is one of the most contentious and highly publicised environmental issues faced by an Australian government during the last decade. Coronation Hill is the site of a former goldmine on a former pastoral lease about 300 kilometres south-east of Darwin. In 1984 a mining consortium identified a further commercial gold, platinum and palladium deposit at the site. The consortium obtained a mining lease that was subsequently frozen by Commonwealth government inclusion of the area in the Kakadu Conservation Zone in 1987. The Conservation Zone, reduced to just 47.5 square kilometres around Coronation Hill in 1989, was available for mining exploration, but not for mining. The Commonwealth government undertook that by 1992 it would choose between allowing mining, incorporation of the zone in the surrounding Kakadu National Park, or division of the land between a mine and the park.

The use of the Conservation Zone for a goldmine was perceived as an environmental problem for two main reasons. First, a mine would create a risk of fuel or toxic chemical contamination of the immediate area and, more importantly, of the nearby South Alligator River, which supplies the wetlands of Kakadu National Park.[3] Second, mining would impose costs on the traditional Aboriginal owners of the area, due to the cultural and spiritual values associated with Coronation Hill and its vicinity.[4] To inform the government about these matters, and the trade-offs between such costs and the value of mining and associated regional development, the question of how the land should be used was referred to the then Resource Assessment Commission (RAC) for inquiry. The RAC report, delivered in April 1991, was non-committal about whether the possible costs of contamination to users and lovers of the natural environment could exceed the benefits from mining. However, the RAC accepted evidence from a majority of Aboriginal representatives and anthropologists that, due to the spiritual significance of the area, a mine would impose major costs on the traditional Aboriginal owners.

The RAC report probably played only a modest role in government decision making. From 1989 to 1991 the Commonwealth government was energetically lobbied by the mining industry and the Northern Territory government (for mining) and by environmental and conservation organisations (against mining). Major Aboriginal organisations, in particular the Northern Land Council representing Aborigines in the Top End, were against, but some Aborigines in the region favoured mining. The Cabinet split between 'pro-development' and 'pro-environment' ministers. In June 1991, with the Australian economy in decline and his party leadership under threat, Prime Minister Hawke decided against mining, giving his commitment to support Aboriginal spiritual and cultural beliefs as the reason.

8.8 Assessing government planning for Coronation Hill

The breadth and vigour of the public and political debate over mining at Coronation Hill suggested that, whatever the final government decision, some people were likely to see themselves as suffering costs to which they had not consented. The mining consortium companies, having been granted a mining lease prior to declaration of the Conservation Zone, saw the denial of mining as an uncompensated taking of their property. They have been pursuing a legal claim for compensation since 1992. Environmental organisations argued strongly that mining could impose substantial costs on tourists and on Australians who want to preserve the natural environment. Most Aboriginal representatives argued that it would desecrate or damage sites of spiritual and cultural significance to local Aboriginal people. Thus we have a classic environmental problem, with property rights not clearly specified, each side claiming that an adverse decision would deprive them of rights to which they reasonably believed they were entitled, and each side therefore seeing the other as attempting to impose costs on them (loss of mining profits for the consortium, damage to the natural environment and losses of spiritual and cultural values for the opponents) without agreement or compensation.

Where property rights are disputed, it is usually impossible to achieve universal satisfaction with any decision. But if the planning *process* is seen as fair, disappointed parties are generally willing to live with it. Consistent with the discussion of social coordination systems in Section 2.4, a reasonable definition of a fair planning process is one that generates accurate information and where decision makers have clear incentives to consider the interests and values of all affected parties. Market exchanges under conditions that approximate the assumptions of the ideal market are generally seen as fair in this sense. Recall that in our ideal avocado market described in Chapter 3, would-be buyers and sellers are forced to reveal true values, and have to pay or to accept prices acceptable to their exchange partners.

Was the decision-making process used for Coronation Hill fair in the above sense? If so, it would be likely to improve social coordination in the long run, and thus reduce or eliminate the environmental problem, so that no significant group of Australians would be seriously upset over the use of Coronation Hill. Our four questions for assessing democratic planning are a way of testing the fairness of planning for Coronation Hill. We ask them in turn. Remember that our objective is to demonstrate the assessment of governmental environmental planning using Coronation Hill as an example; with limited information from the public record, we cannot be sure of the correct answers to the four questions. And we have no way of knowing whether the final decision on Coronation Hill was the correct decision, given unavoidable uncertainty about the facts.

How important are non-rival and non-excludable goods, where government coercive power is an advantage in overcoming free riding? Suppose that the mining consortium has a permit to mine at Coronation Hill. If the values people attach to avoiding mining damage to the South Alligator environment, and to protecting cultural and spiritual values associated with the Coronation Hill area, exceed the net returns from mining, why won't market deals prevent mining? Preservation of the local environment and of Aboriginal cultural and spiritual values are non-rival and non-excludable goods. If the environment and its cultural values are preserved for one person, they are preserved for all, and exclusion is impossible for people who value the site without visiting it. So the beneficiaries of preservation are unlikely to contribute much to achieve it. Why should someone who wants to prevent mining at Coronation Hill pay when his or her contribution will have an insignificant effect on the benefit he or she receives? Most would probably prefer to free ride on the contributions of others.

So, if there are no community rules to prevent mining, it is likely to go ahead even if the total value attached to preservation of the environment and Aboriginal spiritual and cultural values exceeds the net returns from mining. This is where government's coercive power to permit or restrict or deny mining in the interests of the community is an advantage.

What are planners' sources of technical and value information, and what are the incentives of these sources to tell the truth? To determine the trade-offs between mining and preservation at Coronation Hill, the planners needed to know, first, the outputs and inputs of the proposed mining operation, where the outputs include not only gold and other valuable minerals, but also the final condition of the Coronation Hill site and the South Alligator environment. Second, they needed to know the values of the minerals produced and the machinery, labour and other inputs used in mining, and the values of the changes in the environment and in the spiritual and cultural significance of the site. Of these, only the values of minerals, mining inputs and some downstream tourist services, such as tours and accommodation, which may be affected by mining, are signalled in markets.

Identifying the value people attach to preservation of the South Alligator environment is especially difficult. The value of preservation is non-excludable. It is not signalled in the market. Further, people won't all agree on what is being or should be preserved, or on what is 'natural' or 'unspoiled'. Most must value its existence as they imagine it based on films, articles and pictures.

The Coronation Hill planners could not obtain the above information from voting at the preceding 1990 general election, where large numbers of issues were bundled together in the party platforms. Even a single-issue referendum on Coronation Hill would have revealed nothing about the intensity of individual voters' feelings for or against mining.

The planners were provided with considerable technical and value information from government agencies with no obvious incentives to distort information. The Australian Bureau of Agricultural and Resource Economics (ABARE) provided information about the present net value of the mine and site restoration, using project information supplied by the mining consortium (which was subject to verification by government mining experts). The Office of the Supervising Scientist of Kakadu National Park (OSS) reported on the possible nature and extent of environmental damage due to the mining project. Such estimates of future production conditions, damage, benefits and costs are uncertain. No one can know future gold prices or the future environmental impact of an escape of mining residues into the South Alligator River. Faced with such uncertainties, the planners needed to adopt some method of weighting the benefits and costs of possible alternative scenarios by the estimated probabilities of their occurrence. They also had to choose a rate of discount to convert future benefits and costs to their equivalent present values. (Discounting of future benefits and costs is discussed in Chapter 9, and decision making under uncertainty in Chapter 16.)

The provision of information to the planners was coordinated by the RAC, which conducted an inquiry into the use of the resources of the Kakadu Conservation Zone designed to consider the values of all mining and preservation alternatives. The RAC was a purely advisory body, which accepted submissions from any interested party. It was created by the Hawke government in 1989 to establish, for resource use debates and decision making, 'a transparent, consistent process that can be used by all interested parties as a means of having their information and their views impartially and independently assessed and considered by Government'. In addition to the ABARE and OSS reports and evidence presented by interested parties, the RAC engaged consultants to carry out research into a variety of aspects of the natural environment, cultural significance and resource values of the zone, and the possible impacts of mining at Coronation Hill. It commissioned an appraisal of the significance of the Coronation Hill area to Aboriginal people, and undertook an interview survey of 2500 Australians in an attempt to determine the environmental costs of mining. All of this evidence was summarised and evaluated in the RAC's final report to the government. The report listed seven options for the government, ranging from permitting mining at Coronation Hill and elsewhere in the zone to a ban on all mining and exploration. The RAC did not recommend any option, stating that 'the weighting of economic and environmental considerations and of Aboriginal views . . . involves value judgements that only the Government can make'.[6]

The RAC was an innovation in planning natural resource use designed to increase, and reduce the distortions in, the information supplied to planners. The RAC inquiry process had at least two advantages. First, because the RAC held open hearings, with all submissions and evidence on the public record and subject to direct challenge by the commissioners and other parties, distorted information was less likely to be supplied and accepted. Second, ease of access and the equal standing of all those appearing before the Commission was likely to encourage a greater variety of interested parties to provide information to planners, including people who otherwise would not bother. This would be so if interested parties perceived that the inquiry process lowered their costs of information provision and increased its possible impact.

The RAC inquiry process did not eliminate other, less public, forms of information provision, such as demonstrations and personal lobbying of planners, but it lowered their credibility. In the case of Coronation Hill, the inquiry process assured the planners that the major information deficiencies were due to unavoidable uncertainties about environmental impacts of mining, people's valuations of environmental preservation and the spiritual and cultural attributes of the site.

The RAC was abolished in 1994, indicating that the government of the day did not believe that the benefits of the information obtained in inquiries justified the costs involved. The multidisciplinary nature of RAC inquiries and reports, and the RAC's consequent wariness of value judgments leading to definitive policy recommendations, probably contributed to its demise.

What are the incentives of the elected planners to act in the interests of all their constituents? In the case of Coronation Hill the decision makers were the Prime Minister and Cabinet members. We cannot know for sure what their objectives were at the time. Our assumption is that they would each act to maximise the net benefit to themselves from the decision reached. What were they most likely to be trying to maximise?

In the case of the Prime Minister, there are two plausible motivations for his final decision—one moral and one political—and they reinforce one another. First, the Prime Minister had previously demonstrated strong personal support for Aboriginal land rights, on moral grounds. Second, at the time of the Cabinet decision on Coronation Hill, the Prime Minister's leadership of the Labor Party and the government was under challenge from Paul Keating. Bob Hawke had declared that he would lead the party to the next election, but the election was still two years away, allowing plenty of time for a new leader to settle into the job. Thus Mr Hawke's short-term task was to convince the parliamentary party that he was the best person to lead the party in the next election. This was in a situation where both his popularity and that of the Labor Party were low, due mainly to a depressed economy, so that the prime interests of many members (their parliamentary seats) appeared in jeopardy within two years. Thus it is plausible that the Prime Minister's major objective was to maximise his parliamentary party support over the next two years, by demonstrating decisive leadership and obtaining electoral support in the suburbs where most marginal seats were located. Remember that few suburban jobs and incomes are directly dependent on mining, and that environmental and Aboriginal causes have substantial suburban support. This logic is strengthened when we recall that the left faction of the parliamentary Labor Party was almost unanimous in supporting his leadership, and the left faction was opposed to mining.

What about other members of the Cabinet? Cabinet members' prime interests would obviously vary. With a high proportion of safe seats in Cabinet, it is unlikely that getting re-elected was the first priority for many. The re-election of the government would be a priority, but was two years away. Remaining in—or advancing in seniority in—Cabinet in the short term seems a more likely prime objective for many. This would tend to lock diehard Hawke supporters, and diehard Keating supporters, into supporting the position of their candidate, regardless of their own views or the views of their constituents. However, this would not explain the positions taken by those who were non-partisan in the leadership contest, or by those strong minds who attempted to draw their own conclusions based upon the evidence presented to them.

How might a Cabinet independent best assure a continued Cabinet position regardless of who was Prime Minister? Most plausibly by performing well in his or her current portfolio. A minister is most likely to perform well when he or she gets on well with the departmental bureaucracy and its industrial and community clientele. This is more likely if the minister generally sees eye-to-eye with the department and the clients, both on the facts of the case and on values. And of course ministers are selected, and self-select, on this basis. Thus we would expect to find ministers with responsibility for the primary resource industries, commerce and other development-oriented portfolios supporting mining, and those with responsibility for the environment and Aborigines opposing mining. The Cabinet did in fact split along these lines.

The above logic would not necessarily explain the vote of every one of the thirteen Cabinet members. Media reports of the time suggested that the Prime Minister may have been in the minority. Whatever the true position, the Prime Minister succeeded, arguing that, on moral grounds, Aboriginal spiritual and cultural values took precedence over the possible gains from mining. (It is possible that the Cabinet decision was taken on the understanding that some other issue would be decided in a manner favourable to mining and development interests, as compensation for their loss on Coronation Hill. This horse trading would ensure that many more Australians would be reasonably satisfied with the government's overall performance by the time of the next election.)

If valid, the preceding argument suggests that in the case of Coronation Hill, the political decision makers' objectives were more closely aligned with those of certain interest groups than with the objectives of the electorate as a whole. In other words, the political decisions on Coronation Hill were likely to involve externalities for many Australians outside the favoured interest groups. The Prime Minister, the key decision maker, appears to have had strong incentives to satisfy the interests of antibusiness and environmental constituencies in the unions and in suburbia. Cabinet members had incentives to put the interests of their departments and departmental clients ahead of the broad community interest. Most, if not all, had incentives to concentrate on the political and economic impacts of the decision within the next two years, since few Cabinet members could be confident of their circumstances beyond the 1993 election. This was despite the fact that, for voters, most of the real impacts of a decision to allow or prohibit mining, such as additional employment and incomes in the Top End, and possible environmental damage, would occur beyond 1993.

A different but supporting argument, this time applicable to all individual members of the Cabinet and the government, relates to the perceived risk attached to the major options, to permit or to ban mining at Coronation Hill. The main benefit from mining—the expected increase in national income due to mining—was relatively well defined, except for future minerals prices. The nature of and/or values attached to the major costs of mining—possible environmental damage within the South Alligator catchment, the destruction of Aboriginal sacred sites and disruption of the culture of traditional owners—were highly uncertain. Moreover, the ban on mining is reversible, but many of the mine's impacts on the environment, sacred sites and Aboriginal culture are likely to be irreversible. Thus risk-averse politicians would be likely to act cautiously by favouring the environmental status quo, avoiding the uncertainties associated with current mining, while not totally eliminating the possibility of future mining. If you were a Cabinet member of the time, which would rate more highly with you: being responsible for the loss of $50–100 million of national income over the next 12 years, or the likelihood of being responsible for the destruction of local Aboriginal culture, plus the risk of being responsible for major damage to parts of the Kakadu environment? (The pros and cons of a cautious approach to environmental change in the face of major uncertainty are discussed in Sections 16.7 to 16.9.)

What are the incentives of public service planners to act in the public interest as defined by elected representatives, and to implement plans at least cost to the community? Public service bureaucrats appear to have had two major roles in relation to Coronation Hill: providing technical and value information prior to the political decision and implementing the decision to preserve the local environment by incorporating the Conservation Zone into Kakadu National Park. It is unlikely that the researchers in ABARE, the OSS and the RAC had incentives to intentionally distort information. Most of these people are professionals subject to peer review of the evidence made public in the RAC inquiry. Clear flaws or biases in their work would have damaged their professional careers. The same arguments apply, but with less force, to departmental staff preparing non-public briefs for their respective ministers. In these cases, there would be greater tension between the incentive to put forward the best possible argument for the department (and its clients) and the public servant's professional judgment. However, the public inquiry process limited the scope for self-serving submissions from departments, as it did for parties appearing before the RAC.

As in the case of politicians, risk-averse bureaucrats would probably have judged that the adverse outcomes of mining were more uncertain and therefore potentially more damaging (to the bureaucrats) than the economic losses associated with not mining. Thus bureaucrats may have favoured the environmental status quo.

If the Cabinet decision means only that there will be no goldmine at Coronation Hill, implementation is easy. If, to keep faith with environmentalists and traditional Aboriginal owners, it is supposed to mean that the local environment really will be preserved within Kakadu National Park, the Commonwealth government, through the National Park Authority, will have to enforce its rights to control land use in the area. Coronation Hill is relatively remote from the main tourist centres and park headquarters. Excluding tourists, hunters and amateur prospectors who might damage the local environment and its spiritual and cultural attributes may be quite costly. It may also be a low priority with National Park Authority decision makers, given a limited budget and the remoteness of Coronation Hill from the main tourist destinations in the park. Both the federal ministers who are directly responsible for the actions of public servants and most of the citizens who care about preservation are located far from Kakadu. So it is not clear that, in prohibiting the mining project, the government created the conditions necessary to preserve the attributes of the local environment upon which the decision was based.

8.9 A lesson from Coronation Hill

If market and government decision makers were perfectly informed and responded to all the concerns of others, people's unhappiness about the use of Coronation Hill could be quickly eliminated, whichever system was used to coordinate resource use. But we do not live in such a perfect world. Information is costly for buyers and sellers; it is also costly for planners and the citizens they serve. Where market property rights are not precisely defined and enforced, market decision makers neglect costs that they impose on others. Where planners' rights and responsibilities to citizens are not precisely defined and enforced, planners neglect costs that they impose on others. The environmental problems at Coronation Hill, or any other serious environmental problem, cannot be eliminated by prescribing and implementing *the* correct solution, involving the creation of markets or of planning arrangements. We have to trade off the environmental gains against the increased costs of information and enforcement. What we face is a search for the most satisfactory mix of two unavoidably costly (and therefore imperfect) signalling and incentive systems.

Discussion questions

1. Government's power to coerce individuals to pay for environmental improvements and to restrict people's use of scarce natural resources is a blessing for the environment. True, false or uncertain? Explain why.
2. The political process is superior to the market process in expressing the 'will of the people', because in markets the poor get far fewer 'dollar votes' than the rich. True, false or uncertain? Explain why.
3. The European Community's Common Agricultural Policy provides subsidies and price supports for European farmers, encouraging conversion of existing grasslands, woodlands and wetlands to agricultural use. Why do European farmers continue to receive this sort of government assistance, at the expense of the far more numerous food consumers, taxpayers and environmentally-concerned citizens of Europe? Why do you think that subsidies are not restricted to the neediest farmers?
4. From the point of view of providing correct information and incentives to decision makers, what are the most important differences between market planning of resource allocation, and central planning by elected representatives and salaried bureaucrats?
5. Elephant poaching for the illegal ivory trade resulted in sharp declines in elephant numbers in Kenya and Tanzania in the late 1980s. What are the likely problems with a central planning approach to management of elephant herds, i.e. relying on Kenyan and Tanzanian politicians and bureaucrats to carry out the wishes of tourists and lovers of wildlife the world over for preservation of the herds?

III

DECISION-MAKING TOOLS

9

Decision making over time

The beneficial and harmful consequences of use of the natural environment frequently stretch far into the future. Whenever a private or public decision maker compares the values of alternatives which stretch through time, such as logging or preserving an area of forest, he or she confronts two problems. One is how to weight present versus future benefits and costs. The second is ignorance about the nature and likelihood of the future consequences of today's actions. Long-delayed consequences also pose an ethical issue. Since markets and planning cannot incorporate feedback from people yet to be born, what should be the rights of current decision makers to take actions which may harm future generations?

9.1 Comparing values over time

Preservation of endangered species increases the genetic resources available to future generations. Tree clearing and flood irrigation early this century have increased the salinity of Murray–Darling waters, harming current users of river water. Lead residues at industrial sites can affect the health of site users decades later. How do we weigh benefits and costs arising at different points in time?

As explained in Chapter 3, future benefits and costs are valued less than identical present benefits and costs. Why do we prefer to receive a $1000 lottery prize now, rather than in a year's time? For one or both of two reasons.[1] The first relates to people's consumption preferences;

we prefer avocados or a new dress today to an identical amount of avocados or an identical dress next year. The second relates to people's ability to increase their future productivity (and hence pay lenders interest in the future) by investing in themselves or in productive equipment. For example, if you receive the $1000 now, you can pay for a training course that makes you more productive, and so increase your earnings a year hence.

Comparisons of benefits and costs arising at different points in time are made by discounting future values to their equivalent value in present dollars. To see how this is done, consider the familiar compound interest formula used to calculate the growth in the value of investments. If the rate of interest is 5 per cent per year, $952 invested for one year will grow to $\$952(1 + 0.05) = \$1000$ in one year, and to $\$952(1 + 0.05)^2 = \$1050$ in two years. The compound interest formula $(1 + r)^t$ is used to calculate the amount to which a present sum will grow over t years compounded at the interest rate r per year. Discounting works in the opposite direction, from future values to present values. Thus, if the interest rate or *discount rate* is 5 per cent, the present value of the lottery prize payable one year from now is $\$1000/(1 + 0.05) = \$952$. More generally, the present value of $1000 to be received t years from now is $\$1000/(1 + r)^t$.

The present money value (*PV*) of a stream of money benefits (*B*) and costs (C) arising between the present (time zero) and t time periods into the future is calculated as:

$$PV = B_0 - C_0 + (B_1 - C_1)/1 + r + (B_2 - C_2)/(1 + r)^2 + \ldots +(B_t - C_t)/(1 + r)^t$$

where the subscripts refer to time periods and the discount rate is r per time period.

Remember that the *B*'s in the present value formula represent people's willingness to pay for goods or resources, and the C's represent people's costs of, or willingness to accept compensation for, providing those goods or resources. The formula allows these values to be compared across time, providing that all the *B*'s and C's and the appropriate discount rate are known.

To see how this works, consider decision making in a commercial venture where money estimates of benefits and costs are likely to be available. Imagine a uranium mining venture in northern Australia that currently has the opportunity to extract all the remaining commercial ore and close the mine. Suppose that the mining company can sell the uranium to France at the low world price indicated by d_0d_0 in Figure 9.1(a)(i) with current marginal costs of production s_0s_0 (assumed to be progressively rising as the rate of extraction increases). Assume that the mine executives have the alternative of closing down the mine for ten years and then extracting and selling the uranium. The executives confidently believe that, due to greenhouse-driven switches away from carbon-based fuels and to improvements in mining technology, in ten years the inflation-corrected dollar price of their uranium will have

Figure 9.1 Calculating the present value of uranium sales

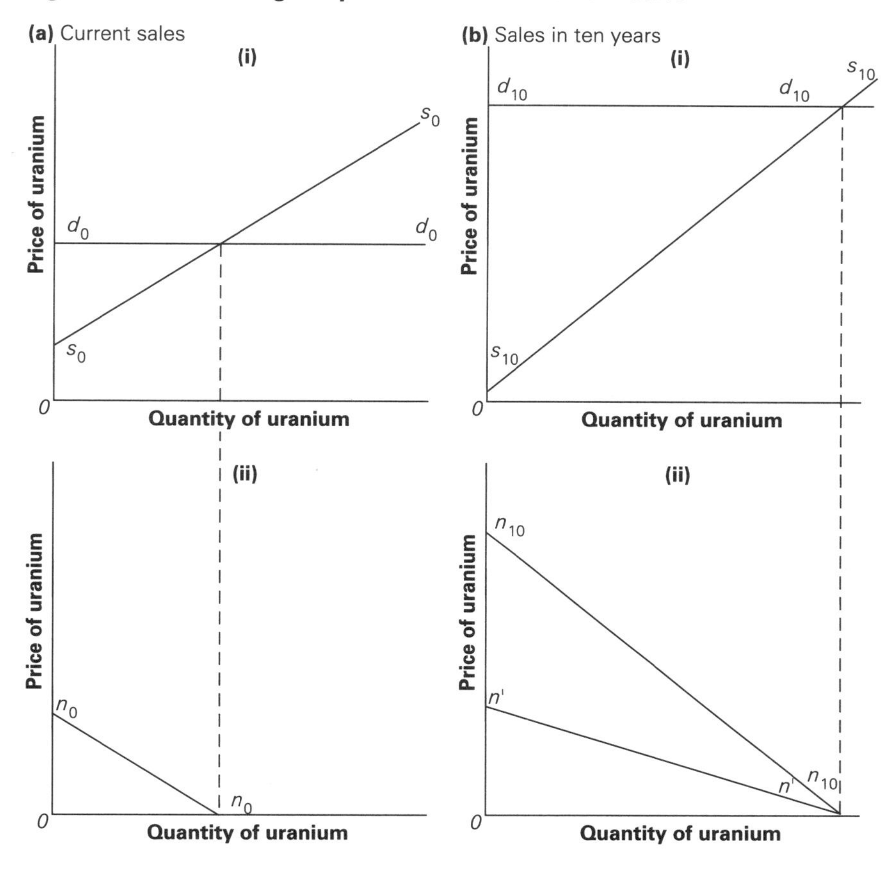

doubled, to $d_{10}d_{10}$ in Figure 9.1(b)(i), and their marginal costs of production will have fallen to $s_{10}s_{10}$.

Net benefits from mining now and in ten years are obtained by deducting the marginal costs from the demand curve—in this case from the price of uranium. Figures 9.1(a)(ii) and 9.1(b)(ii) show the corresponding marginal net benefits curves (n_0n_0 and $n_{10}n_{10}$ respectively) for the two alternatives, in current dollars and dollars ten years hence. If the discount rate is 10 per cent per year, a dollar that the mine earns in ten years time is worth $\$1.00/(1+0.10)^{10} = 38.5$ cents today. Figure 9.1(b)(ii) also shows the present value of the marginal net benefits curve ($n'n'$) obtained after the net returns from uranium sales in ten years time have been discounted. It is these present values of net benefits from future sales that must be compared with the net benefits from current sales. In our simple example, if the present dollar value represented by the area n_0n_0O in Figure 9.1(a)(ii) exceeds the dollar value represented by $n'n'$O in 9.1(b)(ii), the mine executives will maximise company net benefits by mining and selling uranium now. If it is less, they should defer mining for ten years. If the choice is not either/or—that is, if it is economic to mine some of the uranium now and some in ten years time—the executives must compare the present values of marginal net benefits from the two periods in order to decide how much to mine now and how much ten years hence.[2]

9.2 Two problems: ignorance of the future and choosing a discount rate

In reality, mining executives cannot know future uranium prices and costs of production. Future uranium prices depend on, among other things, technological and political developments affecting future energy demands and the future availability of uranium and alternative energy sources such as oil, coal and solar power. Future costs of producing uranium also depend on technological and political developments, such as innovations that lower processing costs and the extent of community and union opposition to commercial exports of uranium. Uncertainty about these and other future developments is unavoidable; nevertheless, decisions about mining now and in the future have to be made.

Ignorance about the nature of possible future events and the chances that particular events will come to pass may be even greater when actions have major consequences for both the economy and the environment. We saw in Chapter 4 that combined economic–environmental systems are very complex, and thus the consequences of some actions affecting the environment are practically impossible to predict.

It is possible to deal with uncertainty by default: to continue as is, believing or hoping that events will turn out for the best. After all, conscious planning for an uncertain future is costly in terms of time, money and nervous energy. Also, some decision makers put little effort into planning for the more distant future because they have high discount rates—they attach little or even no value to benefits and costs arising more than a short time ahead. But such attitudes are generally unacceptable where decisions are made on behalf of or have consequences that affect others—family members, shareholders, and present and future citizens. Private and public decision makers who must justify resource use decisions to others are expected to explain how they deal with uncertainty.

Decision makers also have to choose a rate of discount, weighting present relative to future values. The appropriate rate of discount is clear in principle; it should reflect the decision maker's personal opportunity cost of tying up resources over time. This will be based on some combination of that person's preference for goods today versus goods in the future (a terminal cancer sufferer would have a very strong preference for present over future goods) and his or her opportunities for productive investments that would make the person better off in the future (an inventor who has a device that he or she believes will be very profitable in the future will promise lenders a high rate of interest to finance the development of the invention). The willingness of these and others to pay for loans would create a demand for loanable funds. At the same time, people who have ample funds today and a preference for or need to augment their future resources will be willing to accept compensation for loans, creating a supply of loanable funds. Assume an ideal market for loanable funds analogous to our ideal avocado market in Chapter 3 (this would require, in particular, zero costs of negotiating loans, zero inflation and equal riskiness of all loans). Then the market rate of interest would be the same for all borrowers and lenders, all would therefore face the same opportunity cost of funds, and the discount rate would be the same for all.

In practice, market exchanges of loanable funds may be too costly, in particular because it can be costly for the lender to enforce his or her property right to the borrower's future income or assets. So individual discount rates differ. Also, since many interest rates coexist in the market at any time, and market interest rates incorporate allowances for the costs of negotiating loans, for inflation and for the riskiness of particular loans, the appropriate discount rate is not always obvious.[3] And in cases such as possible global warming and possible extinction of species, where consequences of current decisions may affect people yet to be born, there is the ethical issue of externalities across generations. If markets and planning cannot incorporate feedback from the yet unborn, should decision makers have unrestricted rights to take actions that may harm people in the distant future?

In the remainder of this chapter and in Chapter 16 we discuss the social coordination arrangements that decide the weighting of future values versus present values and determine actions under uncertainty. Who has the right to decide discount rates? Who has the right to choose between alternative courses of action under conditions of uncertainty? What are the sources of information and the incentives of the responsible decision makers? We discuss discounting in this chapter, and the problems posed by uncertainty in Chapter 16.

9.3 Decisions on discounting[4]

Rights to decide discount rates go hand-in-hand with individuals' or democratic planners' rights to decide the use of assets over time. Whenever a decision maker compares the values of alternatives that stretch through time, such as alternative policies to deal with stratospheric ozone depletion or nuclear wastes or endangered species, he or she must decide the relative weights of present and future benefits and costs.

Consider the case of private property rights that extend through time, such as fee simple titles in land (i.e. estates that are inherited without restriction to a particular class of heirs). The owner of a farm, so long as he or she causes no harm to his or her neighbours, can discount future returns as he or she chooses. The farmer can pursue immediate returns by overgrazing or overcropping, decreasing future productive capacity and returns. Or the farmer can choose to increase future productive capacity by sowing new pastures, applying fertilisers and planting trees. In either case, if there are markets for all attributes of the land, the anticipated consequences of these decisions are signalled to the farmer in the land market, as the value of the farm changes to reflect changes in its future productivity.

The farmer's decisions about the use of the land affect not only himself or herself, but also the heirs of the estate, who will in due course inherit either the land or the proceeds of its sale. The farmer's personal rate of discount of future values compared to present values will allow for any value he or she attaches to future incomes received by the heirs. But the heirs do not necessarily have any say in the decisions that may harm or benefit them. Inheritable private property rights include the right to value dollars received by one's heirs.

Market signals to private decision makers will incorporate the individual discount rates of affected parties. Suppose that a mining company operating on land whose surface rights are held by farmers has to make market deals to compensate the farmers for disruption to farm production during the life of the mine. The dollar values of compensation asked by the farmers will depend on the rates at which individual farmers convert anticipated farming losses in the future back to their present dollar equivalents. Compensation deals, and the resulting signals to the mining company regarding the profitability (and hence the geographic extent and economic life) of the mine will be affected by the individual discount rates of both the mining company and the affected farmers.

Where there are no property rights or markets, as discussed in Chapter 5, the individual discount rates of those affected cannot influence market-based decisions. Suppose that the above mine will result in the disappearance of a population of brolgas (native cranes) that frequent the farms. Assume that the presence of the brolgas is a non-rival and non-excludable good for the people of the region. As explained in Chapter 6, present and future values of the brolgas are unlikely to be correctly signalled to the mining company because of free riding. The losses and discount rates of crane lovers who choose not to contribute to crane protection will play no part in the mining company's decision.

As explained in Chapter 8, government coercion can increase provision of non-rival and non-excludable goods, now and in the future. In attempting to achieve the efficient amounts of mining, farming and brolgas, government planners have to value the present and future alternative uses of the land, and compare values arising at different points in time. Assuming that the planners have the power to enforce their decisions on land use, it is now the planners who decide the relative weights of present and future benefits and costs.

Unlike private right-holders, government planners are supposed to weight present and future dollars according to overall community preferences. But, as discussed in Sections 3.11 and 8.4, elected and bureaucratic planners may act in their own short-term interests, which may mean acting to maximise votes or career prospects. Thus, in choosing a discount rate, the signals and incentives that the planners experience may be biased towards the interests of those who can organise effective lobby groups at least cost. For example, if conservation groups are very effective in mobilising public support in marginal electorates, politicians may place a higher value on wilderness preserved for future generations than would be supported by voters as a whole.

9.4 Bequests to the future as non-rival and non-excludable goods

If planners choose a discount rate based on individuals' preferences expressed in markets, is this rate appropriate for the community as a whole? We can reasonably assume that individuals wish to leave bequests in cash and kind to their own descendants, but to the extent that they wish others in future generations to be better off, they may prefer to free ride on the generosity of other members of the present generation. If bequests to the future in general are non-rival and non-excludable goods, they will be inadequately provided in the absence of government coercion. In other words, if intergenerational charity extends beyond individuals' own families, but is subject to free riding, government should give a higher weight to future values than do individuals.

Intergenerational charity doubtless does extend beyond the family for many people who belong to environmental and aid organisations, but it is far from clear that the average Australian family is prepared to go without to provide either more cash or more natural resources to Australian families of the future (who, on past performance, will be better off than today's families). However, even if we grant that government should discount future values at a lower rate than markets, it is not obvious that it can or would. First, suppose that the government increases taxes on the average family, in order to transfer extra assets to future generations. This attempt to raise our collective provision of assets for the future will be undermined if families reduce their private bequests in response. Second, it is not obvious that elected and bureaucratic planners who have the rights to determine discount rates would in fact weight future values more heavily than private decision makers. It depends on the planners' objectives and the political signals that they receive. More importantly, a commitment by today's planners to set aside additional assets (such as compensation funds or mineral deposits or old-growth forests) for future generations will rarely bind tomorrow's planners. Future planners generally have the rights to renege on such commitments. If voters perceive this, they are likely to weight the future *less* heavily in their political choices than they do in their private choices, the exact opposite of the intergenerational charity argument for government intervention explained above.

9.5 Ethics and discounting

Future generations do not vote with ballots or dollars, and so they cannot influence the discount rate chosen by current planners or individuals. Is it ethically acceptable that members of the current generation have unrestricted rights to choose the discount rate affecting members of future generations?

Although future generations cannot represent themselves in market and political bargaining over discount rates, it is in principle possible to define and enforce their property rights by imposing constraints on the actions of current decision makers. For example, in *Blueprint for a Green Economy*, David Pearce and his co-authors suggest that sustainable development requires that future generations have the right to an undiminished stock of natural assets. Consequently, groups of projects undertaken by a decision maker should be subject to the constraint that the stock of environmental assets as a whole must not decrease.[5] Thus a private company planning a ski resort that would reduce stocks of alpine flora and fauna might only obtain approval for the project in conjunction with another project that provided offsetting increases in the same or other environmental assets, for example a flora and fauna conservation project.

The principle of defining and enforcing the environmental property rights of future generations is clear enough. How governments would define and enforce such rights in practice is anything but clear. If future generations' rights to (say) the preservation of populations of plants and animals were precisely defined at the level of every farm and development site, both the costs of enforcement and the restrictions on current production and consumption activities would be great. If rights were defined only in the aggregate, say maintaining the stock of environmental assets for the state of New South Wales, there is the daunting problem of determining the unit values of each of the natural resource components of the aggregate stock. To appreciate why values are necessary, suppose that over time the area of native forests in New South Wales increases, but the water along the coast becomes more polluted; the amount of one environmental asset rises, and that of another falls. We cannot say whether the people of New South Wales are better or worse off as a result without determining the relative values that they attach to extra forests and cleaner (or dirtier) sea water. Also, recalling the discussion of the incentives of public decision makers in Section 3.11, they are likely to have weak incentives to enforce the rights of future generations. The police, courts and other public officials responsible for enforcing such rights are subject to political feedback from the current generation, but not from future generations, so that enforcement activities may be biased against the future.

There are two more serious problems with the rights-based approach to meeting our ethical responsibility to future generations. One, implied above, is the increase in government coercive activity required when government is required to define and enforce an additional set of property rights in natural assets, on behalf of an additional group of right-holders. Since definition and enforcement are costly activities, and mistakes can occur, it is possible that the exercise of these additional powers may leave both current and future generations worse off.

The second serious problem is that, unlike the property rights of the current generation, what is protected today is not necessarily what the future right-holder gets or wants. As discussed in Chapter 4, the complexity of economy–environment interactions over time is such that we cannot be sure of the long-term consequences of maintaining environmental assets today, let alone whether our descendants would prefer such environmental protection to the alternative futures forgone when we restrict current resource use.

Discussion questions

1. High discount rates encourage more rapid use of exhaustible natural resources. True, false or uncertain? Explain why. Is your answer the same for renewable biological resources? Why or why not?
2. What factors will the owner of a stand of timber consider in deciding whether to harvest it now or reserve it for milling some years in the future?
3. It has been argued that, when the consequences of actions are very destructive and very long term (e.g. extinction of species) so that a huge number of people yet unborn are affected, the appropriate discount rate to use is zero (any positive discount rate over very long periods turns catastrophes into trivial present values). Do you think this is sensible or not? Explain your answer.
4. Are the interests of future generations in natural resources likely to be better accounted for by markets (representing the separate decisions of large numbers of individuals) or by governments, acting as stewards of natural resources for future generations? Discuss the factors determining which allocation system best serves our descendants' interests.

10

Cost–benefit analysis of environmental changes

Economists are frequently called on to advise private and public decision makers on alternative uses of natural resources—to advise on improvements in social coordination. The economist comes to such tasks armed with certain disciplinary assumptions and tools of trade, such as cost–benefit analysis. In order to realistically assess economists' advice, it is important to understand how economists, and their tools, deal with the problems of costly and imperfect information about the economy and the environment. Cost–benefit analysis is the economist's standard technique for evaluating projects from the point of view of the community. Its advantages are that it requires a comprehensive listing of all benefits and costs, wherever they fall, and incorporates clear principles for valuation of benefits and costs in money terms. Its limitations are that it involves strong behavioural and moral assumptions and that it requires precise technical and value information to provide clear guidance to decision makers. In the absence of such information, in particular where there are no markets to value project outcomes, the result of cost–benefit analyses are often controversial.

10.1 An example: preservation of forests in south-eastern Australia

The native forests of south-eastern New South Wales and of East Gippsland, Victoria, together referred to as the 'South-East Forests', have been logged for sawn timber since last century. The use of these forests has been hotly disputed since 1970, when the Japanese company Harris-Daishowa commenced export woodchipping at Eden, on the New South Wales south coast. Previous selective logging had attracted little opposition. Export woodchipping, involving clear-felling of native forests adjacent to the Sydney–Melbourne highway and the coastal resorts, with the chips shipped to Japan for further processing, was strongly opposed from the very beginning.[1] Public concern about woodchipping is partly due to the clear-felling, but also results from the economics of the combined harvesting of sawlogs and pulplogs; the utilisation of inferior logs for woodchips makes it commercially feasible to extend logging into areas that would be unprofitable to log for sawlogs alone. Such concern led the Victorian government to prohibit direct use of logs in woodchipping; East Gippsland supplies only sawmill wastes to the Eden chip-mill.

Increased conflict over the management of Australian forests, including the South-East Forests, led to an RAC inquiry into options for the use of Australia's forest and timber resources.[2] The inquiry included an economic analysis of the costs and benefits of alternative uses of the South-East Forests.[3] This research illustrates the use of cost–benefit analysis to evaluate alternative uses of natural resources, and techniques used to value goods and services not priced in markets.

In 1990, of approximately 1.4 million hectares of state forests in the south-east, 600 000 hectares were listed on the Register of the National Estate, declared by the Australian Heritage Commission to be of high conservation value. Of the National Estate forests, 470 000 hectares were protected from logging in National Parks and other nature conservation reserves. The RAC's economic analysis focused on alternative uses of the remaining 130 000 hectares of National Estate forests, which were available for logging under the policies of the forest management agencies in New South Wales and Victoria.

Streeting and Hamilton, of the RAC, used cost–benefit analysis to examine the impact of changing government policy to protect the remaining 130 000 hectares from logging.[4] Table 10.1 reproduces their list of the major costs and benefits of a cessation of logging. The principal costs of the policy change are the value of sawlogs and pulplogs forgone and the economic and social costs of additional unemployment in the region. The principal benefits are the cost savings due to reduced logging and milling and various benefits of forest preservation, including recreational benefits, the scientific and educational benefits of species and ecosystems not adequately preserved elsewhere, and ecological benefits such as maintenance of water quality and prevention of soil erosion.

Table 10.1 is divided into two parts. Streeting and Hamilton obtained the values of the items above the line from markets and information about the Budget costs of unemployment. Table 10.2 reproduces their 'best guess' estimates of these benefits and costs. They estimated that, if the environmental benefits and the social costs of

Table 10.1 Major economic benefits and costs of cessation of logging in the National Estate

Benefits	Costs
Capital and labour cost savings: • forest management[a] • log harvesting • log transport • sawmilling • chip-mill	Value of forgone sawlogs and pulplogs Economic cost of unemployment
Avoidance of damage to natural environment Maintenance of recreation and tourism opportunities in the National Estate Reduced activity by logging trucks: • avoidance of road pavement damage • reduction in air and noise pollution Maintenance of forest suitable for non-timber commercial uses (e.g. bee-keeping)	Social cost of unemployment Loss of opportunities for some recreation and tourism in the National Estate

Notes: a Includes management costs associated directly with harvesting. The ongoing cost of managing the forest for commercial timber production is assumed to be equal to the cost of managing the forest for preservation.

Source: Mark Streeting and Clive Hamilton, *An Economic Analysis of the Forests of South-Eastern Australia*, Resource Assessment Commission, RAC Research Paper no. 5, AGPS, Canberra, 1991, p.61.

unemployment were ignored, ceasing logging in the National Estate forests would cost the Australian community $11 million in 1992 dollars. This estimate of net dollar costs involved the following assumptions:

- Logging ceases in 1992.
- Benefits and costs of the decision extend from 1992 to 2040.
- Benefits and costs falling after 1992 are converted to 1992 dollars using a discount rate of 7 per cent.
- Harvesting and mill throughput and technology are unchanged throughout.
- The prices of sawn timber, woodchips, and harvesting and mill inputs, including labour, are all unaffected by the reduction in output.
- Half of the displaced timber workers remain unemployed for three years, at a Budget cost of $400 per week.

Streeting and Hamilton dealt with uncertainty about their assumptions by conducting a sensitivity analysis. Their analysis was repeated using different combinations of higher and lower prices of sawn timber and woodchips, higher and lower log-processing costs, and higher and lower estimates of unemployment and of its Budget costs. When the combi-

Table 10.2 Results of economic evaluation: cessation of logging in National Estate areas (*$ million, net present value*)

	Discount rate		
Benefits/costs	4 per cent	7 per cent	10 per cent
Benefits (present values):			
Cost savings in bush operations:			
sawlog production, NSW	6.8	5.3	4.2
pulplog production, NSW	16.0	12.4	9.9
sawlog production, Victoria	18.0	11.6	8.4
Cost savings in sawmills and chip-mill	11.9	8.3	6.3
Total benefits	52.7	37.6	28.8
Costs (present values):			
Value of forgone sawlogs	44.5	30.3	22.6
Value of forgone pulplogs	19.7	15.2	12.1
Economic cost of unemployment	3.2	3.1	2.9
Total costs	67.4	48.6	37.6
Net present value	**-14.7**	**-11.0**	**-8.8**
Benefit–cost ratio	**0.78**	**0.77**	**0.77**

Note: Analysis was undertaken over the 49-year period 1992 to 2040.

Source: Mark Streeting and Clive Hamilton, *An Economic Analysis of the Forests of South-Eastern Australia*, Resource Assessment Commission, RAC Research Paper no. 5, AGPS, Canberra, 1991, p.72.

nation of assumptions most favourable to forest preservation was used, the cessation of logging involved a net cost of $3 million. The combination of assumptions most favourable to continued logging increased the net cost to $19 million.

The items below the line in Table 10.1 are not valued in markets. Many of the benefits of forest preservation are non-rival and non-excludable; since beneficiaries cannot be excluded, they have little incentive to signal values using dollars that could be spent on other goods. The personal and social problems resulting from losses of employment in forestry depend on individual and community attitudes and adjustment options. It is extremely difficult to place dollar values on such costs.

While not attempting to measure the social costs of unemployment due to a cessation of logging, Streeting and Hamilton did estimate the costs of a government compensation package equivalent to that announced in September 1991 for timber workers affected by the cessation of logging on Fraser Island in Queensland. Assuming that 110 forestry jobs would be lost in the south-east, the cost of a three-year compensation package, discounted at 7 per cent, was $43 million 1992 dollars.

10.2 Estimation of non-market values of the South-East Forests: the survey and the travel cost study

The RAC used two techniques in attempting to estimate money values attached to preservation of the 130 000 hectares of the National Estate forests subject to logging.[5] The techniques were the travel cost method and contingent valuation.

The survey. The information required for the two techniques was obtained from a mail survey of a sample of 5000 voters in New South Wales, Victoria and the Australian Capital Territory. The survey began by asking respondents to look at a map that identified the areas of the South-East Forests in conservation reserves and the areas currently available for wood production. The latter areas included 130 000 hectares of National Estate forests, which were separately identified as potential conservation reserves. For these potential reserve areas, the questionnaire described the ecological, social and economic effects of the two options, wood production and forest preservation. The interviewees were asked four types of questions:

1 the nature and costs of their most recent trip within the previous year (if any) to the potential conservation reserves;
2 their choice between wood production and preservation, given that preservation would cost them a money sum nominated in the questionnaire (explained as due to a combination of higher timber product prices and higher government charges to fund conservation);
3 their attitude to the environment in general;
4 their sex, age, education, income and occupation.

The mail questionnaire yielded just over 2500 useable responses, a 50 per cent response rate. One-quarter of the respondents (605) had visited the identified areas during the previous year.

The travel cost study. The travel cost method was used to value the benefits obtained by visitors to the potential conservation reserves. Although no fee was charged for spending time in the forest, visitors did have to incur the costs involved in journeying to and from the forest. These travel costs were used to estimate an imputed demand curve for visits to the forests—a relationship between hypothetical entry fees and total numbers of visitors from New South Wales, Victoria and the Australian Capital Territory. This procedure involved several steps. First, the travel costs of the respondents who had made recreational visits to the forests were used to estimate a relationship between travel costs and the number of visits per thousand persons from different zones of origin. Then, by assuming that visitation rates would respond to increases in entry fees as they would to increases in travel costs, the RAC obtained estimates of how visitation rates from different zones would fall as an entry fee was introduced and progressively increased. Finally, multiplying the total population of each zone of origin by the relevant visitation rates, the RAC calculated the imputed demand curve—the relationship between entry fees and total numbers of visitors from all zones. The area under the imputed demand curve provided an estimate of the total amount per year that New South Wales, Victorian and Australian Capital Territory residents were willing to pay for time spent in the forests at the time of the survey: $950 000 per year according to the RAC's calculations. This was equivalent to $8.90 per visitor per year. (The assumptions involved in the travel cost method, and its limitations, are discussed in Section 1.4.)

10.3 Estimation of non-market values of the South-East Forests: the contingent valuation study

In the contingent valuation section of the mail survey, respondents were questioned about their willingness to pay to preserve the forests in the potential conservation reserves, given the two alternative use options described in the questionnaire. The study illustrates the steps involved in contingent valuation (CV).[6] (The assumptions involved in contingent valuation, and its limitations, are discussed in Section 11.6).

Sampling the population of likely beneficiaries The RAC assumed that the people who would benefit from the preservation of the South-East Forests were the people of New South Wales, Victoria and the Australian Capital Territory. Survey questionnaires were mailed to 5000 randomly chosen voters in the relevant electorates. The numbers of voters sampled in New South Wales south coast, Gippsland and Australian Capital Territory electorates were eight times the numbers in other electorates.

Describing the environmental alternatives involved The RAC's mail questionnaire included a map identifying potential forest conservation reserves. Table 10.3 reproduces the RAC's (hypothetical) descriptions of the alternative futures of these areas used for wood production or for conservation; CV practitioners commonly call these 'scenarios'. Respondents were also reminded that the designated areas of forest are only part of the Australia-wide forests that they might wish to pay to see preserved—that other forests might also require part of their conservation 'budgets'.

Describing the payment mechanism and eliciting values The questionnaire offered respondents a referendum-style choice between the wood production and conservation options. Respondents were told that the choice of the conservation option would cost them a specific sum of money each year, for two reasons:

Table 10.3 Alternative scenarios for contingent valuation of the South-East Forests

ALL RESPONDENTS PLEASE READ THE FOLLOWING

Please look at the map again.

We are now going to ask you some questions about what you would like to see happen to the forests in the striped areas shown on the map.

The Resource Assessment Commission is considering two options (A and B) for the future of the forests in the striped areas on the map:

We would like to know which of these options you prefer.

(Option A) **Wood production**	(Option B) **Conservation reserves**
• Half of the striped area would consist of stream-side reserves, wildlife corridors and other reserves for the protection of plants (including trees), animals, soils and streams.	• The striped areas would be placed in conservation reserves.
• The remaining half of the striped area would be used to grow trees for wood production.	• Some parts of these forests have never been logged.
• Each year a different 2% of the wood producing parts of the striped area woud be logged.	• Trees in these areas would not be logged in the future.
• After logging the forest would be allowed to regrow ready to be logged again in the future.	• These forests would be placed where more could be discovered about Australia's natural environment.
• The wood-producing areas would have a mixture of trees ranging in age from seedlings to 80-year-old trees.	• The forests would have a mixture of trees of different ages, ranging from seedlings to trees over 150 years old.
• Some rare and endangered species, both plants and animals, living in the forests may have their habitats disturbed by logging.	• These forests would continue to provide a habitat for a large range of plants and animals, some of which are rare and endangered.
• The forests would no longer be places where little human disturbance has occurred.	• The forests would remain as places where there has been little human disturbance.
• Current job opportunities in the local region would be maintained.	• Fewer job opportunities in the local region would be available.

Source: Resource Assessment Commission, *Forest and Timber Inquiry—Final Report*, AGPS, Canberra, 1992, vol. 2B, p. U25.

1 reduced timber production could raise the prices of timber products such as house frames and paper (note that item 1 is inconsistent with Streeting and Hamilton's assumption of unchanging prices of timber products, listed in Section 10.1); and
2 government charges could be increased to pay for conservation of the reserve areas.

Respondents were then asked to make a choice between wood production and conservation. The specific sum nominated varied from $2 to $400 annually across the sample. By varying the dollar sum and recording the corresponding proportion of respondents choosing conservation, the RAC was able to estimate a relationship between the probability of choosing the conservation option and the dollar cost of conservation to the individual voter. The RAC was then able to estimate the maximum preservation cost per voter that would attract the support of 50 per cent of voters—in other words, the maximum preservation cost consistent with success of a referendum on government action to preserve the forests.

Respondent attitudes and personal characteristics The survey respondents were questioned about their attitudes to other current environmental issues, about whether the government should do more to protect the environment, and about their age, sex, education, income and occupation. Answers to these questions were used to check the representativeness of the survey sample. They also were used to check the likely validity of the estimates of voters' willingness to pay. For example, one would be sceptical of individual statements about willingness to pay to preserve forests if voters' choices between wood production and conservation were uncorrelated with their attitudes to environmental issues in general, or if poorer voters were willing to pay more than richer voters.

The results The surveyed voters' median willingness to pay to preserve the potential conservation reserves was estimated at $22 per person per year, approximately three times that of the travel cost study. The difference between the travel cost and CV estimates is consistent with expectations. The travel cost method measures only values associated with direct use by visitors; contingent valuation measures both the value of direct use and values unrelated to use, such as values attached to the continued existence of the forests and to maintenance of options for their use in the future. Since the number of visitors to the forests is a fraction of all voters in New South Wales, Victoria and the Australian Capital Territory, the RAC's survey results suggested that the non-use values of the forests dominate their recreational values.

10.4 The costs and benefits of forest preservation

Table 10.4 puts all of the above estimates on a comparable basis. Assuming that recreational visits to the forests are valued the same each year from 1992 to 2040, the present value of recreational use is $13 million in 1992. Assuming that the same contingent valuation values apply each year, and that half of the 7.1 million voters in New South

Table 10.4 Economic benefits and costs of a cessation of logging, 1992–2040 (*$ million 1992*)

Benefits:	Production cost savings	38
	Preservation value of forests:	
	• Use plus non-use values	1075
	• Recreation use value only	13
Total benefits:	• Including use plus non-use values	1113
	• Including recreation use value only	51
Costs:	Forgone timber value	46
	Economic cost of unemployment	3
	Costs of unemployment compensation package	43
Total costs		92
Net present value	• Including use plus non-use values	1021
	• Including recreation value only	–41

Sources: Mark Streeting and Clive Hamilton, *An Economic Analysis of the Forests of South-Eastern Australia*, Resource Assessment Commission, RAC Research Paper no. 5, AGPS, Canberra, 1991; Resource Assessment Commission, *Forest and Timber Inquiry—Final Report*, vol. 2B, Appendix U, AGPS, Canberra, 1992.

Wales, Victoria and the Australian Capital Territory are willing to pay the median combined use plus non-use value of $22 dollars per year (which includes the recreational use value) for preservation, and the other half are willing to pay nothing, the overall present value of preservation is $1075 million. Preservation values arrived at using contingent valuation dominate both the commercial and social costs of ceasing logging and the value of recreation in the forests. According to the costs and benefits in Table 10.4, the non-recreational benefits of preservation would only have to exceed $41 million, equivalent to $3 million per year or 42 cents per voter per year, for preservation to be the preferred option.

The RAC did not undertake a comprehensive cost–benefit calculation like that illustrated in Table 10.3. The travel cost estimates were judged of limited usefulness because they measured only past recreational values, and ignored non-use values. The CV estimates were regarded as experimental, and not directly comparable with the market-based benefits and costs. In particular, the RAC was unwilling to extrapolate the per-person median willingness to pay to the whole population, although it seemed willing to accept the sample median preservation value of $22 per year.[7]

Why were the RAC commissioners apparently wary of using cost–benefit analysis and, in particular, of non-market valuation techniques? To understand their possible concerns, we review cost–benefit analysis in this chapter, and non-market valuation in Chapter 11.

10.5 Cost–benefit analysis

Cost–benefit analysis is a formal procedure for evaluating private or public actions from the point of view of the *community*, as opposed to the point of view of the individual. It involves:

1 measuring the gains and losses to individuals, using money as the measuring rod for those gains and losses;
2 aggregating the money valuations of the gains and losses of individuals and expressing them as net community gains or losses.[8]

Recall, from Section 3.3, economists' assumption that individuals compare costs and benefits of proposed actions, and act in ways calculated to yield the greatest net benefit. In cost–benefit analysis economists aggregate across individuals to obtain estimates of net benefits for the community as a whole. For example, consider the costs and benefits of a cessation of logging in all National Estate forests in the South-East, listed in Table 10.1. Different individuals affected by the cessation of logging would experience different costs and benefits. Individual sawmillers would lose the value of the sawlogs no longer harvested, while gaining the resultant savings in harvesting, transport and sawmilling costs. Some forestry workers would lose employment and income. Bushwalkers and birdwatchers would benefit from forest preservation. Four-wheel driving enthusiasts could suffer because of reduced road access to the forests. Such individual costs and benefits are aggregated in the cost–benefit calculation in Table 10.4, to produce an estimate of the net benefit of a cessation of logging for the community as a whole.

In the case of logging in the South-East Forests, different interested parties want different outcomes. Cost–benefit analysis is bound to be contentious because it presupposes a community planner who decides what course of action is best for the community after aggregating costs and benefits across people, *without any guarantee of compensation to the losers*. This raises a number of questions. One, already discussed in Chapters 3 and 8, concerns the motivation of the planner. Cost–benefit analysis implicitly assumes that planners are benevolent, concerned only to maximise net benefits to the community as a whole. A second question is what is the relevant community? In the RAC analysis of the South-East Forests, it was taken to be the people of New South Wales, Victoria and the Australian Capital Territory. Why not all Australians, or just people living in south coast New South Wales and East Gippsland? How wide the net is cast usually depends on the level of government making the decision, and the outcome may vary depending on whether the decision is made by local or state or Commonwealth planners. One part of the economist's job is to ensure that the benefits and costs of the proposal include *all* gains and losses experienced by individuals in the relevant community.

The final question about cost–benefit analysis is how are community members' costs and benefits identified, valued and aggregated? Here the economist is guided by a set of established principles for the conduct of cost–benefit analyses.[9]

1 A proposed 'project' is a proposal to modify the environment, sacrificing some valuable resources and goods to obtain other goods. For example, a cessation of logging in the National Estate forests of the south-east will produce the gains and losses listed in Table 10.1.
2 Cost–benefit analysis compares the value of the additional outputs due to the project with the value of the inputs sacrificed. This is equivalent to comparing people's valuation of the world with the project (a cessation of logging, in the preceding analysis) with their valuation of the world without the project.
3 Individual preferences regarding project outputs and inputs are what counts; in other words, what the individual wants is judged good for the individual.
4 Individual valuations of outputs and inputs are measured, wherever possible, by observing the amounts of money individuals are willing to pay for a benefit, or the amounts they are willing to accept to bear a cost. That is, they are measured by areas under individuals' willingness to pay (demand) and willingness to accept (supply) curves, as explained in Section 3.5. These valuations depend on more than the individual's preferences; they also depend on the person's income, which is in turn dependent on his or her property rights. Thus cost–benefit analysis incorporates the judgment that individual preferences should be weighted according to the existing distribution of income, which in turn implies acceptance of the existing societal distribution of property rights. (Recall the first of the assumptions in Section 3.7).
5 Individual willingness to pay and willingness to accept curves are aggregated by adding quantities of outputs or inputs across individuals to yield market demand and supply curves, as explained in Section 3.5. Areas under the market curves then measure the changes in aggregate value associated with additional outputs or sacrifices of inputs. The addition of dollar values of gains and losses across individuals implies that an extra dollar yields the same amount of satisfaction or welfare for each individual, regardless of his or her identity or personal situation. (Recall the second of the assumptions in Section 3.7). Thus, in analysing projects, cost–benefit analysis ignores the distribution of gains and losses across people.

Figure 10.1 illustrates the measurement of values using market information. The cessation of logging reduces sawlog and pulplog output. What are the values of these sacrifices, borne by timber producers? Suppose that the reduction in sawlog production, from Q_{s0} to Q_{s1} in Figure 10.1(a), causes a slight rise in the domestic price of sawlogs, from P_{s0} to P_{s1}, due to a downward-sloping domestic demand for sawlogs. The sacrificed value of sawlogs in money terms is measured by the area $S^0S^1Q_{s1}Q_{s0}$. In the case of pulplogs, suppose that the reduction in

Figure 10.1 Measurement of values using market information

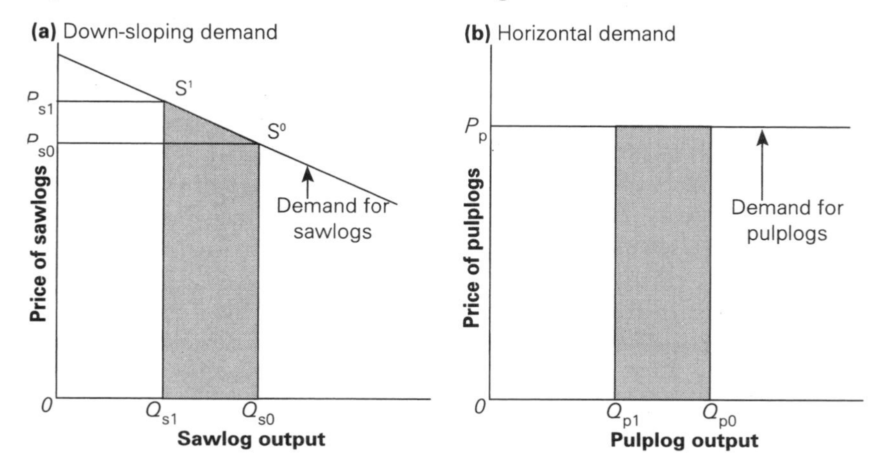

production, from Q_{p0} to Q_{p1} in Figure 10.1(b), has no effect on world woodchip prices and hence on log prices. The money value of pulplogs sacrificed is their price, P_p, multiplied by the output reduction, $(Q_{p0} - Q_{p1})$.

Cost–benefit analysis performed in accordance with the above principles does not produce morally correct decisions—unless the above procedure for identifying, measuring and aggregating people's gains and losses is morally superior to any alternative procedure.[10] This seems unlikely. Some people will disagree that social choices should be based solely on individual preferences. Some will object to weighting people's preferences by the existing distribution of income, and many to the assumption that an extra dollar yields the same satisfaction to a rich person and a destitute person.

Morality aside, some important gains and losses are extremely difficult to identify and value in money terms, for example the value of flora and fauna preserved by a cessation of logging in National Estate forests, and the personal and social costs of the resulting rural unemployment. Thus cost–benefit analysis does not replace political judgment; it is an aid to decision making by planners.

How are the results of cost–benefit analyses presented? Media reports and political arguments often boil the results down to a single number, downplaying or omitting discussion of the behavioural and moral assumptions involved in use of the technique, and of the problems of measuring some values. The single number will be either a *net present value* of the project, or its *benefit–cost ratio*. The former is the present money value of the benefits minus costs of the project anticipated over time, calculated according to the present value equation in Chapter 9. The benefit–cost ratio is the present value of project benefits divided by the present value of project costs, that is:

$$PV_B / PV_C = \{B_0 + B_1 / (1 + r) + \ldots + B_t(1 + r)^t\} / \{C_0 + C_1 / (1+ r) + \ldots + C_t / (1 + r)^t\}$$

Table 10.4 provides illustrative calculations of net present values and benefit–cost ratios. It also illustrates the profound differences in these numbers that can result from the use of techniques to attach money values to goods and services not priced in markets. In such circumstances, economists' techniques for eliciting money values where no markets exist are bound to be controversial. These techniques are the subject of the following chapter.

Discussion questions

1. Placing monetary values on some environmental gains and losses is unacceptable, because some environmental outcomes are a matter of ethics, not dollars and cents. True, false or uncertain? Explain why.
2. When making decisions using cost–benefit analysis, all the community planner needs to know is the final result of the analysis, either the net present value or the benefit–cost ratio of the project. True, false or uncertain? Explain why.
3. In what important ways does a social cost–benefit analysis of a major project—such as a new railway or mine or pulp mill—differ from a private business investor's assessment of the same project?
4. The RAC Forest and Timber Inquiry commissioners chose not to rely on the results of the travel cost and contingent valuation analyses of non-market values of the South-East Forests. Suggest possible reasons for the commissioners' scepticism about the results of these valuation techniques.

11

Valuing the environment

Since money is the most widely-accepted basis for comparing valued alternatives, where markets are absent planners often look for unpriced values related to people's marketplace behaviour. Where people use a natural resource, economists may obtain money values for unpriced environmental benefits or costs by examining the amounts sacrificed on related market goods, such as the costs of travel to and from national parks. Where no use is involved, as when people value the existence of a site or a species, values may be obtained by asking people what they would be willing to pay for preservation of the asset concerned. The first technique commonly involves high data collection costs, the second, high costs of survey and questionnaire design. Thus such valuation techniques are sometimes too costly to be worthwhile inputs into planning.

11.1 Bases for valuations of unpriced changes in the environment

Where markets are absent, due to prohibitively high costs of property rights and of markets (as explained in Sections 5.5 and 5.6), social coordination is commonly the task of government planners. Democratic planners require information about people's private valuations of unpriced changes in the environment, say the value of flora and fauna preserved by a cessation of logging. Information can be obtained from a variety of non-market sources (listed in Section 3.11), such as voting, opinion polling and lobby groups. Recall the method used by the RAC to value the social costs of unemployment among timber workers following reduced logging in the South-East Forests, outlined in Section 10.1. The RAC based its estimate of the social costs of unemployment on the Commonwealth–State compensation package for the timber workers affected by the cessation of logging on Fraser Island, which appears to have resulted from political negotiations between the parties concerned. However, as discussed in Section 3.11, such non-market valuations may be poorly related to interested parties' true willingness to pay or to accept monetary compensation for environmental changes. Non-market value signals may not register intensity of preferences, and can often be distorted at little or no cost to those responsible.

Planners allocating scarce public resources rely heavily on money values where markets exist, as illustrated in Table 10.2. Thus planning decisions commonly incorporate the ethical judgment implicit in cost–benefit analysis, that individual valuations should depend on the existing distribution of income and rights. Further, planners comfortable with market valuations are likely to believe that individuals' (private) willingness to pay or to accept compensation for environmental changes are appropriate valuations of non-marketed benefits and costs, and that non-market valuations are likely to be distorted. Thus it is unsurprising that planners often look for unpriced values with some basis in, or connection to, people's marketplace behaviour.

11.2 Categories of benefits and values

In designing techniques to elicit money values for unpriced benefits or costs of environmental changes, economists need to identify how the changes affect people. (Recall, from Chapter 1, that economics is concerned only with the valuation of preferences held by *people*.) For example, if people enjoy unpriced bushwalking in a forest, they can be questioned about behaviour relevant to that particular site; if the forest absorbs rainfall and protects local communities from flooding, it may be possible to estimate the resulting savings in flood damage; where people simply get satisfaction from the continued existence of the forest, there may be no relevant individual behaviour or community expenditure savings to serve as evidence regarding their valuation of the existence value of the forest. These distinctions suggest three useful categories of unpriced benefits and corresponding values:

1 direct-use benefits/values
2 indirect-use benefits/values
3 non-use benefits/values

Consider each of these in turn, as they apply to benefits derived from the South-East Forests. (Table 20.1 provides another list of benefits derived from forests.)

Direct-use values These are based on conscious use of environmental assets in consumption and production activities. For example, the South-East Forests provide fishing, hiking and timber. Many direct uses are both rival and excludable, and thus can be valued in markets; one person's timber harvesting reduces the amount of the resource available to others. Some direct uses, such as fishing and removal of ferns, are rival but non-excludable because the costs of fencing off or monitoring the valuable resource are prohibitively high. Other direct uses, such as hiking, which do little to modify or destroy the natural resources concerned, are non-rival and non-excludable, at least at low levels of use; as the intensity of use rises they may be rival due to overcrowding. Only the latter categories of direct uses are unpriced.

Indirect-use values These are based on the contributions of natural resources to human life support, discussed in Section 1.3. Humans are dependent on forest ecosystems, which capture the sun's energy, affect climate and cycle essential chemical elements and water, as described in Figure 1.1. The life support benefits of the South-East Forests would include carbon fixation, water purification, watershed protection, soil formation and the decomposition and assimilation of wastes. These benefits are non-rival and non-excludable; aside from the absence of markets, most people are unlikely to fully understand them, let alone attach specific values to them. Yet if the forests were cleared, these indirect benefits would change, probably imposing costs on the community, for example higher water treatment, flood control and erosion control costs and/or additional private health, flooding and erosion costs.

Non-use values These involve no tangible current interaction—no production, consumption or life support linkage—between the environmental asset and the people who benefit from it. The benefit is derived solely from information, specifically knowledge that the environmental asset exists or will exist in the future, for the individual concerned or for others. Such benefits may be based on the possibility of tangible, but as yet unknown, benefits for people in the future. For example, it may turn out that preserved ecosystems yield chemical compounds useful for treating human diseases. However, there is no such tangible interaction at present. Thus many Australians value the remaining undisturbed forests of the South-East regardless of their direct or indirect usefulness to people. Such benefits are non-rival and non-excludable; if the knowledge is available to one person, it is available to many, so that the aggregate benefit from such knowledge can be large. Recall that the RAC's contingent valuation study of willingness to pay to preserve the forests suggested a median willingness to pay among voters of $22 per year, based mostly on non-use values.

Economists have yet to reach a consensus on a classification of benefits and values derived from natural environments, in particular on categories of non-use benefits and values. *Total economic value* is the sum of all use and non-use values. Depending on the author, non-use values (sometimes called *passive use values*) may include *option values*, *quasi-option values*, *existence values*, *bequest values* and *stewardship values*.[1]

Among non-use values, the environmental economics literature most commonly distinguishes option values and existence values. The *option value* of the South-East Forests is the value attached to maintaining future options to use, learn more about and simply enjoy the existence of the forests and their constituents, where we are presently uncertain of both the future usefulness of the forests and of particular resources they contain (and hence of our future requirements for the forests and their constituents) and of their future availability.[2] It stems from our imperfect knowledge of our future preferences, of future technology, of the identity of some forest species, of the functioning of forest ecosystems and of future environmental changes. In these circumstances, risk-averse individuals may be prepared to pay today to preserve the forests as a form of insurance against an uncertain future.[3]

The *existence value* of the South-East Forests is a value people attach to the forests unrelated to their actual or potential use. Existence value is based on the satisfaction which individuals experience simply from knowing that the forests exist, for themselves and for others. Note that existence values are not necessarily unique to natural resources. People could derive satisfaction from the continued existence of the logging industry, logging communities and/or the logger's way of life, despite having no commercial or social connection with the timber industry. Existence values can increase the total economic value of development options, as well as of environmental preservation.

However useful the distinctions between different types of non-use values are in concept, they are currently of little practical value, since available estimates of unpriced non-use values cannot distinguish reliably between them.

11.3 Searching for signals of value: techniques for valuing unpriced goods[4]

Direct benefits The most straightforward way to obtain some direct-use values is to *create means of exclusion*, either 'fences' or surveillance backed by enforcement of payment for use or penalties for misuse. Once exclusive property rights can be defined and enforced, direct use of a good can be provided for and valued in a market. As discussed in Section 6.2, the economic feasibility of this approach depends on the state of technology, resource costs and the value people attach to the resource in question. Thus advances in electronic transmission and receiving technologies and increases in the amounts people are willing to pay for televised entertainment have led to the creation of direct user-pays markets for satellite and cable television. The growth in the values Australians attach to our native fauna has enabled the marketing of access to Australian wildlife. Warrawong Sanctuary, outside Adelaide, which preserves rare native animals inside Australia's first fox-proof and cat-proof fence, is a profitable private tourist attraction.[5] However, exclusion cannot generate indirect or non-use values.

If exclusion is not feasible, but people make direct use of unpriced resources, their market behaviour can sometimes be examined to reach conclusions about their implicit willingness to pay to use environmental assets. Many valued environmental assets can only be enjoyed if the beneficiary is willing to purchase related marketed goods; for example, it is necessary to spend money and time travelling to see penguins at coastal rookeries. Similarly, continual access to scenic views or low levels of air pollution can be 'purchased' by paying higher prices for land and houses, and less polluted working conditions by accepting lower wages. Thus examination of people's travel costs or variations in property prices and wages can reveal willingness to pay for the related environmental assets. These *revealed preference* valuation techniques are discussed below, the travel cost method in Section 11.4 and the use of property prices in Section 11.5.

If there are no particular marketed goods whose purchases are closely related to the unpriced good in question, it is impossible to estimate willingness to pay from market data. For example, visitors to an urban area where air pollution has been reduced cannot express their satisfaction via their bids for property or a job. In such cases economists have two main valuation options: *contingent valuation* and *construction of artificial value estimates* from distinct sets of data on environmental changes and market values of those changes.

The *contingent valuation* technique involves asking the beneficiaries what they would be willing to pay for the environmental benefit in question, based on a hypothetical description of the terms under which it will be provided. For instance, a sample of urban visitors could be asked about their willingness to pay higher transport charges in return for a carefully described improvement in urban air quality. Contingent valuation is discussed in Sections 11.6 and 11.7.

Economists can *construct artificial estimates* of the value of environmental changes by using market prices to value medical, biological, engineering, or other estimates of the effects that specified changes will have on output and/or the resource savings resulting from those effects. For example, the human health benefits of reduced urban air pollution could be estimated by combining medical estimates of the resulting changes in incidence of disease and/or medical resource savings with market values for work time lost, medical treatments, etc. The benefits of reduced pollution for urban structures could be estimated by market valuations of engineers' estimates of savings in maintenance requirements.

When constructing artificial estimates of the values of environmental changes, the economist depends on the advice of medical experts, biologists, engineers, etc. for information about the form of the benefits—the relationship between environmental changes, output changes and resource savings. It is essential to remember that such estimates are not based on the judgment or responses of those who actually benefit from or suffer from the environmental changes. To be credible, such constructed value estimates need to be backed by evidence that people perceive and would respond to the environmental changes in the manner assumed.

Suppose that people do perceive and respond to environmental changes in the manner assumed in the constructed estimates. For example, when urban air pollution falls, the major consequences are a rise in working time and a fall in medical treatment for respiratory illnesses. Then constructed value estimates should understate people's true willingness to pay for the environmental change. An urban commuter whose health improves due to lower air pollution not only gets more work income and/or has lower health costs; the person also enjoys his or her better state of health and the clearer views. Such non-marketed contributions of environmental improvements are not included in the constructed estimates of value. These estimates cannot reveal non-use values, where there are no tangible output changes or resource savings to measure. Also, if people perceive the impacts of environmental changes differently from the experts, the relationship between constructed values and willingness to pay depends on the difference in perceptions. For example, ordinary citizens may perceive the electromagnetic fields generated by high-voltage transmission lines as being more harmful than do experts. If so, the constructed estimates of the value of underground transmission will be lower than the value based on citizens' expressed willingness to pay for undergrounding.

Indirect-use benefits People use the natural environment's life-supporting services, such as oxygen production by living plants, water purification and waste dispersal and assimilation, without having to pay for or even be aware of the nature and value of those services. In the absence of understanding of the indirect life-support benefits of natural ecosystems, it is impossible for most people to make considered estimates of their willingness to pay to preserve these services. However, it may be possible to construct artificial estimates of indirect-use benefits based on calculations of the market costs of public and private actions to replace or mitigate the loss of environmental services such as water purification, waste assimilation and flood protection. Assuming that the community would in fact make such expenditures, the cost savings resulting from ecosystem protection are the best available means of estimating the indirect benefits derived from intact ecosystems, since it is impossible to estimate the community's aggregate willingness to pay for those benefits. Here again the economist is dependent on information supplied by technical experts, and on evidence or assumptions concerning the relationship between community or private willingness to pay and the constructed estimates.

Non-use benefits Where the benefit from an environmental asset is derived solely from knowledge of its current or future existence, the benefit is non-excludable. An individual beneficiary has little incentive to reveal the value he or she attaches to such knowledge, and little incentive to contribute to preservation of the asset, except in the unlikely case that this individual believes that he or she is the only person who values it, or that his or her valuation of it is large in comparison to the aggregate of all other beneficiaries' valuations. In these circumstances, apart from observing people's contributions to specific environmental preservation causes, contingent valuation is the only option. Non-use values can only be revealed by asking non-users what sacrifices they are prepared to make to ensure the continued existence of environmental assets.

11.4 The travel cost method

The travel cost method used to estimate the recreational value of the National Estate forests of the south-east was described in Section 10.2. The technique involves two major assumptions. First, visitors to a site respond in the same way to changes in travel costs as to a change in an entry fee. Second, people in all locations respond in the same way to a given change in the aggregate cost of use (travel cost plus entry fee). Suppose that the aggregate costs of a weekend visit from Canberra (mostly entry fee) and from Sydney (mostly travel cost) to the same South-East forest were each to rise from (say) \$50 to \$75. The travel cost method assumes that the annual visitation rate per thousand persons in Canberra and Sydney would be the same before the rise in access costs and the same (but lower due to the higher cost of a visit) after.

Several questions follow. What is being explained, numbers of visits or number of days spent on site? What if incomes and recreational preferences differ between Canberra and Sydney? What if people who visit the forests also visit beaches or friends or relatives en route? Travel costs time as well as money; is the visitor's time valued, and if so, how? Leaving out time costs will lead to underestimation of the value attached to a site. On the other hand, if people enjoy the trip for its own sake, or visit relatives en route, money costs alone could overstate the value attached to the site. The same is true if the visitor has only a slight preference for the particular site visited over some other site, say a Sydney skier's preference for Thredbo over Perisher; the extra value attached to skiing at Thredbo is far less than the aggregate cost of travelling from Sydney.

The above concerns can be dealt with at least partially by collecting additional information from survey respondents. For example, visitors' income and age can be included, along with aggregate travel costs, as factors explaining differences in visitation rates between Canberra residents and Sydney residents. People can be asked about the main purpose of the trip, and its characteristics; those who travel point-to-point by the fastest route and travel at night are less likely to enjoy the trip for its own sake. Wage rates may indicate the opportunity costs of some travellers' time. Questions about substitute recreation sites and their characteristics may help to determine the incremental value of particular sites, as opposed to the value of a group of substitute sites. There is a substantial literature on the travel cost method, which discusses these and other estimation problems.[6]

There are at least three other significant problems with the travel cost method of identifying use values. First, it cannot identify values for people who do not have to travel to the site, yet people who attach high values to outdoor activities like sailing or skiing may live closeby for that very reason. Second, it measures values attached to sites in the past, which may not always be a good guide to their future values. Most importantly, as in the case of the South-East Forests, the common reason for estimating the use value of sites is to value the consequences of possible degradation of existing sites, such as logging or mining in undisturbed forests, or pollution of beaches. Yet the travel cost method works best when we simply want a use value for an existing site as is, with the characteristics of the particular site and substitute sites held constant. Such values can only be a basis for inferences about how people would adjust to a changed situation, for example what alternative activities they would take up if a particular forest was logged, and the resulting costs.

11.5 Differences in property values: the hedonic pricing method

Hedonic pricing methods of valuation are based on the theory that the value of an asset is derived from the valued characteristics of that asset. Consider two otherwise identical houses on adjacent suburban streets, one subject to freeway noise and the other in a quiet location. The valued characteristics of the houses, which determine their prices, will include features of the properties themselves (number and size of rooms, type of construction, size of garden etc.), access features (distance to the central business district, public transport, hospitals etc.) and environmental features (crime rates, noise, air pollution, views etc.). Purchases/sales of the houses effectively involve exchanges of rights to enjoy, or suffer from, all these characteristics. Assuming numerous well-informed potential buyers for each house, and other would-be sellers of similar houses in the area, the difference in the prices of the two houses should be solely due to the difference in traffic noise.[7]

In practice, no two properties are likely to be identical in every respect except a single characteristic, such as noise level, proximity to a local park or view. Thus, to estimate the effect of a single environmental characteristic on property price, it is necessary to collect sale prices and detailed descriptions of all significant property characteristics for large numbers of properties. The analyst then estimates a mathematical equation designed to explain property price variations in terms of variations in property characteristics, that is, a mathematical function of the general form:

House price = f (property features, access features, environmental features)

Assuming that the estimated equation successfully explains almost all of the variation in house prices, it will provide an estimate of the effect of (say) a one-unit change in noise level upon the value of a house in the study area. Combined with data on house numbers and values, and engineers' estimates of changes in noise levels, this estimate may be useful for valuing the benefits/costs of projects that reduce/increase traffic noise, for example freeway screening, lower speed limits and freeway extensions.

Hedonic value estimates based on property sales represent the present value of all future benefits from a one-unit change in the value of the relevant environmental variable, such as noise. If we are interested in the annual value of a change in the environment, property rental values are relevant.

Hedonic property value studies require large amounts of data, not only market sales prices but also all significant features of each property (or neighbourhood, if the estimation uses average prices and characteristics for neighbourhoods). It is especially important to identify features of the property that have been added to mitigate or complement the environmental feature of interest, for example noise insulation that raises property value but was installed because of proximity to a freeway. The estimates obtained are only based on the true values people attach to environmental characteristics if the property market approximates the description of an ideal market in Chapter 3: property rights in houses are defined and enforced and transferable at low cost, each potential buyer is competing with numbers of other would-be buyers of the same house, each potential seller is competing with would-be sellers of similar houses, and all know the characteristics of each house on offer. As explained in Section 3.6, only under these conditions will the market price of a particular house reveal the buyer's and seller's true willingness to pay and to accept compensation for that house, and for the accompanying noise level. Given the infrequency and costs of housing transactions, in particular the very high costs of obtaining precise information about all house characteristics, and the possibility that people's perceptions of noise levels may be inconsistent with engineers' measurements, estimated equations for house prices may yield incorrect values for traffic noise changes.

Due to their large data requirements, hedonic property value studies almost always produce a single estimate of people's willingness to pay or to accept compensation as the environmental variable of interest (such as noise level) changes. Suppose that planners are contemplating a freeway extension, and that they conduct a hedonic pricing study in the relevant suburbs to estimate the costs of the increased traffic noise. The resulting estimate of the cost of extra noise is only a reliable guide to the future costs of freeway noise if we can reasonably assume that the cost of extra noise changes little as noise levels rise, and as the circumstances of individuals (such as their incomes) change.

11.6 Contingent valuation: communication problems

Contingent valuation (CV) overcomes the absence of a market for an environmental good or service by asking the assumed beneficiaries what they would be willing to pay for the environmental benefit in question, supposing that the benefit was either marketed or provided by a government project funded by themselves and other taxpayers. Unlike the travel cost and hedonic pricing methods, it does not rely on beneficiaries' behaviour revealing values in the marketplace. It is the only valuation technique that can estimate non-use values. The steps involved in the use of CV to estimate combined direct-use and non-use values of the National Estate forests of the south-east were described in Section 10.3.

The South-East Forests study involved a mail questionnaire and a referendum-style choice between wood production and conservation. CV interview techniques and hypothetical payment mechanisms ('payment vehicles' in CV terminology) vary. Where CV studies involve face-to-face interviews, scenario descriptions provided to respondents may include photographs and artists' representations of hypothetical environmental changes; the RAC's earlier CV study of the valuation of mining impacts at Coronation Hill in the Kakadu Conservation Zone employed these descriptive devices.[8] As an alternative to the referendum-style payment mechanism described in Section 10.3, respondents could be asked directly about their maximum willingness to pay (WTP) to preserve the forests, or minimum willingness to accept (WTA) money compensation for sacrificing the forests to wood production. These values are frequently obtained in a 'bidding game'—the interviewer either increases the sum involved starting from a very low value (WTP), or reduces the sum starting from a very high value (WTA), until the maximum acceptable payment or minimum acceptable compensation is reached. These bids could then be aggregated across respondents to estimate WTP or WTA over the group of beneficiaries as a whole.

Referendum-style payment mechanisms and WTP questions are the norm in CV studies, for reasons explained below.

Controversy over the validity of CV estimates of environmental values is a consequence of its use to estimate non-use values where no checks using market values are possible, and of the very large estimates of preservation values produced in some CV studies. For example, in the RAC's study of mining versus preservation at Coronation Hill in the

Conservation Zone adjacent to Kakadu National Park, the median willingness of Australians to pay for preservation was estimated at $600 million per year, or $4–5 billion 1991 dollars over 10 years.[9] This is more than 50 times the estimated net present value of the mine. It is also about 16 times the annual recreational value of the entire Kakadu National Park, as estimated using the travel cost method.

We discuss the problems of communication between CV interviewers and respondents in this section, and other problems of the technique in Section 11.7.

The Bungle Bungle Range of northern Western Australia, with its striking black-and-orange-striped sandstone domes, has recently become familiar from television advertisements. Most of the range is now contained in Purnululu National Park. Suppose that the Western Australian government is presented with a proposal for a major mining project in the Bungle Bungles, and commissions a CV survey of Australians to estimate the value of preserving the range as is. Since only a minority of Australians have visited or plan to visit the area, the CV value estimate is likely to be dominated by non-use values. Suppose that I am selected to be interviewed, and like the majority of respondents, I know nothing of the Bungle Bungles except the pictures I have seen on television. The interviewer describes mining and preservation scenarios and a payment mechanism, and then asks me what I am prepared to pay, either in contributions to a trust fund or in extra taxes, for preservation of the Bungle Bungles.

The survey is designed to identify my 'true' valuation of the preservation alternative. Will I in fact answer the question the government and CV analyst want me to answer? The possibilities for miscommunication are extensive. I may refuse to take the exercise seriously, either because I cannot be bothered, or because I do not believe that the government is capable of ensuring the outcomes described in the scenarios, or because I believe that the proposed method of payment would never be implemented. My response may be to simply 'think of a number'. The same may be true if I find it too demanding to understand the scenarios and judge their impact on me personally. (This is one reason for preferring referendum-style questions, where the respondent is given a specific cost of the conservation option, to direct questions about willingness to pay. Providing a dollar value reduces the effort required of the respondent.) Alternatively, because it is difficult to describe simply the effects of mining on the physical and biological environment of the range, I may interpret the impacts of mining as either more or less damaging than the survey designers intended, and thus state my willingness to pay to achieve or to prevent outcomes not envisaged by the CV analyst.

My moral stance may also impede communication. If I have moral objections to what I view as the commercialisation of nature, I may refuse to nominate a dollar value, even though I feel strongly that the range should be preserved as is. Similarly, if I believe that Australians have a 'right' to the range as is, I may refuse to nominate my willingness to pay because I believe that the miners should compensate me for mining, rather than my bribing them not to mine.

Another problem of interpretation of my answers arises because the interviewer does not know how I perceive my proposed spending on preserving the Bungle Bungle Range versus other claims on my total budget. I presumably have a rough proportion of my total budget that I am prepared to allocate to 'environmental protection', as opposed to other budget categories such as food, clothing, housing and transport. Besides the Bungle Bungles, other actual or potential expenditures from my 'environmental protection budget' may include contributions to help preserve tropical rainforests, to reduce woodchipping of native forests, to protect rare native fauna and flora, to reduce the salinity of inland waterways and so on. In deciding my willingness to pay, I may or may not carefully balance all such competing claims on my limited funds. The interviewer may have difficulty in determining whether I am taking account of other environmental causes or blowing most of my environmental preservation budget on preserving the Bungle Bungles. In the latter case, my willingness to pay to preserve the range may appear implausibly large in relation to my total income.

I may also respond to the CV survey by distorting my willingness to pay, either deliberately or unconsciously, for various reasons. First, if I believe that my answer could influence the final government decision, I may deliberately overstate or understate the true dollar figure, depending on my judgment about the marginal benefits versus marginal costs to me of lying. For example, if I want the Bungle Bungle Range preserved, but do not believe that preservation will have any effect on my after-tax income, I am likely to exaggerate my willingness to pay. Distortions may also occur for less strategic reasons. I may simply want to impress the interviewer, or I may misinterpret the survey as an opinion poll on whether the Western Australian government should do more to preserve the natural environment of the Kimberley. In the latter case a conservationist might signal strong agreement by reporting a high willingness to pay, and a supporter of mining strong disagreement by reporting that he or she is unwilling to pay anything.

By way of contrast, interpretation of respondents' answers is less of a problem where actual expenditures are involved. Suppose that I recently visited the Bungle Bungles on a four-wheel-drive safari tour from Kununurra or Broome. I revealed myself as a beneficiary of the range, and at least part of my valuation of a two or three day visit to it, when I paid for the tour. My decision to take the tour, and my willingness to pay $500–$1000 for it, were based upon the information I have collected from my travel agent or other sources about the features of the range, mode of travel, meals and accommodation provided and so on.

Critics of CV have discussed all of the above communication problems, and other possible limitations of CV studies, at length.[10] CV supporters argue that the communication problems can be greatly

Table 11.1 Guidelines for contingent valuation surveys

1	Avoid mail surveys, which have high non-response rates and eliminate the possibility of asking respondents the reasons for their choices.
2	Provide precise and understandable descriptions of the environmental alternatives (scenarios).
3	Ensure that the payment mechanism is realistic for respondents; hence a referendum format is appropriate for estimating non-use values, where most respondents realise that a market is not feasible.
4	Pretest questions, preferably in in-depth discussions, to ensure that respondents understand and accept scenarios and payment mechanisms.
5	Remind respondents of substitutes for the environmental attribute in question, and of other claims on their limited funds.
6	Avoid CV studies of environmental alternatives where the situation is so unfamiliar or the scientific and behavioural complexities are such that many respondents will be sceptical of the outcomes described in the scenarios.

Sources: J.W. Bennett and M. Carter, 'Prospects for contingent valuation: Lessons from the South-East Forests' *Australian Journal of Agricultural Economics*, vol. 37, no. 2, Aug. 1993, pp. 79–93; Paul R. Portney, 'The contingent valuation debate: Why economists should care' *Journal of Economic Perspectives*, vol. 8, no. 4, Fall 1994, pp. 3–17.

reduced, if not completely eliminated, by appropriate survey methods and questionnaire design.[11] Table 11.1 lists CV survey guidelines and the reasoning behind them.

Strict adherence to the guidelines in Table 11.1 will substantially raise the costs of, and reduce the use of, CV studies. Remember that CV is the only technique for estimating non-use values. The value of the extra (dollar-denominated) information about non-use values has to be weighed against the extra costs of supplying it.[12] The interested reader must consult the sources cited, and make up his or her own mind.

11.7 Contingent valuation: other problems

The RAC's CV study of the South-East Forests, described in Section 10.3, illustrates other problems with the technique.

Identifying beneficiaries: the geographic extent of the 'market'

Because non-use values measured in CV studies are non-rival and non-excludable, aggregate non-use values are obtained by multiplying sample average or median values by the assumed numbers of likely beneficiaries—in the case of the South-East Forests, the people of New South Wales, Victoria and the Australian Capital Territory. If the true beneficiaries come from a smaller geographic area, this assumption could cause an overestimate of aggregate value; if from a larger area, an underestimate.

Is WTP or WTA the appropriate measure of value? By analogy with the market transactions discussed in Section 3.5, where individuals' property rights were clear, maximum willingness to pay (WTP) to preserve sections of the South-East Forests is the appropriate measure of value if a survey respondent is acquiring the right to perserve forest that is presently available for logging. Alternatively, minimum willingness to accept compensation (WTA) is appropriate if planners are considering opening for logging forest that is presently preserved in the community interest.

CV studies almost always ask respondents about WTP, regardless of whether an environmental gain or an environmental loss is involved. This would not matter if survey estimates of WTA and WTP were similar. In practice, surveys typically reveal that WTA is two or three times as large as WTP. There are a number of possible explanations. WTA values are not constrained by incomes, as are WTP values; if I wish to signal a strong desire to preserve the South-East Forests, there is no upper boundary on my WTA, but my income restricts my WTP. Also, WTA makes it easier for me to exaggerate my desire for preservation; my estimate cannot be checked for consistency with my income. Finally, people may feel differently about things depending on whether they count those things as part of their rights and possessions. The prospect of losing a forest to which I believe I have a right may seem to me more important than the prospect of gaining the identical forest to which I believe I have no right. Of course, just because I believe that I have a 'right' to have a forest preserved does not mean that the right exists in law, or that it is consistent with other people's (e.g. forestry workers') perceptions of their 'rights' in the same forest. Thus the valuations obtained in CV studies are not necessarily consistent with a particular clear distribution of property rights, as is usually the case for marketed goods.

CV analysts commonly justify their preference for WTP measures on the basis that the estimates are more reliable, for two reasons. First, WTP reduces the opportunity for strategic distortions of values. Second, WTP estimates are less variable because respondents are more familiar with government imposts to finance environmental changes than they are with compensation for such changes. However, WTP studies are likely to underestimate respondents' true costs of losing environmental assets that they have grown to think of as 'their own'.

In the extreme case, mentioned in Section 11.6, an individual who believes that he or she has a right to have an environmental asset preserved, and values that asset highly, may refuse on principle to specify his or her willingness to pay for preservation.

Contingent scenarios mean the sample is unrepresentative of the population[13] Section 10.3 and Table 10.3 described the hypothetical scenarios and payment mechanisms provided to respondents in the South-East Forests study. The RAC's WTP estimates were contingent on this information, which was not available to the rest of the population of likely beneficiaries—the people of New South Wales, Victoria and the Australian Capital Territory. With the survey sample thus unrepresentative of the population, it seems inappropriate to extrapolate the survey WTP figure to the population, and the RAC avoided doing so directly.[14]

Remember that the democratic planner's task is to acquire information about people's values of unpriced changes in the environment. If CV is used to simulate a referendum on preserving an environmental asset, as was the case for the South-East Forests, how appropriate is a technique that provides decision makers, at substantial cost, with more and different information than they would have when going to the polls? Or should the planners only base decisions on the results of CV studies if they are prepared to undertake the very substantial cost of providing the same information to the whole population? CV involves more than person-to-person communication and sampling problems; it also raises questions about the nature of the public decision-making process for unpriced environmental changes, to which we now turn.

11.8 Unpriced values and planning: is some value better than none?

However estimated, unpriced values are inputs into the democratic planning process, along with the other information sources listed in Section 3.11. As such, they may or may not carry weight with planners. Remember, from Section 3.12, that planners may be more interested in the perceived impacts of their decisions on their electoral and bureaucratic career prospects than on the actual dollar outcomes, let alone the dollar outcomes generated by the various techniques for valuing unpriced goods. Thus, in judging the usefulness of the valuation techniques, we have to consider more than the estimation problems outlined in the preceding sections. We also need to know whether and how their results are used in public decision making. In the case of the South-East Forests, as pointed out in Section 10.4, the RAC commissioners who funded the travel cost and CV studies concluded that they were not a reliable basis for policy decisions.

Sinden, reviewing environmental valuation in Australia, identifies two distinct benefits of valuation studies: to assist in particular environmental choices, and to promote better general understanding of environmental problems and values.[15] Planners are primarily concerned with the former; in addition to the studies of the South-East Forests, the RAC study of mining at Coronation Hill and a CV study of the value of forests on Fraser Island were designed to provide input into contentious policy decisions. However, none of these were given significant weight in the reports that followed, or in subsequent government decisions.[16] It appears that, in Australia to date, valuation studies have been more important in promoting understanding of environmental values.

Elected planners are more likely to take account of valuation studies if they are able to tailor methods/scenarios/questions to their own interests; this is one of the roles of political party pollsters. This points to a potential problem with the above valuation techniques, one encountered with policy-oriented cost–benefit analyses. Where analysts are less than dispassionate for commercial reasons, analyses might be designed as much to shape voter opinion as to obtain valid estimates of unpriced values. The potential risks of manipulation are greatest with CV studies of non-use values, where there are no values revealed in market transactions divorced from the valuation study.

Many environmental economists argue that, where planners must make choices, some economics-based estimates of non-use benefits of environmental resources are better than none, because none can mean implicit valuation by planners shielded from public scrutiny.[17] This can only be true if valuations are performed in ways that minimise estimation problems, and if the benefits from the resulting value information exceed the costs incurred in the valuation studies. The first of these conditions is likely to be violated if political or bureaucratic planners manipulate the design or conduct of valuation studies out of self-interest. The second will be violated if planners take no notice of the valuation results.

CV is inappropriate when the environmental issues involved are unfamiliar or complex. For example, if many people do not comprehend the contribution of forest ecosystems to purifying water, or of reductions in ozone-depleting chemicals to the ozone layer, it is impossible to provide survey respondents with sufficient information for rigorous CV estimates of the values of such services. Even if respondents can be adequately informed, they then become unrepresentative of beneficiaries as a whole.

The use of the above valuation techniques in public decision making raises important questions about the way democracy functions. Of the techniques, the construction of artificial value estimates based upon the judgments of experts appears most, and CV least, consistent with the way most planning decisions about resource allocation are made. This is not a criticism of CV; on the contrary, CV estimates of people's values appear more consistent with the ideals of liberal democracy than reliance on experts. However, if CV is applicable to environmental planning, why not to planning decisions about other important unpriced goods, such as police services, pensions and airline safety standards? If it is not applicable to such goods, what features of the environment justify the use of a different public decision process? So far economists have paid little attention to the broader implications of the use of CV for the planning processes discussed in Sections 3.11 and 3.12.

Discussion questions

1. Attempts to value non-marketed environmental goods and services in money terms are objectionable, since they imply that all values in social decision making are reducible to dollars. True, false or uncertain? Explain why.
2. If an endangered species is unknown to almost all people, the economist has no way of estimating the value of the species. True, false or uncertain? Explain why.
3. It is impossible to measure the value of a neighbourhood park to the local community, since the services of the park are not sold and there are no costs of travelling to the site to serve as proxies for the value of its services. True, false or uncertain? Explain why.
4. In 1993 the Queensland government decided to buy parts of the Starcke property, on the coast of the Cape York peninsula, from a private developer. The 250 000 hectare property includes rainforest, mangroves, wetlands and pastoral land.

   Suppose that the Queensland government wishes to obtain a dollar value for the benefits of preserving the wilderness areas of the property. List the types of benefits which Australians and foreigners are likely to obtain from preservation of the Starcke property. In each case:

   (i) Explain whether the value of the benefit is signalled by prices in markets and, if not, why not.
   (ii) Suggest appropriate techniques which might be used to attach dollar values to various benefits of preservation.
5. According to survey reports, the lyrebird population in Sherbrooke Forest in the Dandenong Ranges east of Melbourne declined from about 120 in 1975 to about 60 in 1989, due to the activities of foxes and domestic dogs and cats. Suppose that the Department of Conservation and Natural Resources wants to know whether the costs of fencing the forest to exclude feral predators would be justified by the benefits to Melbourne residents. The Department undertakes a survey of Melbourne residents, where they are asked how much they would be willing to pay to save the lyrebirds. What questions would you want to ask before deciding whether or not you believe the survey results?

12

Monitoring changes in economic–environmental systems

A country can cut down its forests, erode its soils, pollute its aquifers and hunt its wildlife and fisheries to extinction, but its measured income is not affected as these assets disappear. Impoverishment is taken for progress.[1]

National accounting statistics do not monitor the performance of a country's combined economy and environment. The national accounts only record interactions between the economy and the environment if they generate money incomes for people. Use and degradation of natural resources which are available to all because exclusion is too costly, such as fish stocks or the waste assimilation capacity of streams, goes unmeasured despite the dependence of the economy on natural resource inputs, waste assimilation and ecological life support. However, suggested 'Green' national accounting measures may not improve national economic–environmental management. If we do not know how the combined economy–environment works, a single number for 'Green' national product provides no guidance for planners attempting to avoid adverse environmental impacts of economic growth.

12.1 The neglect of natural resource depletion in national accounting: Nauru

The Pacific island nation of Nauru has one significant domestic source of income, the phosphate rock derived from bird droppings that originally covered almost all of the 21 square kilometre island. Most of the original deposits have been mined and shipped to Australia and New Zealand for the manufacture of phosphate fertilisers. Since independence in 1968, the export of phosphate rock has earned up to $100 million per year, giving Nauruans annual incomes comparable to those of Australians. The phosphate deposits will be exhausted in the next few years. Nauruans' incomes will then be determined by the returns from phosphate-financed investments—Australian and Japanese shares, real estate such as Nauru House in Melbourne, remittances from Nauruans living and working overseas and local investments in tourism, agriculture and fishing. It is not just current income that is of concern, but also the level of income that Nauru's resources can sustain into the future.

If Nauru were a private company, its managers would be subject to clear signals about its economic circumstances and strong incentives to maintain income flows into the future. The company accounts would explicitly recognise the exchange of one kind of asset for another—the acquisition of offshore investments in exchange for the run-down of phosphate rock reserves. The market's valuation of the company would depend on the earning capacity of the respective assets exchanged. Suppose that the company's managers used most of the phosphate proceeds to support a lavish lifestyle; then the company's market value would fall as its future earning capacity was seen to decline. The decline in shareholders' asset values would generate pressure for improved asset management to maintain income flows.

The planner–managers of nation–states do not receive such clear performance signals. Almost all countries measure national economic outcomes using the standardised definitions and national accounting procedures prescribed by the United Nations. These measures fail to account for the depletion or degradation of natural resources, such as mineral deposits, forests and previously unpolluted streams, all of which lower future resource productivity and income-earning capacity. Thus Nauru's planners, observing the official accounts, could be misled and forget that the nation has been largely living off a rapidly dwindling stock of naturally occurring capital, and pay insufficient attention to an investment strategy designed to maintain Nauruan incomes once the phosphate is gone.

12.2 National accounting and its limitations

The standard national accounting measures, used worldwide to assess national economic performance, are *gross national product* (GNP) and *gross domestic product* (GDP). For the purposes of environmental monitoring, GDP, which measures the value of the output of all resources *located in* the country concerned, regardless of who owns those resources, is the more appropriate measure. GNP measures the value of the output of all resources *owned by residents*, wherever located. Nauru's GDP values the output of the island's resources. Its GNP includes Nauruans' earnings from local and overseas assets. In Nauru's case, the difference between GNP and GDP will be large, due to the size of the overseas asset portfolio. For most countries, the difference is small.

The national accounting concepts of GNP and GDP were designed to provide economic managers with output and expenditure data for short-term management of the national economy, to reduce unemployment and inflation. In order to reduce national economic outcomes to a single number, GDP calculations require aggregation across large numbers of separate outputs produced in a national economy. This is achieved by weighting quantities of goods and services by their relative market valuations at some point in time.

GDP is not a suitable measure of a country's economic performance, despite its common use for this purpose, because it omits depreciation of created capital, such as buildings and equipment, and also the depreciation of natural capital that results from resource extraction and waste discharges. *Net domestic product* (NDP), which is GDP minus the value of depreciation of created capital, is a more appropriate performance measure.

GDP per capita is even less satisfactory as an indicator of people's welfare or quality of life.[2] GDP measures production, not the consumption of goods and natural amenities that are the basis for a good life. GDP also leaves out the negative values of congestion, pollution and other side effects of production and consumption which reduce people's quality of life. Further, as an aggregate measure, GDP per capita ignores the distribution of personal incomes. It makes no distinction between the welfare of people in countries with the similar average incomes but very different income distributions (e.g. an inequitable oil-rich nation and Australia). GDP fails to record beneficial activities that are not marketed, such as voluntary rubbish clean-ups and unpaid domestic work. Unmarketed government services are valued according to their cost of production, rather than their benefits for people. GDP also ignores the welfare benefits associated with increased personal leisure time in an increasingly productive economy.

12.3 National accounting and changes in economic–environmental systems

Nauru is an extreme case of dependence on a non-renewable natural resource. The omission of changes in natural resource stocks in conventional national accounting is less serious in a rich country such as

Australia, where the major natural resource-dependent sectors—mining, agriculture, fishing and forestry—account for less than 10 per cent of GNP. However, the discussion of Nauru focuses on only one of the four ways in which the economy and the environment interact, natural resources used in production. Remember, from Figure 4.1, that the environment makes four types of contributions to the economy and to people. They are:

1 natural resource inputs (e.g. coal, timber);
2 waste assimilation (e.g. biological breakdown of wastes in streams);
3 natural amenities enjoyed by people (e.g. the Great Barrier Reef);
4 life support services of ecosystems (e.g. soil stabilisation, water circulation).

Comprehensive monitoring of the performance of a country's combined economic–environmental system needs to take account of more than just human use of environmental assets. Some exhaustible natural resources can be recycled and reused. Living renewable resources can be replenished, by natural increase or by human investments in reproduction, pest and disease control, pollution control, and so on. Also, planners need to know the magnitudes of stocks of environmental assets, to help determine whether the particular asset is relatively scarce or abundant. However economists commonly ignore the effects of economic activity on the environment's waste assimilation, amenity and life support capacities, in effect assuming that these services are freely available at levels that exceed any possible human demands for them.[3]

Figure 12.1, a more elaborate version of Figure 4.1, attempts to take account of the relationships just described, except natural increase and magnitudes of stocks. It depicts, as simply as possible, the nature of the economic–environmental system that we need to monitor if we wish to predict impacts of economic changes upon the environment, and of environmental changes upon the economy. The dashed lines indicate possibilities for investments in substitutes for environmental services.

The reader may protest that Figure 12.1 is complex. That is the point of including it. Any attempt to monitor changes in economic–environmental systems is going to be complicated and difficult; recall the discussion of system complexity in Section 4.4.

As in Figure 4.1, the largest and heaviest box in Figure 12.1 represents the environment. The economy, contained in and dependent on the environment, is represented by the intermediate-sized box in the lower portion of the figure. Natural increases in stocks of natural resource inputs, waste assimilation capacity, amenities and life support services are not shown. We can divide the interactions depicted in Figure 12.1, represented by numbered arrows, into two main categories and four subcategories:

1 *Intra-economy interactions* These are wholly contained within the economy box. They include exchanges of goods and resources

Figure 12.1 Relationships between the economy and the environment

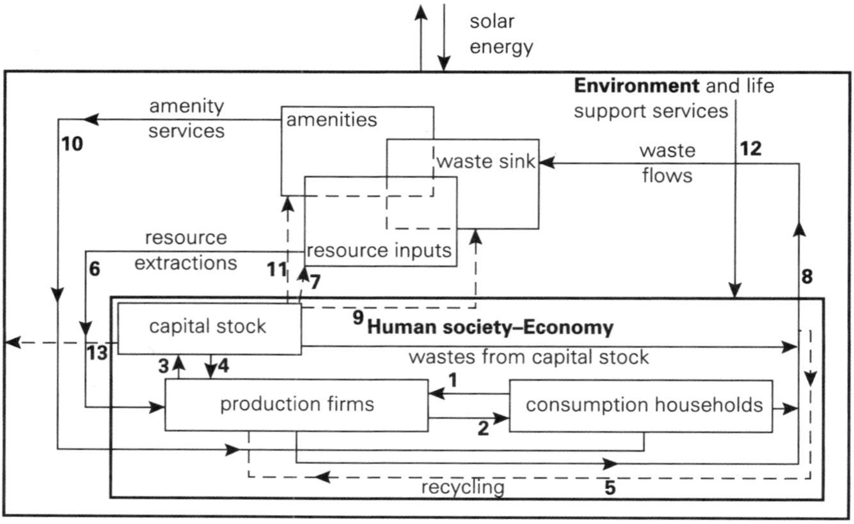

Source: Adapted from Michael Common, *Sustainability and Policy*, Cambridge University Press, Cambridge, 1995, p. 32.

between consuming households and producers (**1** and **2** in Figure 12.1), investment in created capital, for example machinery and education (**3**), use of the services of capital in production (**4**), and recycling of wastes (**5**).

2 *Economy–environment interactions*:

a *Natural input extraction and conservation* Along with the use of natural inputs (**6**), there are investments to conserve or augment natural resource stocks, for example insulating buildings against heat loss and restocking fisheries (**7**).

b *Use and conservation of waste assimilation capacity* Discharges of production and consumption and depreciated capital wastes into the environment (**8**) may be offset by investments to conserve or augment the assimilative capacity of the environment, for example sewage treatment (**9**).

c *Use and conservation of natural amenities* People enjoy natural amenities such as beaches and flora and fauna (**10**). They also invest to conserve natural amenities, for example clean-ups of beaches, and monitoring and penalties to protect rare and endangered species (**11**).

d *Use and conservation of life support services* The biosphere contributes life support services to people, for example plant production of oxygen and biological decomposition of many waste products (**12**). Investments to conserve biospheric life support services include, for example, CFC replacement to preserve the stratospheric ozone layer (**13**).

Which of these interactions are included and which omitted, wholly or in part, in GDP measurement? GDP measurements include the market value of most production for consumption and investment, plus the cost of production of unmarketed government services, plus the value of commercial and government recycling. The value of investment includes investment in conservation of environmental assets, including the cost of government expenditure to conserve environmental assets. In terms of Figure 12.1, GDP incorporates the interactions within the economy box (arrows **1**, **2**, **3**, **4** and **5**), plus investments in conservation of all four types of environmental assets (arrows **7**, **9**, **11** and **13**).

The economy's use and depreciation of natural resource inputs, waste assimilation capacity, amenities and life support services (arrows **6**, **8**, **10** and **12**) are either partially (in the case of extractions of natural resources) or almost totally (in the cases of waste discharges and assimilation, and amenity and life support services) omitted from national accounting. Yet natural input extraction and assimilation of waste discharges generally involve the using up or degradation of these environmental assets, in other words, depreciation of natural capital. On the other hand, the use of natural amenities and life support services commonly involves little or no resource depreciation.)

The national accounts only record transfers between the economy and the environment if they generate money income for people. Thus the extraction of natural resources that are the subject of defined and enforced property rights, and of market transactions (such as agricultural land, mineral deposits, commercial timber and fish) is recorded in national accounts. On the other hand, use and depreciation of non-excludable natural resources are not recorded since they do not generate incomes, depreciation of commercial stocks or expenditures. Such non-excludable, and hence non-recorded, services of the environment include production inputs such as the water that drives hydroelectric turbines, waste assimilation such as decomposition of sewage in rivers, amenities such as coastal scenery and the life support services of ecosystems.

The other major omission from GDP is depreciation of created capital. In national accounting, depreciation of capital equipment, but not human capital, is deducted from GDP to yield NDP. However, because of the difficulties of accurately measuring depreciation, the gross output measure is commonly used in government planning.

Plainly, if countries wish to monitor the ability of their combined economic–environmental systems to maintain living standards far into the future—the notion of sustainability introduced in Section 4.3—GDP will not do. Countries need indicators that incorporate all contributions of the environment to the economy, and all changes in stocks of environmental capital. To this end, some economists and environmental organisations advocate that national accounting procedures be modified to produce 'Green' national accounts.[4] Is this feasible and sensible?

12.4 The problems of moving to Green national accounts

How might national accounting procedures be modified to provide useful information about changes in a country's combined economic–environmental system? Starting with net domestic product, two modifications are widely suggested:[5]

1. *Deduction of environmental defensive expenditures*—expenditures that are specifically undertaken to protect or repair the environment. Defensive activities would include emissions reductions by firms taken to comply with environmental protection laws, extra household cleaning because of air pollution and publicly funded waste collections and clean-ups of lakes and rivers.
2. *Deduction of the depreciation of the stock of natural capital over the relevant time period* The value of depreciation is calculated by multiplying the change in the stock of each environmental asset by an appropriate price, and aggregating across assets. Ideally, depreciation would be calculated for all types of environmental assets—productive resources, waste assimilation capacity, amenities and life support capacity. In practice, attempts to correct national accounts for depreciation of natural capital have concentrated on productive resources directly or indirectly valued in markets, such as minerals, commercial timber and agricultural land.

These modifications would lead to a single number, in addition to GDP, that might be termed *environmentally adjusted national income* or *sustainable national income*. Its implementation would require major increases in resources for national statisticians. Would it be likely to improve the long-term management of national resources?

Consider whether the above modifications are likely to assist in assessing the impacts of clearing of Queensland's coastal mangroves (discussed at more length in Section 19.1) on the future well-being of Australians. Mangrove ecosystems provide all of the environmental services depicted in Figure 12.1: resource inputs (e.g. commercial fish and prawns), waste assimilation (e.g. trapping silt and decomposing effluents), amenities (e.g. flora and fauna enjoyed by tourists) and life support (e.g. storm mitigation). Alternatively, mangrove land can be cleared for recreational development. Possible economic and environmental consequences of mangrove clearing include changes in tourist numbers and total tourist revenue, and in categories of tourists and their spending patterns, and reductions in mangrove-dependent flora and fauna, including commercial fish species. Other possible consequences—increased pollution of coastal and reef waters and increased coastal storm damage—could contribute to further changes in coastal and reef ecosystems and generate new revenues or costs for the region and the nation.

Current national accounting will record short-term consequences of mangrove clearing that register in markets, including the market value of the clearing itself, additional tourist revenues, additional tourist industry costs, reduced commercial fishing revenues and extra government expenditure on pollution control and coastal protection. Is it feasible to go further, to measure and deduct the defensive expenditures and depreciation of natural assets that result from mangrove clearing?

Both identification and valuation of the consequences of mangrove clearing are subject to major problems. Identification of consequences is impeded by imperfect knowledge of the functioning of the economic–environmental system containing the mangroves. There is no reliable way to determine the amount of subsequent expenditures on coastal pollution control and storm protection specifically due to mangrove clearing. Scientific knowledge of mangrove and marine and reef ecology may be inadequate to enable reliable estimates of the long-term impacts of clearing on commercial fish and prawn populations, let alone on the rates of death and replacement of coral in neighbouring reefs, and changes in stocks of mangrove and reef flora and fauna.

Valuation of the consequences of mangrove clearing is impeded by the absence of property rights and markets for most of the direct, indirect, option and existence benefits of mangrove ecosystems. More serious still, many of the consequences of clearing are likely to extend far into the future. Valuation of consequences is also impeded by ignorance of future preferences, technologies and resource costs.

Clearly the suggested modifications to national accounts will not provide information about all the longer-term impacts of human actions on complex economic–environmental systems, as in the case of mangrove clearing. Neither our scientific knowledge, nor our knowledge of the preferences and options of our human successors, which determine future benefits and costs to people, are up to the task.

The mangrove example suggests several problems with environmentally adjusted national accounts. First, recall from Section 4.3 that neither economists nor ecologists understand the functioning of the economic–environmental system. Therefore, contrary to GDP, which is based on widely accepted models of how the economy works, the single number derived from environmentally adjusted national accounts has no reliable economic or ecological interpretation. Environmental accounting could leave us as misguided as hypothetical Nauruan planners who, observing high current GNP figures based on high consumption and low investment, conclude that Nauru's economy is in fine shape.

Further, remember that national accounting figures get used, not by benevolent and logical Mr Spocks, but by individual politicians, bureaucrats, reporters and lobbyists, each with his or her own goals, incentives and perceptions of the functioning of the economy and environment. If there is no consensus, even within individual disciplines, on interpretation of environmentally adjusted accounts, political players will have wide discretion to select the interpretations and policy prescriptions that suit individual agendas.

Second, even if national statisticians knew how the economic–environmental system worked, how could they value all the important outputs of the system (environmental assets as well as produced goods) as required if national performance is to be summarised in a single number? Valuation of the unpriced environmental consequences of human activity is subject to the problems discussed in Chapter 11. The costs of a serious attempt to provide non-market valuations of changes in nationwide stocks of flora and fauna, fresh air and water, specific ecosystems and so on would be prohibitive. More importantly, the non-market valuation techniques discussed in Chapter 11 could give misleading value signals if the values people attach to natural assets misrepresent the long-term ecological significance of those assets. For example, suppose that major new oil reserves were discovered close to the Great Barrier Reef. The current values attached to stocks of oil make no allowance for the possible long-term adverse impacts of oil extraction and transport on the reef environment. It could therefore be misleading to raise Australia's environmentally adjusted income to reflect the readily valued oil stocks while omitting the uncertain and difficult-to-value depreciation of reef assets associated with future oil extraction. The exploitable stocks of natural capital included (oil) are negatively correlated with the stocks of equally important natural capital that are omitted (reef flora and fauna).

Third, it is not clear what the suggested environmentally adjusted accounts measure. Remember, from Section 12.2, that GDP and NDP are measures of national production. Deduction of natural capital depreciation improves monitoring of changes in national production capacity, but deduction of defensive expenditures, which are part of national output, moves us away from a pure production measure.

If we were interested in people's welfare rather than national production, it is still not clear that environmental defensive expenditures should be deducted. The argument for deducting them is that they are undertaken to reduce harm rather than to increase people's well-being. For example, I soundproof my house when the state government constructs a new freeway nearby. Therefore, it is argued, defensive expenditures should not contribute positively to national income. However, the logic is slippery. The national accounts include the extra value of current production due to the time savings gained from the freeway. If the freeway is part of people's actual environment, surely expenditures made to reduce noise do enhance people's well-being, and should be treated the same as other consumption expenditures. It is hard to see any logical reason for distinguishing between expenditures made to reduce harm from freeway noise and those made to reduce harm due to naturally occurring environmental changes, such as severe storms.

12.5 A single number, or state-of-the-environment reporting?

A single-number measure of environmentally adjusted national income cannot be clearly based in economics and ecology, and is subject to diverse interpretations leading to a wide variety of possible policy prescriptions. Is there then any good reason to summarise the state of the national economic–environmental system in a single number? Common believes not. On the contrary, he states:

> The approach which seeks to wrap everything up in a single number . . . runs the risk of fostering the belief that matters are simpler, and better understood, than they are in fact.[6]

The alternative is for planners to rely on a variety of environmental, economic and social indicators to assess national economic–environmental performance. The environmental monitoring involved is commonly termed *state-of-the-environment reporting*. State-of-the-environment reporting simply records physical changes in the quantity and characteristics of selected environmental assets and system outputs, such as forest cover, crop yields, water salinity and populations of selected species.

State-of-the-environment reporting shifts more of the burden of making assumptions about the functioning of a poorly understood economic–environmental system to the planner, rather than less directly accountable national statisticians. Planners, assisted by ecological and economic analysts, have to decide which environmental data should be collected, and how the separate types of information—environmental, economic and social—fit together to yield a picture of the state of the national economic–environmental system. By comparison with environmental accounting, it is relatively easy and cheap to change the set of environmental indicators, and the way the data are combined, to reflect changes in physical, biological and ecological knowledge of the environment, and the corresponding values attached to different environmental assets.

State-of-the-environment reporting faces the same imperfect knowledge and interpretation problems as environmental accounting. For example, scientists or planners might decide, in ignorance, to monitor an ecologically redundant mammal, rather than a bacterium essential to continued functioning of major ecosystems. Conservation groups and the resource-based industries may disagree over the interpretation, and policy implications, of state-of-the-environment indicators. However, separate incommensurable indicators will serve as a constant reminder of the need for better information about system functioning.

Discussion questions

1. Gross domestic product, as conventionally measured, is a misleading indicator of national productivity changes because it omits to measure both the depreciation and conservation of environmental assets that contribute to national output. True, false or uncertain? Explain why.
2. It is better to summarise the state of the national economic–environmental system in a single number reflecting the best judgments of scientists and economists. Political decision makers cannot be expected to interpret complex scientific and economic data in a sensible fashion. Do you agree or disagree? Explain why.
3. If a country depletes or degrades its stocks of minerals, agricultural land, timber, air and water, but compensates by increasing stocks of manmade and human capital (machines, hospitals, high school graduates, etc.), do you believe that it can continue to increase its GDP in the long run? Explain why or why not.
4. What are the respective advantages and disadvantages of state-of-the-environment reporting and Green national accounts as ways of assessing the impacts of the nation's economic growth on the environment?

IV

LOCALISED ENVIRONMENTAL PROBLEMS

15.7 Signalling and incentives under direct controls, marketable permits and taxes
15.8 Taxes and permits when emissions transfers vary
15.9 Who pays for pollution control?
15.10 There is no perfect pollution control system

13

Social coordination in waste disposal and recycling

'It may be garbage to you, but it's our bread and butter.'

(Sign on a garbage truck)

Solid waste disposal and local pollution exemplify environmental problems commonly confined to a modest time frame and a single governmental jurisdiction. In such cases, social coordination can frequently be improved by some combination of market and planning signals and incentives. Solid waste generation and disposal in Australia involves a mixture of market and planning coordination of resource use. It illustrates how, even when environmental damage is readily recognisable and local, resolution of environmental problems may be very costly. Identifying waste disposers and measuring damages due to wastes and enforcing penalties can all require large amounts of private and public resources which have other valued uses.

13.1 Recycling: is it about doing good or making money?

A few people are old enough to remember the postwar years when no one had heard of the environment but most kids collected bottles for recycling because that was one way kids got pocket-money (a halfpenny for beer bottles, and up to threepence for the prized lemonade bottles).

Some kids still collect cans and bottles for the money, but probably more do so because parents or teachers have told them that it is a practical way of caring for our environment. A favourite slogan of the environmental movement is 'Think globally, act locally'. Collecting our containers and newspapers and so on for recycling seems to most children and many adults to be a direct way of doing our bit for the good of the planetary environment.

But are we doing good if the newspapers we faithfully bundle for the weekly collection often end up in rotting piles on an industrial lot in western Sydney or Melbourne? Or if council collections flood the market for recycled containers, drive prices down and put longtime commercial collectors out of business? Is our satisfaction from participating in recycling collections compatible with other people's desires for the materials we collect for recycling, and with the collector's desire to make a living from the trade, pithily expressed by the garbageman? There is no guarantee that it will be unless we have social coordination between the households and industries that supply wastes, the collectors and processors who transport them and convert them to useful products, and consumers like ourselves who desire reprocessed products. Without signals and incentives to coordinate the actions of these diverse groups, there is no guarantee that our keen participation in kerbside collections will lead to any recycling at all, in the true sense of recovery, reprocessing and reuse of used products.

Like the blind men describing the elephant, many of us think of recycling in terms of that part of the total 'product life cycle' with which we are in contact. We pay little attention to the transport, reprocessing and reuse stages of the recycling process, or to the costs of the alternative—extraction of further raw materials from the environment. If we did, we might ask questions such as: Does it pay to spend large amounts of money and energy transporting old newspapers back to paper mills when virgin paper pulp is available from forests close to the mills? What happens to the ink that must be removed and disposed of? Is the quality of the reprocessed paper equal to that of the original newsprint? If not, what is the demand for products based on recycled newsprint?

The whole recycling process is itself only one component of the system determining society's 'solid waste stream', a stream that originates not in mountain ranges but in industries and households when we make production and consumption decisions. These are followed in turn by industry and household decisions about how to dispose of solid wastes, industry decisions about reprocessing of some wastes, and, to complete the cycle, industry and household decisions about use of recycled products. All these decisions need to be coordinated by signals and incentives. If they are not, we are likely to end up with solid wastes in the wrong places, imposing costs on people to which they have not consented, as in the case of the piles of rotting newsprint.

Figure 13.1 The solid waste network

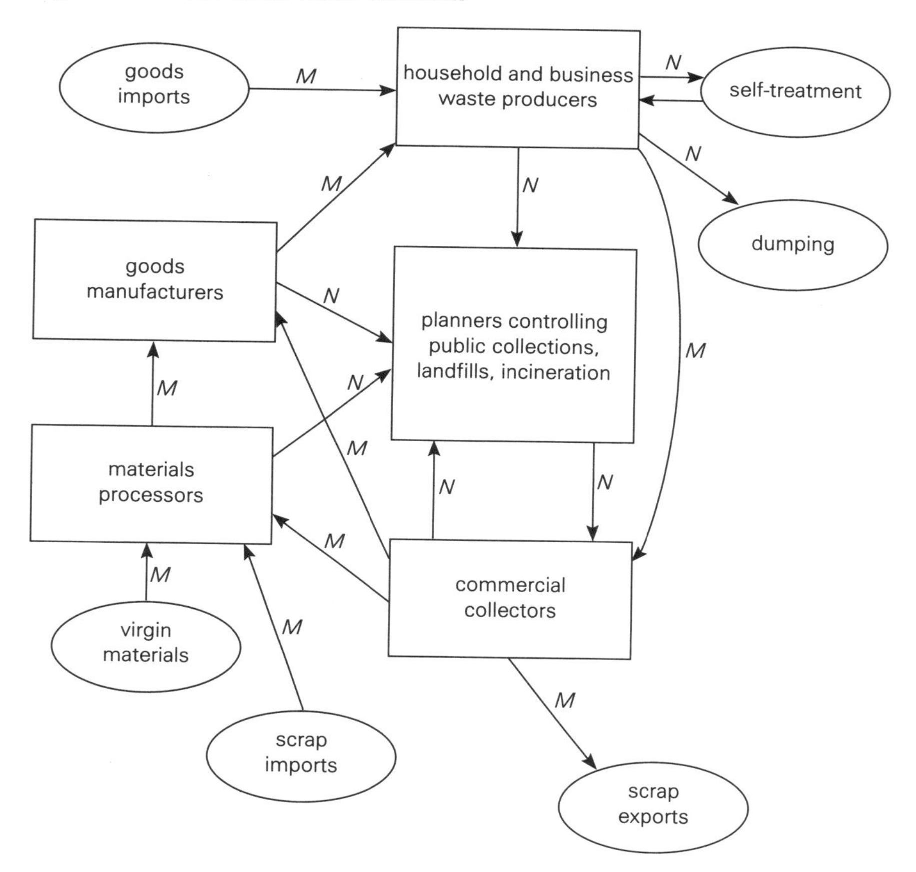

To understand the issues involved in the generation and disposal of solid wastes, including recycling, we must begin by describing the solid waste generation and disposal system.

13.2 The solid waste network[1]

Figure 13.1, adapted from the Industry Commission (IC) 1991 report on recycling in Australia, illustrates the flow of materials, the decisions that determine the amount and disposal of solid wastes, and the choices available to households and businesses. A major choice is between disposing of the wastes into the recycling stream or into the waste management stream (remember, from the law of conservation of matter, that wastes cannot be destroyed, only changed in form).

Each of the arrows in Figure 13.1 represents a transfer of resources or goods or solid wastes between the decision makers identified in the boxes. In Australia, most of the decision makers are private individuals or businesses. The exceptions are landfill and incinerator operators, who are practically always local councils or regional waste management authorities, and councils who organise collection of recyclables, either using their own employees or independent contractors.

In Figure 13.1, the transfers labelled *M* are mainly market exchanges, so that the quantity flows are mainly determined by price signals, and increasing profits or satisfaction from the use or disposal of one's property is the decision maker's major objective. Those labelled *N*, involving collection of recyclables from households, access to publicly owned landfills and incinerators for waste disposal and recyclable collection, and controls on self-treatment and dumping, are largely determined by non-market forces, in particular planners' decisions and individuals' preferences about waste disposal. Planners define and enforce rules on such matters as backyard incinerators, composting, waste dumping and access to public landfills and incinerators. They also determine charges and penalties for legal and illegal disposal, and taxes and subsidies for solid waste production and recycling.

13.3 Determining the extent of recycling: the case of newsprint

The extent of recycling is determined by the choices of the private and public decision makers identified in Figure 13.1: waste-producing households and businesses, waste management planners, recyclables collectors, materials processors and manufacturers of consumer goods. To understand the extent of recycling, we need to understand the alternatives available to each decision maker, and the incentives of and information available to each.

Table 13.1 identifies the separate factors determining the extent of recycling of any particular product, such as newsprint or aluminium beverage containers. The complexity of the table is rather daunting, but it is important for the reader to appreciate the very large number of factors that influence recycling rates. To illustrate the product life cycle and many of the choices involved, we examine the particular case of newsprint.[2]

According to the IC, in the late 1980s about one-quarter of the 650 000 tonnes of newsprint used in Australia was either recycled for use in packaging, or exported. The rest ended up in landfill. At that time, no newsprint was recycled to make new newsprint in Australia. In 1993 Australian Newsprint Mills (ANM) began processing used newsprint at its Albury mill; by 1996 it was using 125 000 tonnes of used newsprint per year, 70 per cent used newspapers and 30 per cent used magazines, mostly collected from Sydney, Melbourne, Brisbane and Adelaide.

Household choices about both their purchases of newspapers and magazines and their disposal of printed materials affect the amount of newsprint recycling. Households (and some businesses) can choose to buy more or less newspapers and magazines. They usually have several alternatives for disposing of used newsprint. They can sort and package newsprint and even deliver it for recycling, at some cost in money and time. They can use the garbage collection service, where available. They can treat newsprint themselves (e.g. backyard incinerators and composting), or they can dump it on other people's land, which is illegal.

For most households the major cost of contributing to recycling is the time it takes, not its money cost. If the household sees newsprint recycling as 'a good thing' in itself, the warm glow that comes with recycling may offset the time and money costs involved. The household estimates its own time costs of alternatives; money costs and returns associated with newsprint disposal are signalled by market prices paid to or by commercial recyclers, or by disposal prices set by planners, including the penalties for illegal dumping. Where the only payment for garbage collection is the annual garbage rate, the money price of garbage disposal of newsprint is negligible.

State and local government planners affect the extent of newsprint recycling when they decide the frequency and methods of garbage and recyclables collection, when they negotiate market deals with commercial recyclers of newsprint, and when they set waste disposal charges. The commercial recyclers commonly provide several services, not only pick-up and transport, but also sorting and storage of newsprint. They include businesses and voluntary organisations such as the Boy Scouts who collect to raise funds. Thus newsprint collection decisions are made by both planners and commercial organisations. Planners are more concerned with budgetary costs and political costs and benefits; commercial collectors are concerned about profits.

Commercial recyclers rely on market prices to signal the profitability of newsprint collections. Australian used-newsprint prices follow closely the export price of used newsprint, which fluctuates widely. In late 1995 and early 1996, the domestic price of used newsprint varied from about $100 per tonne to below $40 per tonne, with the price of magazine paper somewhat higher than that of used newspapers. ANM says that the cost of producing de-inked pulp in Australia is about the same as that of mechanical pulp from softwood forests. Newsprint recycling saves wood and electricity, but uses extra labour, fuel and capital in collection and transport. Newsprint collections from industrial sources such as printing works are generally homogeneous, clean and come in large quantities, whereas household newsprint collections are more heterogeneous, more likely to be contaminated with other wastes and come in small quantities. Thus the per-unit costs of pick-up, transport and sorting will be much higher for newsprint collected from households. Therefore we expect relatively less commercial collection of newsprint from households.

In the long term, reprocessing capacity and reprocessors' use of used newsprint depends on the prices paid for the products they produce,

Table 13.1 Choices determining the extent of recycling

Decision makers	Decision makers' alternatives	Likely objective	Choices determined by	Information sources
Households	• goods producing more or less wastes • recycled v. non-recycled goods • methods of waste disposal: – garbage collection – recycling activity—sorting, delivery, etc. – self-treatment, e.g. composting – dumping—usually illegal	• maximise satisfaction from given budget and time available for acquiring and using goods and disposing of resulting wastes	• money costs of alternatives • time costs of alternatives • consumption preferences • waste disposal preferences	• market prices of goods and of waste disposal alternatives • disposal costs due to planners: – dumping penalties – probability of getting caught
Businesses	• Products and inputs producing more or less wastes • recycled v. non-recycled inputs • methods of waste disposal: – garbage collection – commercial waste disposal – treatment on site, e.g. neutralise acids – dumping—usually illegal	• maximise profits from production and disposal of resulting wastes	• money returns from alternative products • money costs of alternative inputs and methods of waste disposal • business's environmental image	• market prices of products, inputs and waste disposal alternatives • disposal costs due to planners: – dumping penalties – probability of getting caught – costs of litigation
Waste management planners	• frequency and methods of waste and recyclables collection and transport • waste disposal methods (e.g. landfills v. incinerators) and siting • waste collection and disposal charges	• maximise political and bureaucratic career prospects	• budget costs of alternatives • political benefits and costs of alternatives • bureaucratic benefits and costs of alternatives	• market prices of inputs and sites • political signals: voting, lobbying, community meetings, demonstrations etc.

	• environmental damage (incl. dumping), penalties and enforcement			
Collectors of recyclables (businesses, voluntary organisations, councils etc.)	• types of waste collected • extent of pick-ups, transport, sorting, cleaning etc.	• commerical: maximise profits subject to constraints imposed by planners • public: maximise political pay-off	• money costs and returns from alternative collections • political benefits and costs of alternative collections	• market prices of recyclables and inputs—labour, fuel, storage etc. • political signals: voting, lobbying etc.
Materials processors	• types of recyclables reprocessed • mix of reprocessing and virgin materials (where the reprocessor also makes the final product)	• maximise profits	• money costs and returns from reprocessing and from using virgin materials	• market prices of reprocessed materials and reprocessing inputs • market prices of virgin materials—depend on extraction, transport, processing etc. costs
Final goods manufacturers	• mix of reprocessed and virgin materials	• maximise profits	• ability of reprocessed materials to substitute for the virgin materials • money costs and returns using reprocessed v. virgin materials	• market prices of final products, reprocessed materials, virgin materials and other manufacturing inputs

such as recycled newsprint or packaging, on the prices of electricity, labour and other inputs used in reprocessing, and on the price of used newsprint delivered to the factory. Where reprocessors are also manufacturers using the virgin raw material, as is the case for ANM at Albury, they directly compare the prices and processing costs of virgin material—pinus radiata thinnings—with those of recycled newsprint.

In the case of paper, the reprocessed product is not a perfect substitute for the paper pulp from a forest, due to shortening and thinning of paper fibres in reprocessing. To make new newsprint of satisfactory quality, ANM requires 30 per cent of the recycled newsprint to be the higher quality magazine paper. Alternatively, ANM must add a smaller proportion of virgin pulp.

Since reprocessed and virgin paper pulp are good substitutes up to a point, factors affecting the price of forest logs or thinnings are likely to affect the demand for recycled newsprint and the extent of recycling, particularly when expansion of manufacturing capacity is being considered. Thus if government intervenes to alter forest product prices, it may encourage or discourage paper recycling. The IC pointed out that government underpricing of timber could lead to less commercial recycling of timber products, including recycling of newsprint. On the other hand, if consumers prefer 'recycled' products to products made from virgin materials, as may be the case for products such as recycled toilet tissue, higher market prices of recycled products will encourage reprocessing and recycling.

Summarising, the extent of newsprint recycling in Australia is largely determined by the relative magnitude of two sets of costs. On the recycling side, there are the money and time costs of collection, sorting and reprocessing recycled newsprint, adjusted for any reductions in savings because the recycled pulp has to be mixed with some higher quality paper pulp. On the other hand, there are the combined costs of using additional forest timber (including extraction, transport and processing costs) plus the costs of landfill or other disposal of used newsprint.

13.4 Recycling in Australia

Table 13.2, from the IC report on recycling, shows the extent of recovery and reprocessing of products in Australia in the late 1980s. The IC found that substantial recycling was occurring. Recovery rates were one-quarter to one-third of total use for most metals, glass, tyres and for paper as a whole. Recycling is more extensive where the reprocessed materials are good substitutes for new materials, as is the case for metals. Where reprocessed materials can only be used for low value products, as in the case of most plastics, very little is recovered and reprocessed. Glass and papers and packaging recycled from industry

Table 13.2 Recovery and reprocessing of products, Australia[a]

Product or commodity	Quantity recovered ('000s tonnes)	Recovery rate (proportion of consumption recovered,%)	Net quantity of scrap exported (–) or imported (+) ('000s tonnes)	Scrap exported (–) or imported as a proportion of consumption (%)	Quantity reprocessed ('000s tonnes)	Proportion of consumption reprocessed (%)
Aluminium:						
• all scrap	99	31	–51	–16	48	15
• UBC	28	62	–5	–10	23	52
Lead	36	60	–20	–33	16	26
Copper	24	19	+3	+2	27	21
Steel	1616	26	–791	–13	825	13
Tin	<1	37	0	0	<1	37
Glass:						
• reprocessed–all glass	290	25	0	0	290	25
containers	204	24	b	b	204	24
• reused (refillable bottles)	<13	65	0	0	<13	65
Plastics:						
• industrial and commercial	65	50	+2	+3	67	53
• plastic in domestic waste:	1	<1	0	0	1	<1
• PET	<1	3	0	0	<1	3
• polythene	<1	<1	+1	+1	<1	<1
Paper:						
• newsprint	151	24	–48	–8	103	16
• printing/writing	164	22	–48	–6	116	16
• packaging/industrial	720	51	0	0	720	51
Lubricating oil[b]	84	18[c]	0	0	84	18
Organic waste:						
• household	210	9	0	0	210	9
	('000s units)		('000s units)		('000s units)	
Tyres	4000	24[d]	+207	+1	4225	25

Notes: (a) Estimates are for 1988/89 but may refer to different years (both calendar and financial), and are intended only as a guide.
(b) Reprocessing here refers to both rerefining and recycling into heating and other oil.
(c) Of total oil consumption: about half of all lubricating oil consumed is not available for recycling.
(d) Proportion of used tyres recovered in Australia for retreading.

Source: Industry Commission, *Recycling*, vol. I, *Recycling in Australia*, Report no. 6, Feb. 1991, pp. 25–6

are intermediate; they can be used to produce the same product if mixed with new materials.

The bulk of recyclables in Australia comes from industrial sources. The reason was pointed out in Section 13.3: large volumes of clean, homogeneous wastes greatly lower collection and reprocessing costs. To see the difference, compare the rates of recycling for industrial plastics and packaging with those for household plastics and newsprint. Another explanation for the much lower rates of recycling from households is the very different price signals to industry and households. The IC found that in 1990 industries were normally charged $10–40 per tonne for waste disposal at metropolitan landfills, while households paid for garbage disposal through their annual rates. Why bother to separate your stubbies for recycling when the council garbage service will take them away for free?

13.5 Why are planners involved in solid waste disposal?

We saw in Figure 13.1 and Table 13.1 that decisions about the use of recycled and virgin materials are for the most part guided by market price signals. Decisions about solid waste management are mostly made by planners and signalled to households and businesses by planning rules and prices set by planners. Why are the planners involved in solid waste disposal? Do solid wastes create costs that are not signalled in markets?

The market transfers identified in Figure 13.1 involve exchanges of exclusive property rights to resources or consumer goods or recyclables. Buyers' and sellers' actions are coordinated in bilateral market deals backed by the possibility of litigation if property rights are disputed. Government is only required to define and enforce property rights in, and responsibilities for, the goods exchanged, via the legislature, courts and police. (Note that property rights to use and transfer assets involve complementary responsibilities. If I leave the brake off, and my car runs into my neighbour's house, I must compensate my neighbour for the damage. I am also responsible if I misrepresent my car to a buyer prior to its sale.)

Can the production, collection and removal of solid wastes be coordinated by market deals based on exclusive rights in and responsibilities for solid wastes? This would be possible if all households and businesses producing or transporting or treating or storing solid wastes had well defined and perfectly enforced responsibility for any costs that those wastes imposed on others. Then those suffering costs, perhaps a neighbour upset by the smelly compost heap in my backyard, or a rural resident or passer-by upset by a car wrecking yard spoiling a scenic view, could force the offender to pay compensation determined in a market deal or in court proceedings. Government's only role would be to define and enforce rights and responsibilities for solid wastes.

Part of the explanation for the absence of some markets for solid wastes is the high costs of defining and enforcing property rights. Landowners have a legal right to enjoy their land free of other people's garbage, and waste producers have a corresponding responsibility to dispose of wastes in defined ways. Dumping is illegal. However, identifying and penalising illegal dumpers is often costly for both landowners and government. To the extent that people can avoid responsibility for legal disposal of their wastes by dumping, property rights are ill-defined, and waste producers have less incentive to undertake market deals for legal disposal of wastes. But planning has no advantage over markets in this respect. Identification of illegal dumpers is costly, and property rights are ill-defined, under both planning and markets. Therefore households and industries have similar incentives to dump their wastes on others under both systems.

The other important explanation for the absence of markets, and the presence of planning, in solid waste disposal is that some attributes of solid wastes (such as unsightliness and health risks) are non-rival and non-excludable 'bads', making their reduction or elimination non-rival and non-excludable goods. Solid waste disposal became the responsibility of city governments at the time of the Industrial Revolution, as a result of the dangers of solid wastes, especially organic wastes, to the health of urban populations. If household garbage is dumped on a vacant block in the central city or suburbia, there are costs beyond those imposed on the landowner. Numerous neighbours and passers-by suffer from the smell, the unsightliness and, most seriously, the health risks posed by ongoing organic decomposition and the scavenger organisms attracted to the site. Yet the landowner may disclaim responsibility, and each individual sufferer may defer action to solve the problem on the basis that the costs to him or her will be high relative to the benefit from garbage removal; better to wait for another sufferer to act. This is also the common reaction when we see someone littering a public road or park. Why risk an altercation with the offender when our effort to reduce litter will make no appreciable difference to the problem? Government coercion can overcome these free-rider problems by forcing all sufferers to contribute to government provision of solid waste collection and removal. It can also force contributions to public health or pollution control or environmental protection agencies, which monitor and enforce responsibility for wastes on behalf of all.

Recycling, too, can be a source of non-rival 'bads'. For example, newsprint reprocessing can result in discharges of salt and other chemicals and microbial nutrients into the environment.

Where the waste producer can be identified at low cost, for example in the cases of mine tailings containing toxic metals or manure from cattle feedlots, free riding can still impede private efforts to enforce the producer's responsibility for the wastes. Again, government coercion, which forces producers to reduce the non-rival and non-excludable 'bad', may greatly reduce the costs due to such industrial wastes.

For solid wastes, geographic dispersion of non-rival 'bads' is generally confined to a relatively small area. Therefore it is generally appropriate that planning of solid waste collection and removal is undertaken by local governments, reducing the possibility that unaffected parties are required to vote on, or contribute taxes towards, the solutions of particular solid waste problems. If the group of people impacted by solid waste disposal is much smaller than the group who get to choose what to do about it, the unaffected majority will have incentives to vote and lobby against government funding of solutions to genuine local solid waste problems. For example, a city-wide electorate may be unwilling to fund acquisition of more remote landfill sites, despite the problems of dust, smells and scavenging animals experienced by the small proportion of the citizens who happen to live close to the existing landfill.

13.6 Solid waste coordination problems

Can we improve on the present signalling system guiding the disposal of our solid wastes? In its report on recycling in Australia, the IC suggested ways of improving signalling. Let us look at some of the possibilities.

Better definition and enforcement of responsibility for wastes
Well-defined rights and responsibilities for solid wastes would ensure coordination all along the waste stream, either by market deals or by court settlement of disputes. For example, illegal dumpers and those who suffer from the dumping would be identified, the costs measured, and appropriate compensation and/or penalties imposed. Landfill and incinerator managers would either pay neighbours up front to accept the costs of possible damage due to events such as chemical leaching and smokestack pollution, or be liable for court-determined compensation payments after the event.

Unfortunately, such property rights solutions may be too costly at present. As already explained, it may be very costly to identify responsible parties. Such costs may fall with technological advance; for example, chemical or radioactive tracers added to industrial wastes may help to identify firms responsible for transport spills or dumping. However, free riding will still inhibit private action where there are many victims. Also, some forms of damage, such as leaching of chemicals from landfills to neighbouring properties, can take decades to occur, by which time it may be difficult to enforce responsibility, if the responsible party is still around.

There are also institutional barriers to property rights solutions, rules that may restrict liability or compensation payments by solid waste managers. One example is bankruptcy law, which may rule out compensation payments by a business in financial difficulties. Public waste managers may be protected from civil suits to recover damages. Such rules may be in the community interest on grounds that have nothing to do with waste disposal; for example, bankruptcy is believed to be in the interests of creditors as a group. In any case, there are costs associated with changing the rules.

The preceding problems could be overcome by requiring solid waste managers to post bonds in advance, as security against possible future damage to other parties. This is already the practice in a roughly analogous situation in Australia. Mining companies are commonly required to post bonds to guarantee that mine sites will be restored when mining ceases.

At present, enforcement of waste managers' responsibilities is mostly undertaken by government pollution control or environmental protection agencies. This overcomes the free-rider problem discussed in Section 13.5, but not the high costs of identifying those responsible for externalities. And these policing agencies are subject to the usual weaknesses of planners discussed in Chapter 8: they may provide misleading signals about costs due to solid wastes, either because they have inaccurate technical and value information or because the planners' incentives differ from those of the victims of solid wastes they represent. They also have limited budgets to detect and prosecute offenders.

Council charges better reflecting the full costs of waste management The IC found that council charges for garbage are typically too low to meet the financial costs of waste disposal, plus the costs of providing for replacement landfill sites, plus the costs of environmental protection measures at landfills. Yet household and business decisions to dispose of solid wastes in the garbage create all these costs for the community, and households and businesses should receive the appropriate feedback.

Local politicians, looking to the next council election in a year or two, are understandably reluctant to raise garbage charges. By not doing so, they fail to clearly warn garbage producers that landfill space close by, like a low-cost mineral deposit close by, is an increasingly scarce resource whose replacement will be much more costly in terms of site development and/or transport. Consequently garbage producers are less likely to cut down on disposal now, by reducing solid waste output or increasing recycling. As a result, they miss the chance to delay the much higher costs involved in shifting to new disposal sites. Skimping on charges now may also lead to skimping on environmental protection measures at landfills, leading to environmental damage now or in the future.

Whatever the charge for garbage, producers are more likely to act to vary their disposal practices if the charge is clearly related to the quantity of garbage collected. Annual garbage rates, which bear no relationship to the quantities collected from a household or business, provide no reward for reducing the quantity of garbage.[3] Thus the IC recommended quantity-based charges for garbage collection, as is the case when businesses deliver wastes direct to landfills. The cheapest forms of quantity-based charges, implemented in some USA cities, involve charges based on the number of garbage bins a resident is entitled to have emptied, or the purchase of tagged garbage bags where collection and disposal costs are prepaid when you buy the bags. Such charges are relatively costly to implement for the small quantities of solid wastes collected from households and small businesses, and may not be justified where landfill and transport costs are low, as may be the case for smaller rural cities and towns. Another likely cost of quantity-based charging is increased illegal dumping; when the costs of garbage disposal rise, we expect some households and businesses to turn to the alternatives listed in the second column of Table 13.1.

Better coordination of household collections and commercial recycling Recall, from Section 13.3 and Table 13.1, that recyclables collections from households are largely driven by households' attitudes to recycling and planners' decisions on the provision of recyclables collections. On the other hand, commercial collectors and recyclables processors aim to maximise profits. Thus there is no guarantee that recyclables collections will match the amount reprocessed. If households and planners believe that recycling is 'a good thing', the possible results, described at the beginning of this chapter, include stocks of unwanted newspapers and commercial recyclables collectors going out of business at the hands of groups of enthusiastic amateurs.

Figure 13.2, adapted from the IC report, illustrates the nature of the coordination problem and its consequences. In panel (a), the quantity of a recyclable item (say, old newspapers) that households supply to commercial collectors and councils (who in turn try to sell newsprint to the commercial collectors) is determined by several factors:

1. the cost of the least costly disposal alternative (which could be cancelling the paper, putting it in the garbage, composting it or dumping it on a weekend drive. Usually garbage disposal is the cheapest, because people pay for it in their rates, and the papers can easily fit in the allowed numbers of bins);
2. their personal disposal preferences;
3. their costs of bundling and delivering newspapers for recycling, which depend partly on newspaper collection arrangements.

The money cost of alternative disposal is measured vertically. Households' supply of old newspapers per month, S_hS_h, is determined by their preferences for newspaper recycling and their costs of supplying newspapers for collection. The higher the charge for garbage disposal, the more newspapers would be recycled, all other things equal. According to the diagram, because many households get satisfaction from participating in 'recycling', they are willing to supply OQ newspapers per month when there is no quantity-based charge for garbage disposal.

Panel (b) of the figure illustrates market determination of the

Figure 13.2 Household collections and the commercial recycling market

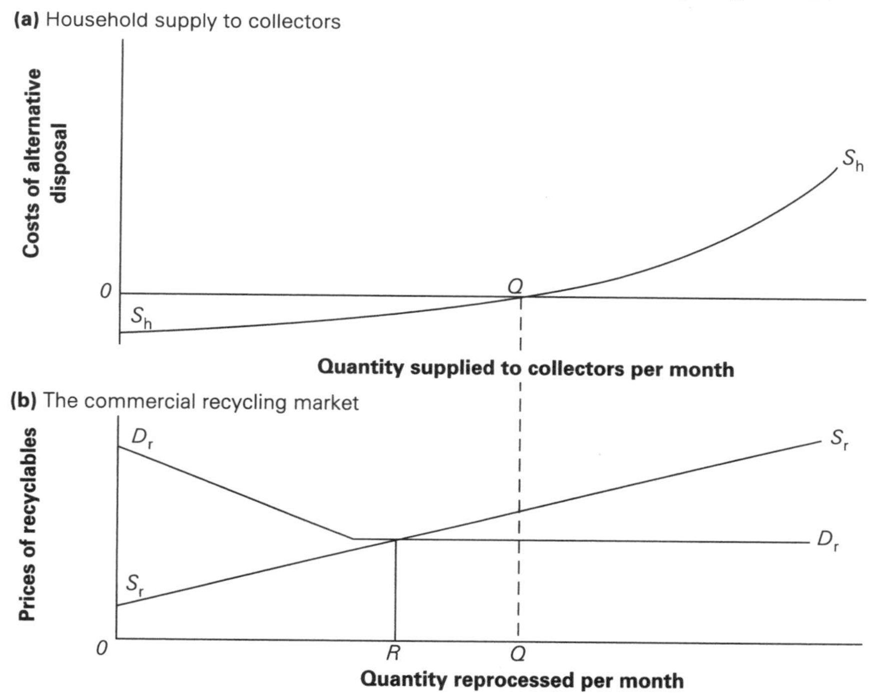

Source: Industry Commission, *Recycling*, vol. 1, *Recycling in Australia*, Report no. 6, Feb. 1991.

quantities and prices of newsprint for reprocessing. The demand for newsprint by reprocessors in Australia and overseas is D_rD_r. Demand slopes down at low quantities, reflecting Australian processors' decreasing willingness to pay for additional quantities. At higher quantities willingness to pay is constant, because Australian exports have no effect on the world price of used newsprint. Willingness to pay depends on the prices of the final products (paper and packaging materials), the prices of competing virgin paper pulp, and on the costs of reprocessing the used material. The supply of used newsprint from commercial collectors, S_rS_r, depends on collectors' pick-up, transport and storage costs, plus their costs of purchases from council collections. If reprocessors and collectors behave as described in the diagram, *OR* newspapers per month will be collected and reprocessed. The cost of the last unit of newsprint to the collector will just equal its value to the processor. Greater amounts of recycling would mean one or both were losing money. In the figure, commercial recycling of newspapers is less than the quantities collected for recycling, and stocks of used newsprint accumulate, with resultant possibilities for environmental damage.

The inconsistency between household collections and commercial reprocessing is unlikely to persist in the long run. Planners who decide on collections will receive feedback via markets (when they cannot sell all the newsprint they collect and have to pay for storage) and via the political process (citizens become concerned about recycling budgets and environmental damage). Feedback would be direct, and coordination of household and commercial actions would occur much faster, if households dealt directly with commercial collectors of newspapers. In this case, households might have to pay collectors to take newspapers, or collectors might pay households, depending on the price of used newsprint, and on how households and collectors divided up the tasks of newspaper transport and storage. If it was not profitable for commercial collectors to regularly visit individual households, households would have to bear the costs of transporting newspapers to collection points, in which case S_hS_h in Figure 13.2(a) would rise substantially. S_rS_r in Figure 13.2(b) might fall, depending on commercial collectors' costs of dealing with large numbers of individual households, rather than just a few councils. In any case, market feedback would more quickly eliminate discrepancies between household collections and commercial recycling.

Market deals between households and commercial collectors provide the required feedback. They also shift the costs of communication, pick-up, storage and transportation of recyclables, presently borne by councils that organise collections of recyclables, to households and commercial collectors. If households and collectors consider these costs too high relative to the costs of disposal to landfill and prices paid by reprocessors, the result will be less collection and less reprocessing of recyclables. Councils that face very high costs of disposal to landfill may still find that it pays to subsidise commercial collections of recyclables, because the council's saving on the costs of extra landfill space exceeds the cost of the collection subsidy. Alternatively, or in addition, councils can impose higher garbage collection charges on households.

13.7 Do we have environmental problems with solid wastes?

Many people, and hence some political and bureaucratic planners, think that we do. Many regard limited recycling by households, diminishing space in existing landfills and limited landfill sites close to metropolitan areas as evidence of such problems. But limited recycling and current landfills approaching capacity do not necessarily mean that private and public solid waste managers are not bearing all the costs of their decisions. It may simply be that landfill disposal is cheap, and recycling is therefore unattractive. Even illegal dumping, which clearly does impose costs on people who have not consented, is not a soluble environmental problem if the costs of policing and penalising dumpers exceed the costs suffered by its victims.

To decide whether we have an environmental problem with solid wastes, we need to apply a two-step test. First, are private or public decision makers responsible for externalities—costs that they do not themselves bear? Second, is it possible to create market or political feedbacks to overcome the problem at a cost less than the cost of the externality? If the answer to both questions is yes, we have a soluble environmental problem; otherwise we do not. In its report on recycling in Australia, the IC was careful to point out that signalling improvements, such as volume-based or weight-based charging for household garbage collection, might be too costly to implement in some cases.

Solid waste externalities do exist. Illegal dumping is not always policed and punished; local planners put off raising garbage charges, raising costs for future ratepayers, in order to get re-elected; households and councils collect recyclables that sometimes end up in unsightly piles in another municipality. The question is, can those responsible be made accountable at an acceptable cost? The answer is not clear without looking at the circumstances of each case. We do know that accountability is costly, whether it involves identifying dumpers, measuring leaching from landfills, measuring quantities of garbage, public reporting of political and bureaucratic decisions, or legal or political actions to penalise decision makers. So it will not pay to identify and penalise every errant solid waste manager.

Private and public accountability can be improved in the long run, as identification and measurement technologies improve, as people are better informed about political and bureaucratic decisions, and as rules restricting liability and compensation payments are modified. But solid wastes illustrate the point that there are no perfect solutions to environmental problems, whether we rely on the market or government planners or some combination of the two.

Discussion questions

1. Recycling of household wastes is far less extensive than recycling of industrial wastes, because consumers care less about the environment than industry. True, false or uncertain? Explain why.
2. Mandatory recycling programs, such as South Australia's requirement for compulsory beverage container deposits refundable at bottle collection centres, could do more harm than good to the environment. True, false or uncertain? Explain why.
3. What do you see as the main factors determining the extent

of recycling in Australia? In Japan? How would you predict that recycling levels would vary between the two and why?

4. The City of Melbourne is trialling a system of weighing individual garbage collections and charging households and businesses according to garbage weight. Discuss the advantages and disadvantages of this system compared to paying for waste disposal as part of your property rates.

14

The economics of pollution control: two parties

If pollution is legal and benefits someone, its benefits to the polluter have to be balanced against costs imposed on pollution sufferers. Thus the socially efficient level of legal pollution is rarely zero. In the simplest cases, involving a single polluter and a single sufferer, differences may be resolved by bilateral negotiation or private legal action, but only after prior determination of whether the polluter has right to pollute or the sufferer has the right to be free of pollution. The question of who has what rights is not an efficiency issue; it is matter of community ethics and justice.

14.1 The benefits and costs of pollution

Pollution must benefit someone, or it would not occur. It is commonly a by-product of legal production and consumption, where the decision maker does not bear all the costs of his or her actions. Once we recognise these realities, it is plain that the benefits of a legal activity to the polluter have to be balanced against costs imposed on those who suffer from pollution.

Such a trade-off between the benefits and costs of pollution was illustrated in Figure 5.2, which is reproduced here as Figure 14.1(a). In Figure 14.1(a), the steel company's benefits from producing extra steel and smoke are the extra profits it can make, the difference between the price of the steel, P_{st} (remember that the company gets the same price for all tonnes of steel it sells) and the private marginal cost of

Figure 14.1 Marginal benefits and costs of smoke pollution

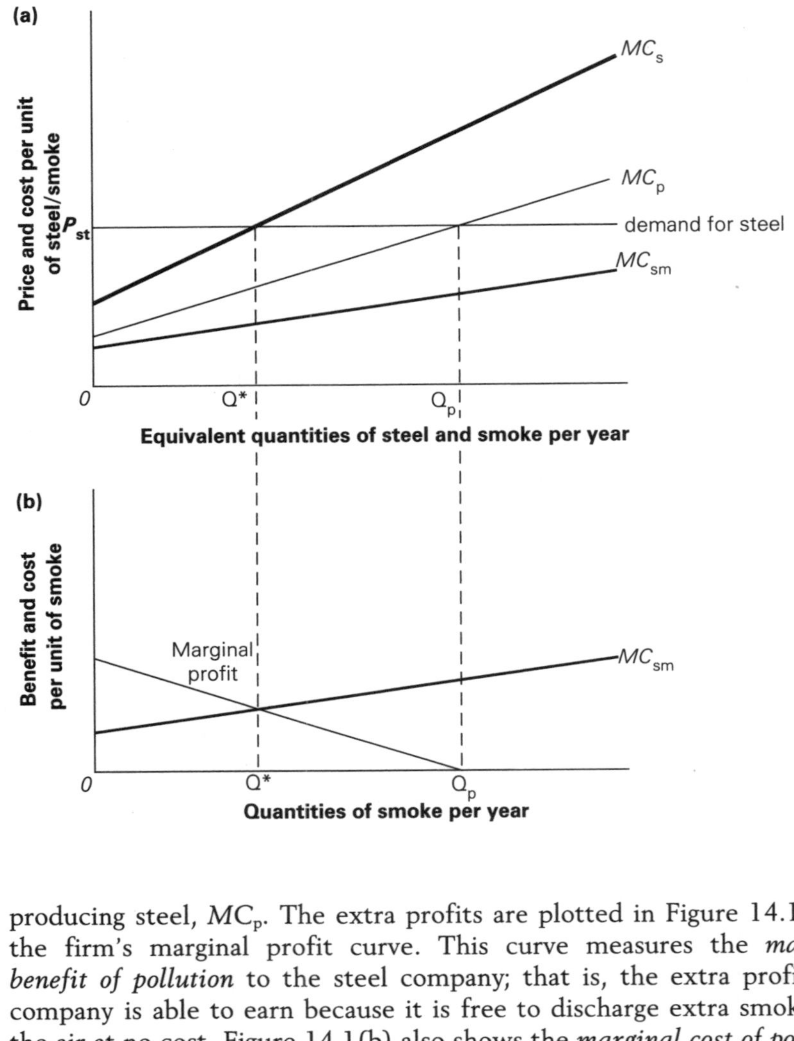

producing steel, MC_p. The extra profits are plotted in Figure 14.1(b) as the firm's marginal profit curve. This curve measures the *marginal benefit of pollution* to the steel company; that is, the extra profits the company is able to earn because it is free to discharge extra smoke into the air at no cost. Figure 14.1(b) also shows the *marginal cost of pollution* to those who suffer from the smoke, which is the MC_{sm} curve from Figure 14.1(a). The net benefits from steel and smoke combined are maximised at that level of steel and smoke output where the marginal benefits of extra smoke (the extra profit from the accompanying extra steel) are just offset by the marginal costs of the extra smoke. This occurs when steel and smoke output is Q*. (Textbook depictions of the efficient level of pollution commonly show the marginal benefits of pollution falling, and the marginal costs of pollution rising, as the level of pollution increases. This is not necessary, provided that the marginal cost curve cuts the marginal benefit curve from below.)

The logic is the same when the polluter is another household, rather than a business, except that the benefits of pollution come in the form of satisfaction to the polluter, not profits. Suppose that students in a flat get pleasure from playing hard rock at high volume. The residents of neighbouring flats dislike loud rock music. Assume that the students' private willingness to pay for loud rock, and the neighbours' private willingness to pay to reduce stereo volume, were somehow known. If the horizontal axis in Figure 14.1(b) measures stereo volume, the marginal benefit of pollution to the students is indicated by their willingness to pay for incremental increases in stereo volume. The marginal cost of pollution to the neighbours is indicated by their willingness to pay for incremental reductions in volume. Students and neighbours combined will be best off when the volume is set at Q^*.

These two examples demonstrate an important point. If pollution is legal and benefits someone, the efficient level of pollution is unlikely to be zero.

14.2 Legal and illegal pollution

Our concern in this chapter is social coordination where pollution is legal. In the absence of specific markets for pollutants such as smoke and noise, how do polluters and pollution sufferers communicate the values they attach to different levels of pollution, and how are polluters motivated to respond to the desires of sufferers and vice versa?

If pollution is illegal, there are no rights to use assets for polluting activities. I have no right to shoot clay pigeons in my suburban backyard, or to fish using dynamite. Industries have no right to dispose of radioactive wastes in Sydney Harbour or the Swan River, or to drill exploratory oil wells along the Great Barrier Reef. In such cases, polluters and pollution sufferers are not permitted to explore the benefit–cost trade-offs along the horizontal axis in Figure 14.1(b); planners set the level of pollution at zero, point O in Figure 14.1(b).

What justifies making some pollution illegal? Remember, from Chapter 2, that there are good reasons to give democratic planners the power to define and enforce property rights and other rules specifying the legitimate uses of assets. Societies can only exist when people agree to observe rules governing the use and transfer of scarce assets. Such rules ban many activities that benefit one person without forcing him or her to bear all the costs imposed on others, for example theft of someone else's stereo or running red lights. If polluting activities threaten social harmony because they impose costs that are widely viewed as unacceptable because of their magnitude or on ethical grounds, they too may be made illegal. For example, some people may argue that we have no right to use nuclear power because the future generations who will have to deal with our nuclear wastes have no say in today's decisions.

Ideally, polluting activities would be made illegal because their by-products or residuals create costs that exceed their benefits even at very low levels of production or consumption. In terms of Figure 14.1(b), this implies that the marginal cost curve is always above the marginal benefit curve, so the efficient level of pollution is at O.

The problem with making polluting activities illegal is that since rights to conduct those activities may no longer be exchanged in legal markets, it is more difficult to estimate the benefits and costs of pollution. For example, if government prohibits incineration of hazardous wastes on environmental grounds, industries' and hospitals' willingness to pay for incineration may remain obscure. This would not matter if democratic planners had accurate information about the benefits and costs of pollution, but we will see in Section 15.2 that this is unlikely to be the case. We also know that pollution control legislation is the subject of energetic lobbying campaigns by industry, labour and environmental interest groups, who have limited incentives to tell the truth. And we saw in Chapter 8 that planners generally have some freedom to pursue their personal goals at the expense of citizens' goals. Thus it is quite possible that the efficient level of some legal polluting activities could be zero, and that of some illegal polluting activities could be positive.

14.3 Pollution involving two parties

The simplest pollution problems to resolve are those between neighbours. Suppose that you install solar panels to heat your pool, and your neighbour grows shade trees in his or her yard. The higher the trees, the cooler and dirtier is your pool, due to more shading and more leaves and other debris in the pool. The trade-off between your neighbour's benefits and your costs from the trees is illustrated in Figure 14.2. The horizontal axis measures the height of the trees. In this case, the marginal benefit of pollution (per time period, say per year) is measured by your neighbour's (private) willingness to pay for extra tree height. The marginal cost of pollution per year is measured by your (private) willingness to pay for reductions in tree height. The efficient level of pollution—the tree height that maximises net benefits to you and your neighbour combined—is Q^*. But your neighbour would obviously prefer Q_m, where the benefits from shade are maximised, and you would prefer Q_l, where your costs of shade and debris are minimised.

What coordination mechanisms can be used to achieve Q^*? To be successful, any mechanism must signal the values attached to different tree heights, and provide incentives for each party to respond to the desires of the other. We consider three alternative coordination mechanisms: private negotiation, court awards of damages and rules determined by planners. The first two depend on existing property rights. The last involves defining and enforcing new property rights.

Figure 14.2 Trade-offs in two-party pollution

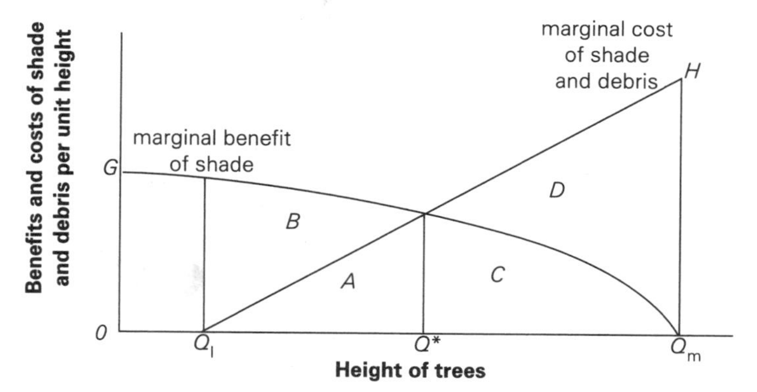

14.4 Private negotiation

The starting point for negotiated deals over the height of the trees, as for market deals over avocados in Chapter 3, is clearly specified individual property rights. If you and your neighbour disagree about whether you have the right to direct sunlight, or whether your neighbour has the right to grow trees as high as he or she wishes, a court must decide the existing rights. The court decision determines who must pay whom to get their way, and negotiation cannot commence until this is clear.

Suppose that the court determines that your neighbour has the right to grow trees to whatever height he or she chooses. Left alone, your neighbour will let the trees reach Q_m. You can only reduce shading and pool debris by paying your neighbour to trim the trees back. Starting from Q_l, the marginal cost curve HQ_l shows the maximum amounts per year that you would be willing to pay for successive reductions in tree height. The marginal benefit curve Q_mG shows the minimum amounts per year that your neighbour would accept for successive height reductions. Assuming that the two of you are honest about these maximum and minimum amounts, you will agree to reduce the height of the trees from Q_m to Q^*. There will be no deals to trim the trees below Q^*, because below Q^*, the amount you are willing to pay for height reductions is less than your neighbour will accept. In reducing tree height from Q_m to Q^*, you save costs equal, in money terms, to the area $C+D$ per year; you pay C of this, plus some of *D*, to your neighbour to compensate for the loss of the area C, the value of the benefits he or she loses because the trees are trimmed. The net benefit to the two of you combined is the area *D*. How *D* is divided between you and your neighbour depends on the size of the payments, which can be anything between your maximum (the marginal cost curve) and your neighbour's minimum (the marginal benefit curve).

Now assume that the court grants you the right to be free of shade and debris. Left to yourself, you will obtain a court order to prevent the tree height exceeding Q_l. Your neighbour can have higher trees and extra shade by paying you to accept more shade and debris. Starting from Q_l, the marginal benefit curve GQ_m shows the maximum amounts per year that your neighbour would be willing to pay for successive increases in height. The marginal cost curve Q_lH shows the minimum amounts you would accept. By the same reasoning as before, the two of you will agree to allow the trees to grow to Q^*; your neighbour will gain, in money terms, the area $A + B$, of which A plus part of B will be paid to you as compensation for extra shade and debris. The overall gain from the deal is the area B, which is divided between the two of you.

Our analysis of the pollution problem between neighbours incorporates two assumptions, and leads to what at first may seem a surprising conclusion, first pointed out by the Nobel prize-winning economist Ronald Coase.[1] We discuss the assumptions first, and then the conclusion that follows from them.

Recall, from Sections 5.4 to 5.6, that with property rights determined, obtaining the information required to conclude market deals—to identify the other party, to measure what is being transferred, and to identify what the other party is willing to pay or to accept—is costly. We have assumed that, in deals between you and your neighbour, these information costs are negligible. In particular, we have assumed that the two of you honestly reveal the amounts you are willing to pay or to accept. In practice, in the case where you have to pay your neighbour to trim the trees, your neighbour could exaggerate his or her marginal benefits from shade, in effect asking a higher price for tree trimming than he or she is truly willing to accept. You could understate your true shade and debris costs, trying to get tree trimming on the cheap. In the avocado market discussed in Chapter 3, such false offers get no takers due to the presence of competing sellers. But you and your neighbour are the only buyers and sellers of the changes in tree heights that affect your two properties. And neither of you has much information about the other's alternatives, which determine individual benefits and costs. For example, your neighbour may not know what it would cost you to heat your pool with gas; you do not know whether lower trees would cause your neighbour to erect shadecloth or to use more sunscreen. If each of you doubts the other's offers, the negotiations may be costly in terms of time and nervous energy, if not money. It is possible that a private deal may seem too costly. And if there is no private deal, the final tree height is very unlikely to be Q^*; it is more likely to be Q_m if your neighbour has the right to decide tree height, and Q_l if the court grants you that right.

We can illustrate the preceding point using Figure 14.2. Suppose that your neighbour has the property right; you have to pay your neighbour to trim the trees. There is no advantage to you in negotiating to move from Q_m to Q^* if you expect the negotiations to cost you more than your share of D, the net gains from the deal.

What is the second assumption implicit in our discussion of Figure 14.2? Remember that the court's decision on who has the property right to decide tree height determines who pays whom. It therefore has some impact on the wealth of yourself and your neighbour. The second assumption is that your's and your neighbour's willingness to pay and willingness to accept compensation, represented by the marginal benefit and marginal cost curves in Figure 14.2, are unaffected by the amounts of money paid and received in resolving the height of the trees.

Given the assumptions that deals are costless and that money transfers between the two parties do not affect their willingness to pay or to accept, the preceding discussion of Figure 14.2 illustrates Coase's conclusion: *the efficient, as distinct from the socially equitable, level of pollution is unaffected by the legal allocation of the property right.* Coase used the example of a cattle-raiser whose cattle destroy a neighbouring farmer's crops. To quote Coase:

> It is necessary to know whether the damaging business is liable or not for damage caused since without the establishment of this initial delimitation of rights there can be no market transactions to transfer and recombine them. But the ultimate result (which maximises the value of production) is independent of the legal position if the pricing system is assumed to work without cost.[2]

Coase pointed out that this is perfectly logical if we ignore the moral dimension of any externality. In a purely technical sense, the 'blame' for an externality such as the shading of your pool arises as much from your sensitivity to debris and water temperature as it does from your neighbour's action in letting trees grow. If you enjoyed pool cleaning and swimming in freezing water, the pollution problem would disappear.

The preceding argument suggests that the question of who is the 'polluter' and who is the 'victim' is not just a matter of resolving which party's action physically impinges upon or invades the other's property. It is also a matter of community ethics and justice. For example, imagine that your neighbour is an elderly lady who has lived there, with her trees, all her life. You have just built on a large lot next door and, knowing the problems of shade and debris, have still chosen to put your pool next to her trees. By your purchase of the lot (possibly a little cheaper because of her messy trees) and pool placement, haven't you effectively consented to bear the costs due to the trees? The physical invasion test suggests that your neighbour is the polluter. If we subscribe to Coase's view that both parties contribute to the problem, consideration of the history of the externality suggests that you are the party whose actions are its immediate cause. It could therefore be argued on ethical grounds that, if you (with the help of the court) succeed in forcing your neighbour to bear the costs of trimming her trees, you are polluting her, by imposing costs to which she has not consented. Thus Coase's arguments cast doubt on the universal applicability of the widely espoused 'polluter pays principle', as long as we define pollution in terms of a physical invasion.

Casual observation suggests that explicit financial deals to solve pollution problems between neighbouring households are comparatively rare. This does not necessarily mean that private negotiations do not resolve such problems; households can exchange a wide variety of goods and services over long periods of time. For example, if you are on good terms with your neighbour, you may let your neighbour use the pool and repair his or her leaking roof in return for allowing you to trim the trees.

14.5 Court awards of damages

If you and your neighbour are not on speaking terms, or find negotiations too costly, there is an alternative coordination mechanism based on individual property rights. After a court has decided which of you has the right to determine the height of the trees, it may go on to resolve the problem by applying a *liability rule*. Liability rules award monetary compensation—damages—to the party whose property has been devalued by pollution. In the present case, suppose that you have a right to enjoy your pool free of shade and debris, and the court decides to award you damages, which must be paid by your neighbour. Since damages are designed to compensate the pollution sufferer for the costs imposed, the court would ideally determine the amount of damages using the marginal cost curve Q_1H in Figure 14.2. This would give your neighbour the correct incentive to cut the trees back to Q^*. At Q^*, your neighbour's annual benefit from the last increment of height is just equal to the extra damage payment he or she must make. The total damages payment per year is measured by the area A in Figure 14.2. In this ideal situation, court application of a liability rule produces the efficient amount of pollution.

Compared with private negotiation, the costs of applying a liability rule—court time, legal fees and so on are high. We would expect damage awards to be mostly confined to situations where the costs of pollution are high and concentrated on one or few parties, for example sparks from faulty power lines setting fire to crops or chemical spills forcing evacuation of businesses. So our tree problem is unlikely to be resolved by a liability rule, unless the case is representative of a class of pollution problems. In that situation, the damages award may cause other tree growers to modify their polluting behaviour in a manner consistent with the court-determined damage costs.

Could the pollution sufferer, the pool owner in this case, ever be liable for damages? Historically, English common law has applied liability rules to award damages against parties whose activities impinge physically on another's property.[3] In our example, your neighbour's actions do physically affect your pool, but your actions have no physical impact on your neighbour's property. In these circumstances, we expect that the tree owner will be liable for damages, if any damages are awarded. However, as pointed out in the discussion of private negotiations, there may be technical and ethical bases for arguing that your action in constructing your pool and demanding trimming of the trees has polluted your elderly neighbour, rather than that her trees have polluted your pool. Courts almost never decide this way. If they did, the appropriate liability rule would involve you paying damages to your neighbour, determined using her marginal benefit curve Q_mG, in return for the court-ordered trimming of her trees.

In order to correctly compensate the pollution sufferer, and correctly signal to the polluter, a court needs to identify the sufferer's marginal cost curve. In the present case, only you know the precise costs you suffer as the trees increase in height. You have an incentive to exaggerate your costs, hoping that this will increase the damages award. So administrative determination of damages by courts is costly, for the same reasons that verification of values in private negotiations is costly, and court awards of damages may not produce efficient amounts of pollution.

14.6 Rules determined by planners

Why might you and your neighbour rely on planning rules to coordinate your activities, rather than negotiating or going to court? Planning rules change your pre-existing individual property rights, rather than use them as the basis for resolving the pollution problem. For example, planning regulations may limit tree heights or pool placement on urban lots. Planners cannot afford to tailor specific new rules for your particular situation; thus planning is unlikely to yield the efficient amount of pollution depicted in Figure 14.2.

Why are zoning and environmental regulations a common response to pollution problems between neighbours? One explanation is the costs of private negotiation and of court proceedings. Another is that planners such as state governments and local councils have the power to change property rights, and you or your neighbour may expect to gain from changing the existing rules. However, changes in individual property rights over trees and pools will be non-rival and non-excludable goods, affecting many others as well as yourselves. You will not have much incentive to pursue such changes unless the benefits or costs you experience as a result of shading and debris are very large. A third possible explanation for reliance on planning rules is the perception that existing property rights in trees and solar access are outdated. With the development and widespread installation of solar collectors and pools with automatic cleaning systems, the costs of tree shading and debris may be much higher than was previously the case. People such as yourself, with the income and leisure to install and enjoy a pool, come to think of access to sunlight and freedom from debris as necessary rights. Politicians, seeking votes, are likely to respond to your desires with new rules. So two-party pollution problems may be the subject of planning rules where the circumstances are common.

The existence of planning rules dealing with buildings, pools, trees, noise levels, domestic animals and so on does not prove that they are crucial in resolving pollution problems between neighbours. You and your neighbour can still negotiate, either accepting or ignoring the planning rules. As mentioned earlier, private negotiations may be concealed within the variety of exchanges between neighbours over long periods of time. The non-market coordination mechanisms listed in Chapter 5—in particular, conventions of neighbourly behaviour, such as consulting about plans for pools and trees—may also help to resolve pollution problems.[4]

Pollution involving two parties is not a major environmental problem for the community. Once property rights are clear, private negotiations or courts can achieve coordination between the parties. The only question is whether the gains from coordination justify the costs of specifying property rights, measuring the damage and haggling to try to reveal true valuations. If not, the parties combined are better off living with the pollution, unless the problem is so common that a planning rule, applied to large numbers of similar cases, reduces the overall costs to the community.

Discussion questions

1. If cultivation causes soil erosion which pollutes rivers and deposits unwanted soil on neighbouring farms, then all cultivation of erodable land must be stopped, since any soil erosion is too much. True, false or uncertain? Explain why.
2. It is not always efficient or fair to require the polluter, defined as the party whose action physically impinges on a neighbour, to provide compensation to the neighbour. True, false or uncertain? Explain why.
3. There are two people who work together in the same office. One enjoys smoking and the other cannot stand smoke. The smoker's marginal benefits (MB) from extra cigarettes per day (in dollars) and the corresponding marginal costs (MC) of the non-smoker are as follows:

Number of cigarettes	MB ($)	MC ($)
1	7	1
2	6	2
3	5	3
4	4	4
5	3	5
6	2	6

(i) From the point of view of the two people combined, what is the efficient amount of cigarette pollution?

(ii) Assume that there is a legal liability rule which requires the smoker to pay for damages suffered by the non-smoker. Carefully explain how negotiation between the two parties can lead to the efficient amount of pollution.

(iii) Suppose that it is possible for the smoker to rent an adjacent office, so that the non-smoker would not have to breathe smoke at all, for $10 per day. Does this change the efficient outcome? Why or why not?*

4. Explain why negotiation between you and your next door neighbours over use of your backyard incinerator has a good chance of yielding the most efficient level of local pollution. What are the main impediments to social coordination between you and your neighbours?

* Based on a similar question posed by Steve Polasky, Boston College, reproduced in Ed Tower, *Environmental and Resouce Economics*, Eno River Press, Durham, 1990, p.133.

15

The economics of pollution control: many parties

Serious pollution commonly involves many polluters, many sufferers, and mixing of emissions from different sources. Because many suffer from the same emissions and private property rights and market negotiations would be prohibitively costly, better social coordination depends on planners. However, because of the numbers of polluters and sufferers and the variety of possible adjustment alternatives, planners have to set aggregate pollution targets in substantial ignorance of the costs and benefits of reduced pollution. Once a target is set, signals and incentives for polluters may involve either 'command and control' regulations or taxes/marketable permits which put a price on emissions. While taxes and permits utilise the knowledge of firm managers to minimise pollution control costs, governments generally prefer the familiarity and flexibility of regulations.

15.1 Pollution involving many parties

Private negotiation and court action cannot resolve problems if a pollutant imposes non-rival and non-excludable costs on many people. This is usually the case when the pollutant is dispersed in air or water, by electromagnetic radiation or by wildlife movements. Remember the case of acid rain, discussed in Section 5.7. Even if there is only one source of the pollution, many people downwind suffer from the same emissions. Thus negotiation of the efficient level of acid rain requires a multilateral deal. Either all the sufferers have to agree on sharing the costs of emissions reductions (if the emitter has the property right to use clean air), or they must agree on sharing the compensation payments from the emitter (if sufferers have the right to clean air). As explained in the discussion of Figure 6.1, in almost all cases such negotiations will be impossibly costly.

In these circumstances, as explained in Section 6.4, most pollution sufferers will prefer to free ride. The individual sufferer reasons that his or her contribution to pollution reduction will make little or no difference, but he or she will reap the benefits of reduced pollution regardless. Thus the sufferer may make no contribution at all. There may be little or no private negotiation or court action to provide feedback to polluters. Thus it is left to planners to act to coordinate the activities of polluters and pollution sufferers.

Pollution that causes major environmental problems commonly involves many polluters and many sufferers. Examples include motor vehicle pollution; dust and smoke from construction and industrial sites; salt, phosphates and organic materials polluting inland waters; organic and chemical pollution of coastal waters; and industrial heavy metals contaminating air, water and wildlife. Both households and businesses pollute and suffer. If planners are to achieve efficient levels of pollution, identified as Q^* in Figure 14.1, they must be able to identify the marginal benefits and marginal costs of pollution aggregated over many separate polluters and sufferers. When there are many polluters and sufferers, the true marginal curves are very difficult to identify. This is a major problem for environmental planners.

15.2 Information problems in pollution control[1]

Consider the planners' information problems when many profit-maximising firms, not necessarily in the same industry, all produce the same pollutant. To simplify matters, assume that there are no household polluters, and that the polluting firms are competitive, and thus price takers in their respective product markets. The benefits of pollution are equal to the profits the firms obtain from being free to emit pollutants at no cost to themselves. Figure 15.1 shows the trade-offs involved in different levels of emissions. The marginal benefits of pollution (*MBP*) curve shows the incremental changes in polluters' aggregate profits as emissions rates increase. Read from right to left, the same curve shows incremental changes in polluters' total costs of reducing emissions (their sacrifices of profits) as emissions rates decrease. The marginal costs of pollution (*MCP*) curve measures incremental increases in the aggregate costs of sufferers, both households and businesses, resulting from successive increases in emissions.

Superficially, Figure 15.1 appears the same as Figure 14.1(b), which depicts the marginal benefits and costs of pollution by a single firm.

Figure 15.1 Trade-offs in pollution with many polluters and many sufferers

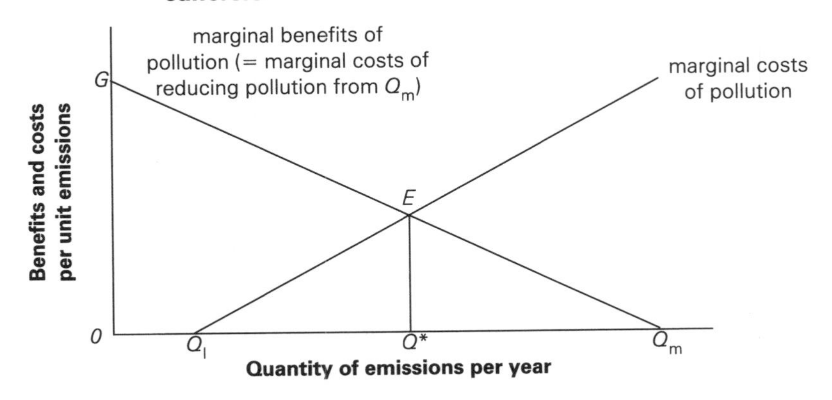

The efficient level of emissions is at Q*, where the profits gained from the last unit increase in emissions are just offset by the additional costs imposed on sufferers. But the profits in Figure 15.1 are aggregated across many firms and come from a variety of products. For example, in a metropolitan industrial area, coal smoke (ash and soot) due to coal burning might come from power stations, steel makers and foundry furnaces. Also, in Figure 15.1 we drop the unrealistic assumption that emissions are proportional to firms' outputs. Firms may reduce emissions by reducing output, but they may also reduce emissions without reducing output, for example by changing their input mix or by installation of emission control equipment. As we shall see, pollution sufferers also have various options for reducing the costs of pollution.

Why are the two marginal curves in Figure 15.1 difficult to identify? Remember that each is the summation of the individual benefit and cost curves, respectively, of all the polluters and pollution sufferers. Let us take the curves in turn, using coal smoke pollution in a metropolitan industrial area as an example.

The marginal benefits of pollution curve What options does a single polluting firm in the area, say a steel maker, have to reduce emissions of smoke particles? We assume that in the past, prior to the introduction of pollution control legislation, the firm's right to emit smoke has not been challenged, so initially it is producing Q_m units of smoke per year. Starting from Q_m, its emission reduction options may include:

- reducing steel output;
- changing production processes, to reduce the ratio of emissions to steel (e.g. using cleaner burning coal, more frequent furnace maintenance);

- installation of emissions control equipment (e.g. smokestack precipitators to remove ash and soot);
- relocating production to an alternative site;
- varying the timing of emissions.

The firm's true *MBP* curve, which, read right to left, is its marginal cost of reducing pollution, is the least costly combination of emission reduction options as the level of emissions changes. For example, the steel maker may begin reducing emissions by maintaining its output, while changing its production set-up to reduce emissions (using cleaner coal, precipitators to reduce smokestack emissions etc.), thereby incurring progressively higher input costs (and thus lower profits) as it reduces emission levels. As a result, the *MBP* curve rises from right to left at a rate determined by the increasing costs of inputs. Once emissions have been reduced to a lower level, it may become cheaper for the firm to cease changing its production set-up and to reduce emissions further by reducing output, in which case the subsequent rises in the *MBP* curve will reflect the combined effects of reduced sales revenue and reduced input costs. Thus the true *MBP* curve for the steel maker is a hybrid; the marginal benefits of pollution are determined by different emission reduction options as emission levels change.

Figure 15.1 shows the aggregate *MBP* curve for all coal smoke emitters in the area, say a power station, the steel plant and several metal foundries. The aggregate curve will be more complex than the steel maker's *MBP* curve, because of the variety of emission reduction options available to the different firms.

A monopolist will produce fewer emissions than a group of competitive polluters with the same costs of production and emissions reduction. This is because a monopoly restricts output to below the level that maximises the net benefits from exchange. In terms of Figure 15.1, where the *MBP* curve GQ_m represents the profit gains from successive increases in emissions for a group of competitive polluters, the corresponding *MBP* curve for a monopolist would begin at G and slope down more steeply, intersecting the horizontal axis to the left of Q_m. Only if the monopolist chose to reduce emissions exclusively by switching production processes, rather than by reducing output, would the *MBP* curves for the group of competitive polluters and the monopolist be the same.

The marginal costs of pollution curve In reacting to pollution, sufferers also have several options. The household or business that suffers costs due to coal smoke can:

- bear the costs;
- alter consumption and production activities to reduce damage (e.g. avoid hanging out the laundry on windy days, switch from cabbages to root crops to avoid ash contamination);
- adopt measures to insulate activities from damage (e.g. individuals

can wear masks, households and businesses can install window screens or air filters);
- relocate.

The true *MCP* curve will incorporate the least costly combinations of these options for each sufferer at each level of emissions.

Emissions transfers between sources and sufferers Another information difficulty results from the usual physical separation between emission sources and receptor points. In the simplest case there is a fixed proportional relationship between emissions at source and emissions received by sufferers, so that the horizontal axis in Figure 15.1 is valid for both source and received emissions. However, the ratio of source emissions to received emissions of a pollutant generally will vary for every source–receptor pairing; it may also vary with environmental conditions. Planners need to know these emission transfer ratios in order to set appropriate controls on emissions sources.

The complexity of the *MBP* and *MCP* curves and emissions transfers creates formidable information problems for planners. In order to identify correctly the two marginal curves and the efficient level of coal smoke Q^*, planners need a huge amount of information about the production and adjustment alternatives of individual polluter firms, the adjustment alternatives and preferences of individual pollution sufferers, and emissions transfers between the two. Also, planners are likely to be aiming at a moving target, because industrial production and pollution control technologies, and product and input prices, change over time. Thus the marginal benefits and costs of coal smoke pollution, and Q^*, are likely to vary over time.

15.3 Planners' responses to the information problems

We explained in Section 15.1 that private negotiation and court action will not achieve the efficient level of pollution when the pollutant is non-rival and non-excludable. So planners cannot identify the *MBP* and *MCP* curves by observing what individuals and courts do. They have to get information about the marginal curves from scientific studies, markets (for products sold by polluters and purchased by consumers, for polluters' inputs and for pollution control equipment), surveys of polluters and sufferers, and the political process. There is no reason to expect that the latter two sources will be accurate. Remember from Chapter 8 that interested parties often have incentives to distort values and technical information supplied to planners. Specifically, polluters have incentives to exaggerate pollution benefits (the costs of reducing pollution), to persuade the planners to set lenient controls on emissions, so long as they do not expect to be compensated for reducing emissions. Thus polluting industries frequently stress the job losses involved in reducing emissions, since elected planners are sensitive to consequences that can alter voting behaviour. (But, as the preceding discussion of the *MBP* curve shows, action to reduce emissions could also increase employment.) Conversely, pollution sufferers have incentives to exaggerate pollution costs, so long as they do not have to pay for the resulting tight controls.

In practice environmental planners have poor information about the costs of reducing pollution and even less about pollution costs. Pollution control agencies in Australia and overseas commonly set aggregate legal levels of pollution based on scientific criteria that do not involve explicit comparisons of these two types of costs. For example, aggregate pollution levels may be set using scientific data from bodies such as the National Health and Medical Research Council, or based on environmental standards adopted by international organisations.[2] Alternatively, aggregate pollution levels may be the result of negotiation between politicians, the pollution control agency, polluting industries and environmental groups. This procedure has the advantage of allowing polluters' costs of reducing pollution and sufferers' costs of pollution to be explicitly considered. It has the disadvantage of permitting distortions of the information supplied to planners.

If the elected and public service planners who determine legal pollution levels put their own personal interests ahead of the community interest, they will be less interested in the true *MBP* and *MCP* curves depicted in Figure 15.1. Then planners' choices of aggregate pollution levels will depend, at least in part, on the relative abilities of polluters and sufferers to reward and penalise the planners concerned.[3]

Aggregate pollution levels set by planners serve no purpose if they are not enforced. This requires continuous or at least regular measurement of emissions, and enforcement of legal or other penalties on polluters who violate set standards. Measurement of air and water pollution often involves expensive monitoring equipment, and legal actions to penalise violators are often expensive and time-consuming. As a result, environmental protection agencies operating with tight budgets sometimes de-emphasise achievement of quantitative emissions targets, and concentrate on prescribing specific technical measures that must be adopted by polluters. For example, coal-burning power stations may be required to use low-sulphur coal, or to install specific smokestack precipitators. New motor vehicles sold in Australia are all required to have specific emissions control equipment.

Prescriptive technological standards have the advantage of being cheap to enforce. Planners have only to check that the required input is being used or the required equipment is installed. But such checks provide no direct evidence about actual levels of emissions of pollutants; this depends not only on the inputs and equipment used, but also on the manner of use, levels of maintenance and so on. Another disadvantage of technological standards is obvious from our previous discussion of the *MBP* curve. Planners, who have no direct interest in businesses' profits, may not choose the least costly emission reduction options.

15.4 Social coordination in pollution control

However the aggregate level of pollution is decided, this is only one part of the environmental planner's task. Having determined an aggregate target for a particular pollutant, the planner must signal this plan to emitters, and provide incentives, either rewards or penalties, for polluters to comply.

There are two distinct approaches to signalling and incentives in pollution control. One is for planners to specify allowable levels of emissions and/or inputs and/or technologies for each individual polluter, together with legal penalties for violations of the specified standards. This is the regulatory or 'command and control' or *direct controls* approach to pollution control, equivalent to the central planning system briefly described in Chapter 2. Individual decision makers are subject to directives, backed by penalties for non-compliance, which reduce their discretion in determining their input, output and emissions levels. A common procedure is for each polluter to be licensed to discharge no more than a specified quantity of emissions per time period. If the environmental protection authority also specifies inputs and equipment to be used, plant operators' discretion is reduced further.

The second possible approach is for planners to put a price on emissions of pollutants, thereby signalling the costs of pollution to polluters, and to allow polluters to respond as they see fit, as would be the case in a market system. One way in which this can done is by imposing *pollution taxes:* taxes per unit of emissions. It can also be done by setting a limit for aggregate emissions combined with a newly created market in emissions permits, and allowing existing and would-be emitters to compete in the market for a limited supply of permits. *Marketable pollution permits* (also termed *transferable pollution permits*) are property rights to use designated bodies of air or water or whatever for waste disposal. Like any property right, they will maximise social coordination in use of the scarce asset if they are exclusive, clearly defined, enforceable—requiring monitoring of use and penalising of violators—and transferable.

With pollution taxes or marketable permits, individual polluters are free to determine their input, output and emissions levels, subject to the tax rates or aggregate pollution levels set by the planners. In each case, polluters have financial incentives to respond by eliminating emissions wherever the benefit gained is less than the cost of the tax or of extra permits.

The advantages and disadvantages of the 'command and control' and 'putting a price on pollution' approaches are clearest when there are many sources of the same pollutant, and planners have to allocate responsibility for emissions reductions across sources.

15.5 Coordinating pollution reductions across many polluters

If there is just one source of a pollutant, for example a mine in a rural area, the planner's task is relatively simple: using either directives or financial incentives, the planner must persuade the mine operator to reduce emissions to the chosen target level. With more than one source, the planner also has to decide on a process to allocate responsibility for emissions reductions among the separate sources. Thus, in the coal smoke pollution example, the planner using direct controls decides the emission levels of the power station, the steel mill and each metal foundry, possibly also specifying input use and technologies. Alternatively, the planner could put a price on coal smoke pollution, and leave all decisions to the polluting businesses. This could be done either by imposing a tax on emissions that the planner believes representative of the costs imposed on pollution sufferers, or by somehow allocating a number of marketable pollution permits equal to the allowed aggregate level of smoke emissions, and allowing the power station, steel mill and foundries to buy and sell permits. (The effects of different allocations of permits on the incomes and welfare of different polluters and pollution sufferers are discussed in Section 15.9.) With a tax or permits each emitter firm would make its own decisions on methods of emissions reduction and its level of smoke emissions, based on comparison of its costs of reducing emissions (the profits it obtains from extra emissions) and its costs of emissions (either the tax rate or the price of emissions permits).

If the planner sets and enforces the aggregate level of emissions, varying the allocation of emissions reductions across polluter firms has no effect on the total costs of pollution. However, because different firms have different costs of reducing emissions, varying the allocation of emissions reductions usually will alter the aggregate costs of reductions. For example, in our coal smoke pollution example, suppose that current emissions levels and the (constant) unit costs of emission reductions are as set out in Table 15.1.

Suppose that environmental planners decide to halve annual smoke emissions, to 100 000 units. If they direct each emitter to halve its emissions, the total cost of the reductions will be $930 000 (50 000 × $5 + 30 000 × $10 + 10 000 × $30 + 10 000 × $8). But, if the planners knew that the power station could reduce emissions more cheaply than any other source (which is unlikely, as already explained), they could achieve the same pollution outcome for only $500 000, by requiring the power station to reduce its emissions to zero. Thus the distribution of reductions across emitters can make a big difference to the aggregate costs of pollution control.

Table 15.1 Coal smoke emissions and cost of emissions reductions

Source	Units of annual emissions	Unit cost of emissions reductions
Power station	100 000	5
Steel mill	60 000	10
Metal foundry *A*	20 000	30
Metal foundry *B*	20 000	8

Requiring the power station to eliminate emissions is the most efficient policy. It achieves the target level of emissions at least cost to the community, and therefore maximises the net benefits to the community from the planners' chosen level of pollution control.

It may be argued that, despite its efficiency, requiring the power station to eliminate emissions while the other polluters continue as before is unfair. The power station operator and electricity users (whose electricity charges will rise) are meeting all the costs of emissions reduction, while the steel mill and foundries get off scot free. The fairness problem can be overcome without compromising efficiency if planners are willing to allow the polluters to negotiate among themselves to settle who is best placed to cheaply reduce emissions. Suppose that the planners feel that fairness demands that each polluter be responsible for eliminating emissions equal to half its current output, but allow each to achieve reductions *from any source*. Thus foundry *A*, which would incur a cost of $300 000 per year to halve its own emissions, is allowed to pay another polluter to cut its emissions by the required 10 000 units. The obvious course of action for the owner of foundry *A* is to negotiate with the power station to reduce its emissions by an extra 10 000 units, at a price that will be somewhere between $5 and $8 (the minimum price that foundry *B*, next cheapest source of emissions cuts, would accept for additional cuts in its emissions). In this way, even though the high-cost emitters do not make the actual cuts, they end up paying for them, and fairness is achieved.

We have just described a (very simple) marketable pollution permit system; each polluter is allowed a specific level of emissions (half its initial emissions, in the example), but rights to emit pollutants are marketable. Polluters who can reduce emissions cheaply (the power station) can increase profits by selling permits. Polluters who find it very costly to reduce emissions (because extra emissions are very profitable—foundry *A*) can increase profits by purchasing extra permits at prices below their costs of reducing emissions. In this way negotiations between polluters result in the desired aggregate emissions reductions being achieved at least cost, whatever the initial allocation of permits by planners.

Pollution taxes also achieve aggregate emissions reductions at least cost. Suppose that planners impose a tax of $9 per unit on all coal smoke emissions. In deciding how to respond, each polluter will compare the profits gained from an extra unit of emissions (its marginal costs of reducing emissions) with the tax rate. Each will eliminate all emissions for which the tax exceeds the extra profits earned. Based on these calculations, the power station and foundry *B* will eliminate their emissions. The total reduction will be 120 000 units. This will be achieved at least cost, since these two sources have the lowest marginal costs of reducing smoke emissions. Note that with a tax, the planners cannot usually achieve a particular emissions target at the first attempt, since they do not know the sources' marginal costs of emissions reductions. In the present case, if they desire a cut of 100 000 units, they can only achieve it by a process of trial and error, involving successive alterations of the tax rate, until the desired level of emissions is achieved.

The example in Table 15.1 oversimplifies in several ways, particularly in assuming constant marginal costs of reducing emissions (horizontal *MBP* curves) for each source, contrary to the logic used to derive the *MBP* curve in Figure 14.1. We now compare direct controls, marketable permits and taxes when *MBP* curves are downward sloping.

15.6 Direct controls versus marketable permits and taxes

Assume that a pollutant is produced by just two firms, which earn different profits from extra emissions; that is, their marginal costs of reducing emissions differ. Panels (b) and (c) of Figure 15.2 show the *MBP* curves for firms *A* and *B* respectively; firm *A* derives modest profits from emissions of the pollutant, while emissions are highly profitable for firm *B*. In the absence of any controls on emissions, firm *A* will produce more units of emissions per year than firm *B*. Panel (a), which measures the total emissions experienced by pollution sufferers, shows the aggregate *MBP* curve for the pollutant, obtained by horizontal addition of the individual *MBP* curves. No *MCP* curve is shown in panel (a), because to do so might suggest that planners could identify the efficient level of emissions, where the aggregate *MBP* and *MCP* curves intersect. In fact, as previously explained, at the outset the planners do not know the individual and aggregate *MBP* curves shown, let alone the aggregate *MCP* curve.

Suppose that the planners select an aggregate emissions target, Q_t in panel (a), using one of the criteria mentioned in Section 15.3. They then have the task of allocating emissions reductions between the two firms. Consider first the case of direct controls. Assume that the planners decide that the firms should get an equal share of the rights to discharge the pollutant. They therefore issue each firm with a permit to discharge no more than $\frac{1}{2}Q_t$ units of emissions per year, shown in panels (b) and (c). Assume that compliance with this condition is strictly enforced.

Figure 15.2 Direct controls versus marketable permits

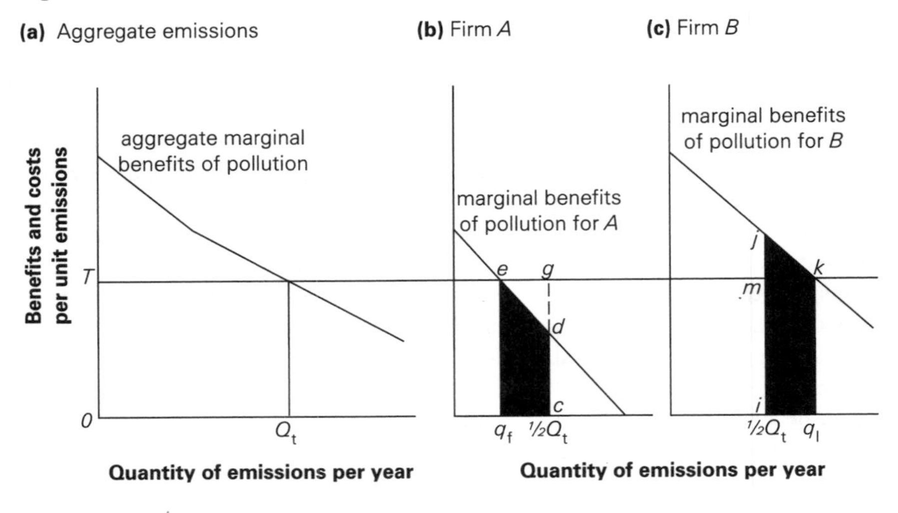

The diagram shows that when firm *A* is meeting its permit conditions by discharging $\frac{1}{2}Q_t$ units of emissions, it derives only \$*cd* of extra profit from the last unit. On the other hand, when firm *B* is meeting its permit conditions, its last unit of legal emissions earns \$*ij*. If the managing directors of *A* and *B* were to meet over drinks at the Melbourne Club, they might well agree that this arrangement does not seem sensible; *A* earns only a small profit from an emission right that would result in large additional profits if held by *B*. But under direct controls, they are prohibited from negotiating a deal to transfer the right to *B*.

Such a market deal would be legal if, after allocating $\frac{1}{2}Q_t$ units of emissions to each firm, the planners allowed market transfers of emissions permits. Then firm *B* would purchase emissions permits from firm *A* at some price between \$*ij* (the maximum price that *B* could afford to pay) and \$*cd* (the minimum price that A would accept). How many permits would change hands? If the managing director of *A* honestly reveals what firm *A* is willing to accept (based on *A*'s *MBP* curve) and the managing director of *B* honestly reveals what firm *B* is willing to pay (based on *B*'s *MBP* curve), and they can easily identify one another, market deals will cost very little. In this case, permits will change hands as long as the marginal profit sacrifices of *A*, which increase as *A* cuts emissions, moving up its *MBP* curve, are less than the marginal profit gains of *B*, which decrease as *B* increases emissions, moving down its *MBP* curve. So the exchange of permits will cease at the emissions levels where the marginal benefits of pollution are equal for *A* and *B*, that is at q_f in panel (b) and q_l in panel (c). If all permits exchange at the same price, the permit price will be OT $(= e_f = k_l)$.

What do the firms gain from the switch from direct controls to marketable permits? Firm *A* loses profits measured by area $cdeq_f$ when it has to cut its emissions from $\frac{1}{2}Q_t$ to q_f, but it gains an amount measured by area *deg* overall, because it gets paid $cgeq_f$ per year for the permits. Firm *B* achieves a net gain measured by area *jkm* after paying $imkq_l$ for permits that increase its annual profits by $ijkq_l$.

What is the cost saving to the community as a result of making emissions permits transferable? The saving comes in the form of profits that are preserved because emissions reductions are shifted to the source (firm *A*) that can do the job most cheaply (i.e. that sacrifices the least profit when it reduces emissions). By shifting permits to *A* from *B*, profits measured by area $ijkq_l$ are preserved at a cost of a sacrifice of profits measured by $cdeq_f$. The overall cost saving resulting from transferability is measured by the difference, the sum of areas *deg* and *jkm* in Figure 15.2.

An alternative way of thinking about the benefits of making emissions permits transferable is to think of the waste disposal capacity of the air as a scarce community resource, like a timber-producing forest or land suited to market gardening, which has to be allocated between profit-oriented businesses. In determining the aggregate amount of a pollutant that can be discharged into the air (Q_t in Figure 15.2) environmental planners are allocating some air to waste disposal, as opposed to its other valued functions, such as human, plant and animal life support and maintaining visibility. The community would presumably like to see the limited waste-disposal capacity of the air used so as to generate the maximum net benefits for the community. A market for pollution permits facilitates the allocation of waste disposal rights to businesses who can most profitably use the air's limited disposal capacity, in the same way that the market for market-gardening land facilitates the allocation of land to the farmers who can use it most profitably.

We can also use Figure 15.2 to show how a pollution tax will achieve emissions reductions at least cost. Suppose that, based on their 'guesstimates' of the costs of pollution, planners impose a uniform pollution tax of \$*OT* per unit emissions on firms *A* and *B*, shown as a horizontal line in the diagram. In attempting to maximise profits net of the tax, the firms will eliminate all emissions for which the tax exceeds the profits previously earned (measured by their respective *MBP* curves). So firm *A* will cut emissions to q_f and firm *B* will cut emissions to q_l, where their marginal benefits of pollution are equated (\$$eq_f$ and \$$kq_l$ respectively). Aggregate emissions will be reduced to Q_t in the same least-cost fashion as was the case for a total of Q_t marketable permits. However, because the planners do not know the *MBP* curves in Figure 15.2, they will be unable to predict the individual and aggregate emissions reductions that occur.

Note that taxes and permits do not produce least-cost emissions reductions where the polluting firm is a monopolist. The monopolist will adjust its emissions to the level at which the tax or permit price intersects the *monopoly MBP* curve. However, at the chosen level of emissions, the true marginal cost to the community of reducing emissions is measured by the corresponding point on the (higher) competitive *MBP* curve. So, in responding to taxes or marketable permits, a monopolist will operate where its marginal cost of reducing emissions is, from the community's point of view, too high.

15.7 Signalling and incentives under direct controls, marketable permits and taxes

Marketable permits achieve least-cost emissions reductions because, unlike direct controls, they make use of polluters' private information about their individual *MBP* curves, which measure polluters' costs of reducing emissions. We have seen that planners generally have little information about these costs. Thus planners are likely to allocate emissions reductions using rules based on some notion of fairness, such as the equal percentage reductions for all sources, or rules designed to minimise overall monitoring and enforcement costs, which might dictate concentrating on the sources with the largest emissions. Our examples suggest that such rules will rarely achieve the desired reductions at least cost.

Once polluters are given transferable property rights to emit pollutants, they find it in their own interests to reveal their true willingness to pay and to accept money for permits. The reason is the same as for the avocado buyers and sellers of Chapter 3, who found it in their own interests to reveal their true valuations of avocados—if permit buyers and sellers do not reveal true values, they may miss out on profitable deals. And in revealing willingness to pay and to accept, permit buyers and sellers reveal information about their private costs of reducing emissions. The discussion of Figure 15.2 shows that, once this information is revealed, polluters as a group can coordinate their activities to achieve planners' target level of emissions at least cost.

Pollution taxes also enlist private information to achieve least cost emissions reductions. Planners do not allocate reductions; instead, each polluter has a financial incentive to reduce emissions to the point where its (private) marginal benefit of pollution equals the tax rate.

Both pollution taxes and marketable pollution permits allow planners to learn more about individual and aggregate *MBP* curves. A tax is represented by a horizontal line in Figure 15.2, its height above the horizontal axis measuring the tax rate. The tax line will intersect the downward-sloping aggregate *MBP* curve for all polluters at a single point. Assuming that individual polluters try to reduce emissions as cheaply as possible, they will be on their individual *MBP* curves. In attempting to maximise profits net of the tax, polluters as a group will adjust along the aggregate *MBP* curve until they reach its intersection with the tax line. Thus, after planners set a tax rate, and have given polluters time to adjust in a least-cost manner, the planners can be reasonably certain that the levels of emissions observed represent points on individual polluters' true *MBP* curves, and the true aggregate curve for all polluters. In this way, by altering tax rates and observing the resulting levels of emissions, planners can gradually identify the true individual and aggregate *MBP* curves (assuming that technologies and prices have not changed in the meantime).

The analogous learning exercise using marketable pollution permits involves successive variations in the allowed aggregate levels of emissions. An aggregate level of emissions is represented by a vertical line in Figure 15.1. In attempting to maximise profits net of the cost of purchasing permits, polluters as a group will end up at the intersection of the aggregate *MBP* curve and the emissions line. By observing changes in the prices of permits and individual sources' levels of emissions as aggregate emissions are varied, planners can again identify true *MBP* curves.

15.8 Taxes and permits when emissions transfers vary

To signal accurately costs to pollution sufferers, taxes and permit allocations should reflect emissions at receptor points; however, taxes have to be imposed on, and permits allocated to, sources. To meet environmental quality standards at receptor points, environmental planners must have reliable information about emissions transfers between sources and receptors.

Where all sources contribute equally to emissions received at all receptor points, as appears to be the case with global carbon dioxide rises, the costs of emissions are identical for all sources. Hence, in the case of carbon dioxide emissions, the same tax rate would apply worldwide, and a whole-world permit market is appropriate.

Where emissions transfers vary for each emitter for each receptor point, it is in principle necessary to set a different tax rate for each emitter–receptor pairing. Alternatively, a separate permit market is required for each receptor point, with permits to be traded according to the sources' respective ratios of emissions discharged to emissions received. These complications are likely to greatly increase the costs of implementing pollution taxes and marketable permits. A common compromise solution to the problem of varying emission transfer rates is to define zones within which a single tax rate, or a single permit market, applies. All emissions within a zone are then assumed to impose approximately the same costs at all receptor points.[4]

Even with small permit markets, atmospheric mixing of airborne emissions, discussed for acid rain in Section 5.9, is likely to make it very expensive to identify and monitor all the sources contributing to pollution in one locality. And even if the main sources of emissions received can be identified and monitored, a small permit market may contain too few potential buyers and sellers of permits to guarantee the result achieved in the ideal avocado market described in Section 3.6, where parties are forced to reveal their true willingness to pay or to accept. Thus, when emissions transfers vary, the community cost savings or profit gains claimed for taxes and marketable permits in the preceding sections may be offset by planners' high costs of collecting accurate information about emissions transfers. Note that planners imposing direct controls on polluters will require the same information; this is an ubiquitous limitation of the central planning approach to allocation of the air's limited waste disposal capacity.

15.9 Who pays for pollution control?

The allocation of property rights to release pollutants into the environment affects the distribution of income and welfare between emitters and recipients of pollution. If polluters have a legal right to discharge pollutants, they will operate at Q_m in Figure 15.1, and sufferers or the community will have to pay them to reduce emissions, by paying subsidies or purchasing emission rights. If sufferers have the legal right to be free of pollution (the 'polluter pays' principle), the starting point will be Q_l in Figure 15.1; would-be emitters must pay sufferers or the community for emissions, either by paying a tax on emissions or by purchasing emission rights.

At starting points on the horizontal axis between Q_l and Q_m, the rights to discharge pollutants are effectively shared between polluters and sufferers. Suppose that somehow planners impose direct controls that yield the efficient level of emissions, Q^*, in Figure 15.1; then polluters and sufferers each bear some costs. Polluters are restricted to emissions of Q^*, rather than Q_m, and lose potential profits represented by the area Q_mEQ^*. Sufferers have to put up with moderate pollution, bearing pollution costs represented by Q^*EQ_l.

Both pollution taxes and marketable permits can, in principle, be structured to achieve a variety of distributional outcomes. In the case of taxes, different allocations of rights to pollute can be achieved by combining different emissions standards with a tax on emissions in excess of the standard and a subsidy on reductions below the standard. In the case of marketable permits, different allocations of rights can be achieved by combinations of free distribution and auctioning of permits.

Direct controls, taxes and marketable permits all raise polluters' costs of production compared to a situation of no controls. If all or a significant proportion of the firms producing a particular good are affected by pollution controls, this will raise the market supply curve for that good. The result is to shift some of the costs of pollution control from polluters to consumers of any good where a significant proportion of production is subject to pollution controls. Suppose that all of the avocado growers of Chapter 3 are subject to controls on the use of sprays that can drift across property boundaries and harm their neighbours. Then the market supply curve for avocados, *SS* in Figure 3.2, will move up and to the left, the market price of avocados will rise, and the quantity of avocados produced and sold will fall. As a result, some of the costs of pollution control will be passed on to avocado consumers, who will find themselves paying higher prices for a smaller quantity of avocados.

The same logic applies to shifting costs of pollution control backwards to those who supply inputs to polluters. If pollution controls and consequent lower profits reduce avocado growers' willingness to pay for (say) nursery-raised trees, the price and quantity of nursery trees sold will fall. So nursery operators will bear some of the costs of the controls.

15.10 There is no perfect pollution control system

Governments in Australia and other rich countries rely heavily on direct controls to reduce pollution. Direct controls have the advantages that their operation is familiar and readily understood by polluters and pollution sufferers, and that their flexibility permits rapid responses to changing environmental conditions and technology. On the other hand, economists criticise direct controls for failing to achieve emissions reductions at least cost. Also, because direct controls maximise the discretion in the hands of planners, they maximise interest groups' opportunities to manipulate controls in their favour, and planners' opportunities to put their personal interests ahead of the community interest.

To date, direct controls' cost-ineffectiveness and susceptibility to manipulation have not been seen as severe defects by policy makers. However, the identification of pollutants whose reduction, if judged desirable, could require major changes in state or national production, income and employment (in particular, greenhouse gas emissions) has stimulated interest in economic instruments that could reduce the costs of achieving given emissions targets.[5] Already Australia's *Ozone Protection Act 1989* (Cwlth) allows market trading in quotas for the production and import of ozone-depleting substances.

Despite the advantages of pollution taxes and marketable permits in achieving pollution targets at least cost, they can be very costly to implement. We saw, in Section 15.8, that if emissions transfers vary for each polluter–sufferer pairing, planners should ideally set taxes for individual emitters, or emissions targets for individual receptor points. And both pollution control instruments may run into political opposition, because the shift of discretion from planners to polluters may be seen as government granting 'licences to pollute'. This is despite the fact that taxes do, and marketable permits may (by auctioning permits), implement a 'polluter pays' approach to pollution control.

Discussion questions

1. Consumers, not industries, are the true guilty parties in cases of industrial pollution. True, false or uncertain? Explain why.
2. If an electricity generating firm sells a steel mill marketable permits to emit 1000 tonnes of particulates per year, the community is no better off, since total allowable emissions of particulates remain the same. True, false or uncertain? Explain why.
3. Compared to direct regulation of the emissions of individual firms, putting a price on pollutant emissions—by means of a pollution tax or marketable pollution permits—allows planners to learn more about individual firms' costs of reducing emissions. True, false or uncertain? Explain why.
4. Lead in the environment is linked to a lowering of children's IQ scores. Do uniform standards across all Australian states and territories regulating the maximum lead content of petrol make economic sense? Explain why or why not.
5. Thinking about the information and incentives of political and bureaucratic planners, suggest reasons why they generally choose direct controls on pollution in preference to pollution taxes and marketable pollution permits.

V

UNCERTAINTY AND WORLDWIDE PROBLEMS

22.3 Major environmental problems involve non-rivalry and non-excludability

22.4 Environmental and economic complexities make errors in environmental policies inevitable

22.5 It is more important to understand processes than to design 'solutions'

16

Social coordination under uncertainty

Major environmental problems commonly involve major scientific and behavioural uncertainty. Scientific uncertainty is due to the complexity of interactions within the environment, discussed in Part I. Behavioural uncertainty occurs whenever the impacts of a single action affect many people. As discussed in Part II and Chapter 15, dispersion in space greatly increases the costs of identifying impacts, as the number of people affected and adjustment options increase. When impacts also extend far into the future, as is the case in this and the succeeding chapters, uncertainty is further increased. Uncertainty has two important consequences for environmental decision making: first, it raises the costs of better social coordination, making societies more willing to put up with actual or potential environmental damage; second, it increases reliance on the perceptions and preferences of the decision maker, in particular, his or her attitude to risk.

16.1 Decisions under uncertainty[1]

Economists commonly use the term *uncertainty* to describe situations where we are ignorant about both the nature of future possible events and their probabilities, or where we know the possible events but not their probabilities. If both events and their probabilities are known, the situation is described as one of *risk*. Thus it would be reasonable to describe an alpine resort operator as facing risk in deciding for how long a period to hire ski-tow staff for the coming season, because past snowfall records are available. On the other hand, when the resort operator is deciding whether to create a new ski slope, there is uncertainty because the operator cannot know all the impacts of damaging the alpine ecosystem.

Our discussion of environmental problems to this point has assumed that the consequences of resource use are known. In reality, our limited knowledge of the functioning of combined economic–environmental systems, discussed in Section 4.4, means that serious environmental problems involve uncertainty.

With uncertainty, a single possible outcome of a decision is replaced by a variety of possible scenarios. For example, when the steel maker depicted in Figure 5.2 produces steel and smoke, possible variations in world steel prices, union wage demands, local weather patterns and the reactions of smoke sufferers, all outside the control of steel company decision makers, mean that the outcomes of steel-making decisions are not certain. Steel/smoke production decisions require a different diagram for each scenario envisaged, and estimates of the probability of each.

How are decisions made under uncertainty? Choices made under uncertainty involve a number of steps on the part of the decision maker. To demonstrate the steps required, we begin with a decision between alternative actions where the possible outcomes and their probabilities are assumed to be known; that is, a simpler problem involving only risk. Reconsider the mining decision problem described in Section 9.1, with the single modification that future uranium prices are uncertain—they could be 'high', 'medium' or 'low', with corresponding effects upon the profitability of the mine in 10 years time. In this case the decision maker is a mining executive, and the steps are:

1 Specify the decision maker's objective or objectives (in this case, to maximise the present value of profits from mining).
2 Define the alternative actions that could be taken (to mine and sell the uranium now or 10 years hence).
3 Identify the possible outcomes of each action, which may depend on events outside the decision maker's control, together with their probability of occurrence. (The executive expects a definite profit outcome from mining now, and three possible profit outcomes from mining 10 years hence. Assume that, based on his or her experience of the uranium market, and advice about technological and economic changes during the following 10 years, the executive judges that the probabilities of 'high', 'medium' and 'low' uranium prices in 10 years time are 0.2, 0.5 and 0.3 respectively.)
4 Calculate the contribution of each outcome to the chosen objective or objectives (in this case, the present value of the profits from mining for each outcome, weighted by its probability of occurrence).
5 Aggregate across outcomes to choose the action that contributes most to the decision maker's objective or objectives. (Which is

Figure 16.1 Decision tree for a mining decision

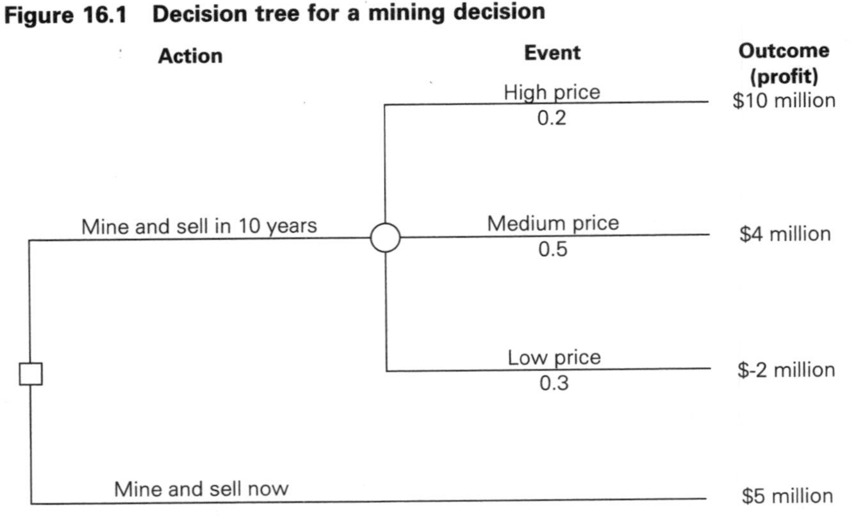

greater, the profits from mining and selling the uranium now, or the aggregate present value of the three sets of possible future profits?)

16.2 Structuring decisions involving risk: decision trees

The mining executive's decision problem is illustrated by the decision tree diagram in Figure 16.1. The decision tree depicts the actions available to the decision maker, the uncontrollable events that can occur and the relationship between actions and events. A *decision point* or *decision node* is represented by a square, with alternative actions branching out from squares; an *event node* is represented by a circle, with uncertain events, not controlled by the decision maker, branching out from circles. The time sequence runs from left to right, with the values attached to the outcomes of the action–event scenarios listed at the ends of the right-hand branches. According to the decision tree, the executive judges that if the uranium is mined and sold now, the outcome is a certain profit of $5 million. If mining is deferred for 10 years, there is a 0.2 probability that the present value of those profits (after discounting) will be $10 million, a 0.5 probability of their being $4 million and a 0.3 probability of a loss of $2 million. The values do not have to be in terms of present values of profits, as depicted in Figure 16.1; they could measure the satisfaction the decision maker gets from each possible outcome of the choice process.

If the problem involves only risk, with probabilities and values of the possible outcomes known, as is the case in Figure 16.1, it is possible, working backward from the right-hand side of the decision tree, to determine which action will maximise the present value of mine profits.

If the choice is to mine the uranium now, the expected profit is $5 million; if the choice is to mine in 10 years, the weighted aggregate present value of the expected future profits is:

$$0.2(\$10m) + 0.5(\$4m) + 0.3(-\$2m) = \$3.4m$$

In this case, the decision tree diagram contains enough information to undertake the five steps listed above, and to choose between the alternative actions. The mining executive will maximise expected profit from the mine by choosing to mine now.

To be useful, the decision tree must correctly represent the important interactions between the decision maker and those parts of the economy and environment outside his or her control. Thus the decision tree will be incorrectly structured if it ignores the fact that the executive can choose to defer mining but lock in future uranium prices by forward contracts that specify delivery of uranium at given prices in 10 years time. The same will be true if the decision tree ignores other uncertain events that can have major impacts on mine profits, for example possible changes in miners' wages, shipping costs and government controls on uranium exports.

Depiction of possible action–event scenarios in a decision tree not only helps in evaluating alternative actions; it also assists the decision maker in identifying additional relevant uncontrollable events (such as government export controls), and additional actions that might be taken (such as forward sales contracts). Use of the tree to make actual decisions requires the decision makers to specify clear objectives and to make judgments about outcomes, values and probabilities for each action–event scenario.

Of course, where the future is uncertain, following a logically correct choice process in no way guarantees that the decision to mine today will actually turn out to be the correct one. A greenhouse-induced switch back to nuclear power generation could see the 'high' price scenario come to pass. Then the executive's choice turns out, with the benefit of hindsight, to result in profits $5 million less than could have been made by deferring mining. But this does not mean the decision itself was wrong, given the information available to the decision maker at the time.

None of the above steps in structuring a useful decision tree are too implausible in our mining example. Business decision makers generally have clear objectives and reasonably comprehensive understanding of the important uncertainties affecting their business and their options for action. They have their business experience and past market quantities and prices as guides to possible outcomes, values and probabilities. Similarly, it may not be too difficult to determine options for action, plausible action–event scenarios and their likely value pay-offs in environmental problems between neighbours, such as the pool shading problem of Chapter 14. In that situation, possible actions might include negotiating compensation for your neighbour for trimming the trees, or installing gas heating for your pool; uncontrollable events might include a variety of possible reactions by your neighbour when you request trimming of the trees. On the other hand, the problems in creating a useful decision tree are much greater when we turn to planners' choices of actions to deal with major environmental problems involving non-rival and non-excludable goods.

16.3 Structuring decisions involving uncertainty: the case of Leadbeater's possum

Consider the problem of preserving the rare Leadbeater's possum, a small arboreal marsupial found only in the wet mountain eucalypt forests of the central highlands of Victoria.[2] The forests supply sawlogs and pulp to the timber industry. They are almost all on public land controlled by the Forests Branch of Victoria's Department of Conservation and Natural Resources. Leadbeater's possum nests in hollows in mature or dead forest eucalypts and feeds on insects and plant saps and gums in understorey wattles. The mountain eucalypts need to live about 200 years before they develop suitable hollows. Clearfell logging on a 50–80 year rotation, the typical timber management regime, therefore eliminates most of the suitable nesting sites. Further, following the 1939 'Black Friday' wildfires, large areas of fire-damaged trees were logged. Thus very little old-growth forest containing suitable nest-trees remains in the central highlands, and as the number of remaining hollow trees decline, it is believed that the survival of Leadbeater's possum is threatened.[3] However there is still much uncertainty about possum behaviour, forest fire behaviour and the responses of the possum to different logging regimes.[4] Thus there is uncertainty about the impacts of alternative forest management policies.

Any decision on use of the central highlands forests will affect multiple objectives, including preservation of Leadbeater's possum and other forest-dependent species as well as maintenance of timber industry income and jobs and the survival of small towns dependent on logging. There is no market for the continued existence of the possum; it is a non-rival and non-excludable good valued by many people. Thus the values attached to the different forest management objectives cannot be reconciled in commercial markets.

Suppose that a senior executive of the Department of Conservation and Natural Resources has the power to decide the use of a block of forest known to contain a population of the possums. An initial, simplistic, depiction of his or her decision problem is suggested by the decision tree in Figure 16.2. The planner has just two options. The first is to continue clearfell logging on the present 80-year rotation, with retention of streamside reserves and unlogged strips as 'wildlife corridors'. The second, based on the results of recent computer simulations

Figure 16.2 Decision tree for forest–possum management problem

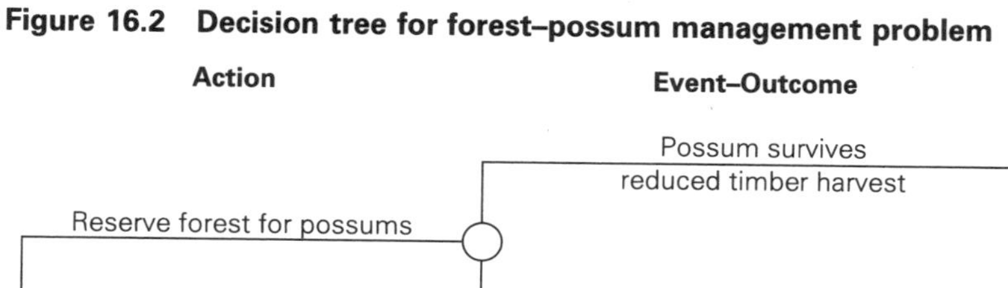
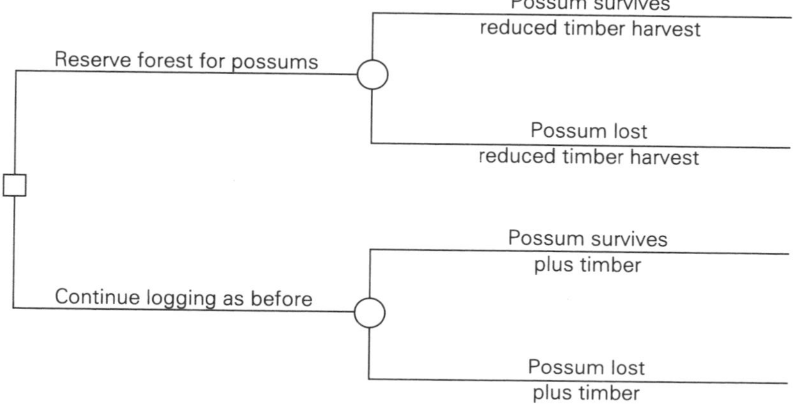

of the results of different logging regimes, would involve preservation of fire-damaged trees and patches of regrowth forest which would otherwise be available for logging, with the aim of creating larger stands of nest-trees in wood-production forests.[5] In each case, the possible events of prime interest are survival or non-survival of the possum population after (say) 100 years, combined with different harvests of sawlogs and pulpwood.

The decision tree in Figure 16.2 provides a starting point for improved depictions of the planner's options and the important events affecting possums and timber production that are outside his or her control. Structuring the decision problem involves identifying additional action–event scenarios that the planner believes are likely to have important impacts on possum survival and timber production. Actions not represented in Figure 16.2 probably would include a variety of logging and forest and wildlife management regimes, plus actions to gather information designed to clarify which events may occur, their outcomes and their likelihoods. Uncontrollable events influencing the outcomes of the planner's decisions might include forest fires, climatic changes (e.g droughts and possible global warming), naturally occurring biological and ecological changes that are not understood and are thus unpredictable (e.g disease outbreaks, predation by other species), and changes in the policies of other government agencies (e.g. Treasury cutbacks in funds for fauna protection and research, granting of mining licences that impinge on the possum habitat). The planner has to judge how many action and event branches to include in the tree.

To make choices about use of the forest and research on possums, the planner has to perform all five decision steps listed earlier. Identifying action–event scenarios covers steps two and three. Reaching a decision also requires clarification of objectives, identification and evaluation of the outcomes of each action-event scenario according to those objectives, and some idea of the relative likelihood of different possible outcomes. (For a decision rule which assumes no information about probabilities of events see Section 21.8.) This is asking a lot in the Leadbeater's possum case. In democratic planning, the planner is supposed to get his or her objectives from elected officials, and ultimately from citizens affected by logging and changes in the possum population. Conflicting values have to be traded off in political decision making. We saw in Sections 3.11 and 8.2 that it is difficult for democratic planners to obtain accurate non-market signals of people's values. It is also difficult for the planner to identify the outcomes and probabilities of different action–event scenarios. This is partly due to a lack of scientific information about the functioning of the forest ecosystem that contains the timber resources and the possums, so that it is difficult to predict natural fluctuations in system characteristics, or the impacts of logging and wildlife management regimes. It is also partly due to the unpredictability of human behaviour that affects the system, including the behaviour expressed in markets such as the market for timber, the political behaviour of politicians and lobby groups, and the behaviour of forest users who may or may not observe the planner's directives regarding forest use.

16.4 Uncertainty and market coordination

Rights to choose actions under uncertainty go with private individuals' and democratic planners' rights to decide the use of assets over time. In the case of public forest in central Victoria, the executive of the Department of Conservation and Natural Resources has the legal right to decide the use of forest flora and fauna. He or she determines the objective or objectives, defines the possible actions and events that structure the decision tree, and decides what information to obtain, preparatory to making a decision. In the case of a private forest, the owner or lessee takes these steps. As always, it is the information and incentives of the decision maker that determine how well he or she coordinates with others.

What are the implications of uncertainty about events and outcomes for social coordination of private decisions? No social coordination problems arise if no one perceives any consequences of a decision that are not subject to appropriate market or non-market feedback to the decision maker. Suppose that steel production, unlike the situation in Figure 5.2, was completely non-polluting. Then it might be the case that, whatever the world price of steel and the wages of steel workers, all the consequences of every possible steel production scenario would be borne solely by the steel company and its employees, each of them subject to market and non-market feedbacks from the other. So, assuming that the company and its employees can negotiate a mutually satisfactory course of action, including workers compensation, full social coordination is achieved, despite uncertainty.

Uncertainty does pose barriers to social coordination through private deals. Market coordination between people is more costly when the parties each recognise a variety of possible action–event scenarios, increasing the amount of information required before a deal can the struck. The possible actions of one party will be uncontrolled events for the other party. For example, the steel company cannot control the cooperativeness of its employees, which affects the profitability of investment in new equipment, and the employees cannot control the company's investment in new equipment. What is controlled and what is uncontrolled depends partly on the parties' respective property rights.

Recall from Chapter 5 that the higher the costs of the information required for market exchange, the less is the chance that markets will exist. Thus the greater the uncertainty attached to some course of action, the less likely it is that market deals will provide decision makers with feedback about all possible benefits and costs they may inflict on others. For example, it is possible to buy maritime insurance to cover the accidental loss of ships and cargoes, but not to cover the health and environmental damage due to the loss of hazardous cargoes such as crude oil and radioactive materials, because of uncertainty about possible consequences in the latter cases. Uncertainty is likely to create similar barriers to the establishment of non-market feedback mechanisms.

Public perception of uncertainties affects community decisions on the allocation of property rights. Going back at least to the publication of Rachel Carson's *Silent Spring*,[6] and stirred by highly publicised examples such as Chernobyl and the *Exxon Valdez* and Melbourne's Coode Island chemical fire, the Western public has become increasingly concerned about the uncertain environmental consequences associated with the use of modern technologies. This has resulted in laws that have diminished private rights and increased government planners' rights, where the public has perceived that new technologies and development projects (e.g. chemical pesticides, genetic engineering, paper mills and power stations) may harm people who would have no say in private decisions.

16.5 Uncertainty and planning coordination

How does uncertainty alter the process of communication between the planner and other interested parties, discussed in Chapter 8?

Thinking of problems in terms of decision trees helps to see the difference in communication requirements. Suppose that the planner responsible for forests and possums must choose between forest use options, perhaps as few as the two depicted in Figure 16.2. In the absence of uncertainty, there would be only a single outcome for each option. This is the situation envisaged in Chapter 8, where the planner's problem was to obtain correct technical and value information from interested parties prior to making and implementing a decision. With events and outcomes uncertain, communication between the planner and interested parties is more complex in several respects. First, the planner and other parties have to communicate information about action–event scenarios and their outcomes for possums and timber production. Second, the planner needs information about the probabilities of possible events and outcomes. Third, if people have objectives related to uncertainty itself, the planner needs information about the values attached to those objectives, as well as to objectives such as maximising the chances of possum survival and maximising profits from logging. We now consider these additional complexities and the difficulties they pose for social coordination under democratic planning.

First, the scientific and behavioural uncertainty involved in forest and possum management means that no one, not the planner nor biologists nor the timber industry nor conservationists, has a clear idea of the outcomes of particular actions. Action–event scenarios and their perceived outcomes will differ across parties, each constructing decision trees oriented to their particular interests. Biologists will emphasise the possible consequences of different forest-use decisions for possum survival, foresters for timber output, and local communities and milling companies for logging and milling returns and costs. Conservationists may envision greater threats to possum survival than most biologists. With different perceptions of possible events and outcomes, interest groups' preferred alternatives may differ for these reasons, as well as differences in objectives and values. Communication of scientific, technical and behavioural information about the possible consequences of alternative forest management regimes may narrow such differences, so that more people are satisfied with the final planning decision. So democratic planners may have a role in communicating information about possible events and outcomes, in addition to collecting the information necessary to make and implement decisions. In Australia, this role is performed, among others, by parliamentary committees, the Productivity Commission, and in the case of the future use of Australian forests, by the former Resource Assessment Commission.[7]

Second, together with information on events and outcomes such as droughts, bushfires, high or low timber prices, mill closures or non-closures and possum survival or non-survival, the planner needs estimates of their probabilities. Such estimates may be highly subjective, but this may also be true of the scientific and behavioural assumptions made in estimating outcomes, and the values attached to outcomes. As illustrated in discussing Figure 16.1, the final choice of an action (a primary branch of the decision tree) will reflect all three components: possible outcomes of that action, their values and their probabilities.

Third, there are objectives related to uncertainty itself. Economists commonly classify individuals or groups as 'risk-averse', 'risk-neutral' or 'risk-preferring'. For example, in the mining decision problem discussed in Sections 16.1 and 16.2, we assumed that the mining executive making the decision was risk-neutral. In other words, the executive's choice between the alternatives was based solely on the expected present value of profits from each, and not on the degree of perceived uncertainty or the range of the possible profit outcomes associated with each. A risk-averse decision maker would favour actions yielding a more certain and narrower range of possible outcomes, and a risk-preferrer the reverse.

Uncertainty-related objectives might plausibly take two forms. One is seeking or, more likely, avoiding, the unknown. Uncertain outcomes may be feared simply because they are unknown, aside from their contribution to other values. In terms of a decision tree, an action branch is less preferred, the more possible outcomes and the more subjective their probabilities. For example, some people may be concerned contemplating the unknown consequences of logging for Leadbeater's possum. Others may be concerned at the unknown consequences of local mill closures, if logging is halted. In either case, if people knew the outcome of policy decisions, they would be able to respond calmly and constructively. The costs of stress based on people's fears of unknown and unfavourable outcomes, stress that occurs whether or not such outcomes eventuate, may be substantial.[8] Needless to say, opponents of particular planning decisions (e.g. permitting logging) will emphasise such unknown hazards. This strengthens the rationale for the planner's communication role. Politically sensitive planning will involve two-way communication of the judgments of 'experts' (biologists, foresters, economists etc.) and of the concerns and values of those directly affected (people dependent on logging, conservationists).

The field of risk communication emphasises the need to communicate what is 'culturally rational', in terms of fitting in with acceptable standards of morality, decency and due process, as well as what is 'technically rational', as perceived by technical experts.[9] For example, 'objective' science may totally discount the possibility of harm to the public from consumption of foodstuffs containing recombinant DNA, while citizens who would be exposed to any such harm see the risk as significant, due to the broader social context in which the decision is made.

Another type of objective related to uncertainty itself concerns the distribution of possible outcomes. People may have preferences concerning the distribution of outcomes resulting from an action, as well as concerning particular outcomes. People who are risk-averse will shy away from choices that appear to combine the possibility of large gains with the possibility of large losses (bidding for the Olympic Games) compared to choices that promise a much narrower range of possible gains and losses (bidding for the Commonwealth Games).

The problem of preserving Leadbeater's possum is typical of major environmental problems involving large numbers of people. Scientific and behavioural uncertainty and multiple conflicting objectives mean that planners have to collect and communicate much more information in seeking to maximise gains to the community as a whole. Planners' specific objectives and decision trees will, at least initially, be imprecise. The greater the imprecision concerning planners' objectives, options, possible events, their probabilities and consequent outcomes, the more difficult it will be to arrive at logical judgments about the action that contributes most to the decision maker's chosen objective, and hence to assess the quality of decision making. In this sense, planners making decisions on environmental policies will have a great deal of discretion—it is difficult to demonstrate incorrect choices when the information about alternatives is incomplete.

16.6 Implications of uncertainty for social coordination systems

Does uncertainty change our earlier assessments of the merits and limitations of markets and democratic planning? Not in any obvious way, because it raises information costs for both market and political decision makers. Both have to process much more information about the possible outcomes and probabilities; both have to consider objectives related to uncertainty itself. Because uncertainty makes market exchange more costly, it increases the likelihood that market feedback will not take account of all possible consequences of decisions. It tends to expand the roles of planners, due to public perception of uncertain harms that are not signalled in markets. For example, this appears to be the case for new technologies such as genetic engineering of commercial plants and animals, and irradiation of fresh produce to reduce spoilage. On the other hand, because uncertainty increases the information requirements of democratic planners and the citizen-voters they represent, it exacerbates the limitations of planning discussed in Chapter 8. So uncertainty increases both market and planning externalities.

As explained in discussing the specificity of property rights in Chapters 2 and 5, in choosing social coordination arrangements, societies must trade off the gains from better coordination of production and consumption decisions against the costs of better coordination in the forms of more precise property rights and monitoring of market and political decision makers. By raising the information costs for both market and political decision makers, uncertainty tips the balance in favour of less precise social coordination arrangements, both private and public. In terms of the diagrams of Chapter 2, uncertainty means that more information has to be signalled between the parties. Thus it becomes efficient for people to put up with less accurate feedback, and more market and political externalities, because of the higher costs of operating signalling and incentive systems.

16.7 Anticipation as a response to uncertainty: the precautionary principle

Uncertainty is most troubling when the possible outcomes of current actions may involve serious *and* irreversible damage to ourselves or to the community. Decisions to undergo risky major surgery have this character for individuals. Figure 4.1 suggests that major decisions on our use of the natural environment may have serious and irreversible consequences for whole communities, if an irreplaceable natural resource or ecological life support system is irreversibly altered. The construction and operation of the Chernobyl nuclear plant is one example. The decline of historic civilisations whose irrigated agriculture succumbed to increasing salinity appears to be another.

The pre-eminent current example of human impacts on the environment that could have both serious and irreversible consequences for humanity is the prospect of global warming due to increasing atmospheric concentrations of the so-called greenhouse gases—carbon dioxide, methane, nitrous oxide, ozone and CFCs—resulting from human activity (see Chapter 17). The serious, possibly life-threatening, consequences contemplated include expansion of the tropical cyclone belts to higher latitudes, extension of deserts due to changes in rainfall patterns, and substantial rises in global sea-level.[10] While it is certain that atmospheric concentrations of greenhouse gases are increasing as a result of human activities, the resulting changes in world climates and costs and benefits for humans, and the costs and benefits of proposed strategies for responding to possible global warming, are not at all certain. There is scientific uncertainty about interactions within the environment, and behavioural uncertainty about people's reactions to climate change. For example, the roles of clouds and oceans in modifying any global temperature changes are poorly understood, as are the abilities of species and ecosystems to adapt or evolve in a greenhouse gas enriched environment.[11]

Given the uncertainty about the greenhouse effect, there is an unavoidable choice between doing something now to avert possible damage and postponing action until there is evidence of how the climate is changing. Pearce, Markandya and Barbier, in *Blueprint for a Green Economy*, call these *anticipatory* and *reactive* environmental policies. They argue that a risk-averse society, one that wishes to avoid large losses from future environmental disasters, will prefer the 'prudent pessimism' of anticipatory policy, for example reducing greenhouse emissions now, before we are certain of the climatic response.[12] In the same vein, the Intergovernmental Agreement on the Environment (IGAE) signed by the then Prime Minister of Australia, state premiers, chief ministers and a local government representative in 1992 included the *precautionary principle*:

Figure 16.3 Greenhouse policy decision tree

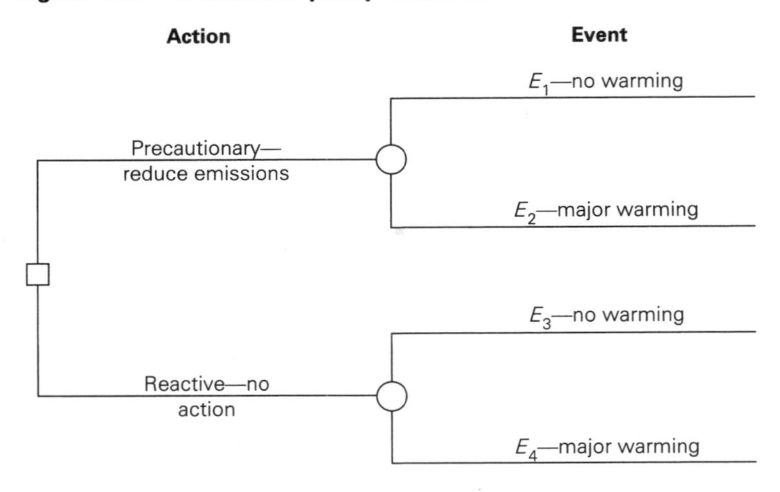

> Where there are threats of serious or irreversible environmental damage, lack of full scientific certainty should not be used as a reason for postponing measures to prevent environmental degradation.
>
> In the application of the precautionary principle, public and private decisions should be guided by:
> (i) careful evaluation to avoid, wherever practicable, serious or irreversible damage to the environment; and
> (ii) an assessment of the risk-weighted consequences of various options.[13]

A precautionary approach to greenhouse warming would involve action to reduce human greenhouse emissions now, despite uncertainty about global warming due to those emissions, and about the costs of any global warming that does occur.

To help clarify the choice between precautionary and reactive policies, consider the simplistic greenhouse policy decision tree depicted in Figure 16.3.[14] The decision maker has just two present options: to act now to reduce human emissions of greenhouse gases that may lead to major global warming within a few decades, or to postpone action, which avoids present costs. Each option is followed by one of two possible climatic events; either major warming will occur or it will not. In the future the decision maker will choose an action tailored to whichever of the specific climatic events E_1–E_4, has occurred.

The decision tree in Figure 16.3 would be correctly structured if it included all possible action–event scenarios. However, in the case of greenhouse gas emissions, this is not the case. While it is certain that atmospheric concentrations of greenhouse gases are increasing as a result of human activities, uncertainty about greenhouse science, future technologies and people's future responses to climate change means that decision makers cannot identify all the possible action–event scenarios, let alone their probabilities.

At first glance, if we believe that people are risk-averse, the precautionary option seems sensible; better to act now to avert possible future disasters that may cause irreversible damage to society. But precautionary policy makes little or no sense if the decision maker is uncertain about possible events and outcomes—if the future disaster cannot be foreseen, or if there is no certainty that precautionary policy will avert it. Choosing precaution over reaction implies that the decision maker *knows something about events and outcomes*, namely that a precautionary policy of reducing emissions now will eliminate or greatly reduce the possibility of severe future damages due to warming. If the decision maker does not know that a precautionary policy will be effective, it is possible that the precautionary approach leads to the worst of both worlds—society incurs the up-front costs of precautionary actions, but still suffers the future costs of unanticipated (given precautionary measures) disasters.[15] Putting it another way, if the chances of very costly climatic changes, such as substantial rises in sea-level, are little affected by costly reductions in current emissions, why incur such costs now? The adoption of precautionary policy is not necessarily consistent with major uncertainty about its outcomes. However, remember that risk-averse decision makers prefer actions yielding a more certain and narrower range of possible outcomes. Precautionary policy may be preferred to reactive policy if it narrows the perceived range of possible events and outcomes.

16.8 The information requirements of precautionary policy

Figure 16.4 illustrates both the possibility of ineffective precautionary policy and the information required for precautionary policy to make sense. It incorporates the same climatic possibilities as Figure 16.3, plus the possibility that precautionary emission reductions may be either effective or ineffective in reducing global warming.

To choose between precautionary and reactive policy, the decision maker needs information about, first, the magnitude of the present costs of precautionary measures, and second, the future consequences, costs and benefits of precautionary and reactive policies. In the greenhouse policy choice depicted in Figure 16.4, both are subject to major uncertainty. In the case of current actions to control emissions, the uncertainty about cost is due to the major economic and social dislocation which could attend severe restrictions on the production of greenhouse gases, in particular carbon dioxide. In the case of future consequences, limited scientific understanding of global climate change and ignorance of future technologies and human responses to climate change cause major uncertainties.

The outcomes of possible action–event scenarios in Figure 16.4 are measured in terms of the (discounted) total costs of precautionary policy and of the climate change itself, less any savings due to future reactive

Figure 16.4 Greenhouse decision tree when policy may be ineffective

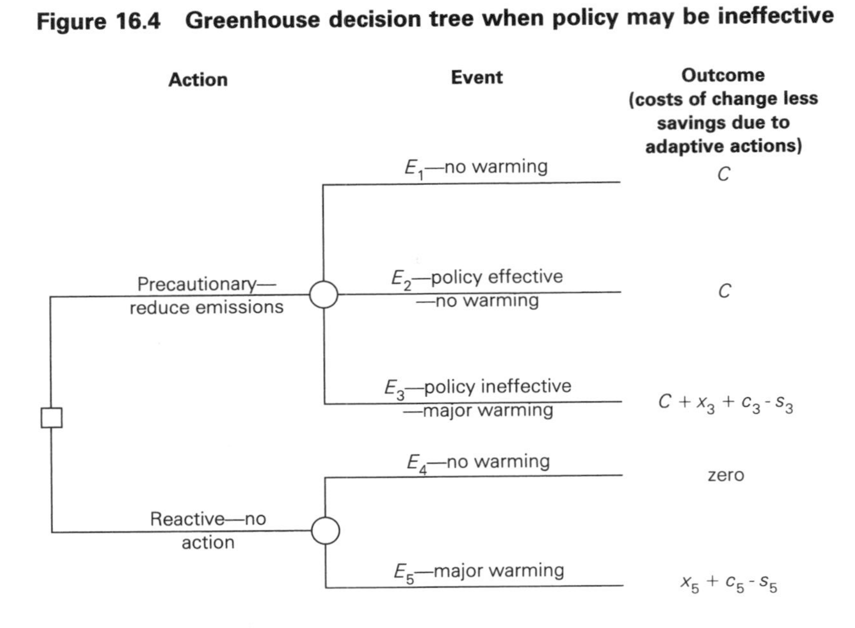

policies. It is assumed that effective precautionary policy leads to a climatic outcome identical to that with no warming, and that ineffective precautionary policy makes no difference to warming. Thus the climatic outcomes are the same for E_1, E_2 and E_4, and E_3 and E_5. The precautionary option of reducing current greenhouse emissions involves a present cost C. If global warming does occur (events E_3 and E_5), the present values of the resulting losses and adjustment costs for unassisted households and businesses are x_3 and x_5. (Future costs will vary between E_3 and E_5, because the prior adoption of precautionary policy will alter future activities, incomes and prices.) The costs of future reactive policies when society acts to adapt to or counter climate change are c_3 and c_5. The savings due to future reactive policies are s_3 and s_5. As explained above, the values of C and of the x's, c's and s's in Figure 16.4 are all subject to major uncertainty.

Assume that the decision maker knows the possible events, $E_1 - E_5$, but not their probabilities or the associated costs and savings. Under what circumstances will the precautionary option make sense? It depends on the decision maker's degree of risk aversion and beliefs about the efficacy of precautionary measures. Assume that the decision maker is so risk averse that he or she cares only about the worst possible outcomes of the precautionary and reactive policies: he or she will choose whichever policy minimises the maximum possible total cost. For the precautionary option of reducing current greenhouse emissions to minimise the maximum possible cost, he or she must believe that the worst of the possible outcomes of precautionary policy, $C + x_3 + c_3 - s_3$, will cost less than the worst of the possible outcomes of reactive policy, $x_5 + c_5 - s_5$. Yet precautionary policy involves the additional costs of current precautionary measures. Thus, if there is a perceived slight chance that precautionary policy may be ineffective, an extremely risk-averse decision maker may choose reactive policy. A less risk-averse decision maker will also give some weight to the other possible events, E_1, E_2 and E_4, and their costs, C, C and zero respectively. He or she is willing to accept a perceived low chance of the high costs of ineffective precautionary policy in return for a perceived higher chance of effective policy which avoids the high costs of global warming: he or she will prefer precautionary policy.

The important point is that in the case of greenhouse, a risk-averse society could make matters worse by adopting precautionary policy. It is commonly believed that precautionary policy, involving the minimisation of currently observable environmental damage, is sensible if people are risk averse. This is almost certainly the case when we are dealing with relatively well-understood environmental systems such as a municipal sewage disposal or local air pollution. In such situuations the decision trees can be fairly precisely structured, and the chances of very costly surprise outcomes of precautionary policy are negotiable. On the other hand, with the major uncertainty characteristic of extensive and complex economic–environmental systems, precautionary policy can be ineffective and can lead to unpleasant surprises, and therefore can be more damaging than reactive policy.

Note that the preceding discussion understates the uncertainty attending greenhouse policy. Contrary to the decision tree depicted in Figure 16.4, decision makers cannot identify all of the environmental–social consequences which may follow the choice of precautionary and reactive policies. Event branches, as well as the corresponding outcomes on the right hand side of the tree, are not clearly specified.

16.9 Precautionary versus resilient environmental policies

Discussing the widespread adoption of health and safety legislation in Western societies, Wildavsky argues for less emphasis on precautionary policies and more on reactive policies—what he calls *resilience*.[16] He argues, first, that widespread adoption of anticipatory policies designed to reduce damage that may never occur reduces the resources available to counter unforeseen dangers. Second, anticipatory policies designed to avoid change limit individuals' opportunities to learn from the experience of change; as a result, people are less able to design cost-effective responses to the unforeseen changes that do occur.

Wildavsky's first argument is an argument for substantial discounting of future values. He argues that the resources sacrificed in implementing anticipatory policies may be productively invested to make people much better off in the future. The discount rate measures the productivity of the forgone investment opportunities. Putting the point in another way, if the income sacrificed by reducing greenhouse emissions now was invested in an environmental contingency fund to deal with future climate changes, it would compound rapidly to much larger future sums. Thus, in making comparisons between today's costs of precautionary policies and the possible future costs of climate change, a substantial discount rate should be used; future values should be weighted substantially less than current values.

Wildavsky grants that precautionary policies may be appropriate in some circumstances. If a future climatic catastrophe is reasonably certain, and if reductions in future greenhouse emissions are reasonably certain to avert severe damage, precautionary action is likely to be the best strategy. But, as already explained, if catastrophes cannot be foreseen, or there is no certainty that precautionary action can avert them, it is better to develop resilience. Resilience is likely to be highly correlated with societal productivity and wealth. Wildavsky points out that 'richer is safer'—modern disasters kill overwhelmingly more people in poor countries than in rich countries.[17]

16.10 The planning bias towards anticipatory policies

Wildavsky argues that democratic planning is biased towards precautionary policies.[18] The possible disasters that we recognise loom much larger in our imaginations than disasters that we cannot imagine or the everyday harms to which we have grown accustomed. Experimental psychologists report that, in case of risks to people where statistics are available, lay persons overestimate the probabilities of events that are widely publicised and thus easily imagined, such as nuclear accidents and shark attacks. Conversely the probabilities of unspectacular and rarely reported events such as strokes and domestic accidents are underestimated.[19] This leads to demands for anticipatory policies to avert the recognisable disasters, with inadequate recognition that those policies may reduce society's ability to respond to the unforeseen and to reduce the incidence of the mundane. In the case of greenhouse, many people perceive the possibility of serious and irreversible damage due to climate changes and rises in sea-level, and demand emissions reductions as a result. They are less likely to recognise that the resulting reduction in community income and wealth can reduce expenditures on other ways of saving people's lives and property, and on the research and development that equips society to handle the unforeseen. For example, the search for a cure for the unforeseen disaster of AIDS is underpinned by research in molecular biology that occurred before AIDS revealed itself.

Private decision makers regularly face choices between precautionary and reactive strategies. We can accept or refuse to undergo immediate surgery to repair constricted arteries or to have a breast removed after inconclusive indications of tumour development. We can insure or not insure our house or car. Does the bias towards anticipatory action apply to private decisions? In general, no. In personal decision making, individuals have strong incentives to acquire symmetric information about the relative outcomes, costs and benefits of precaution and reaction. On the other hand, when a politician or public servant is making the decision, citizens' information gathering and signalling will be influenced by the voter logic explained in Section 8.3. If we do not expect our vote to make any difference to the planner's decision, there is little incentive to collect information beyond that ready to hand, and considerable likelihood that votes will express prejudices rather than judgments based on facts.

Many planners will recognise that both precautionary and reactive policies could lead to major costs for society. If they act to reduce greenhouse emissions, and science later establishes that increased greenhouse gas emissions are either beneficial or cause little damage to humanity, community income has been sacrificed for little or no benefit. If planners postpone action, emissions from human sources may cause climate changes that seriously harm humanity. (The tradeoffs between these alternatives are further explored in Sections 21.7 to 21.9.) However, since the public is more aware of the possible costs of not acting on greenhouse than the costs of acting and getting it wrong, planners' self-interest encourages them to act now. It takes a strong politician or bureaucrat, dependent on the public's votes and its tax dollars, to resist demands to 'do something' by way of precautionary action in response to well-publicised environmental dangers.

Discussion questions

1. When uncertainty is present, decisions that turn out to be incorrect in the circumstances are not necessarily wrong decisions. True, false or uncertain? Explain why. Can you suggest an example involving (say) medical treatment?

2. The more uncertain we are about the future environmental and economic impacts of current resource use, the more we should rely on government planners, who are responsible to the community rather than just to themselves. True, false or uncertain? Explain why.

3. Consider the forest management problem depicted in Figure 16.2. Suppose that there is a change of executive personnel in the Department of Conservation and Natural Resources. The forester previously responsible for planning the use of the central highlands forests containing the possum is replaced by an ecologist. Do you think that the nature of the decision tree is likely to change? Suggest why and how, or why not.

4. The pay-off matrix below, taken from Pearce, Markandya and Barbiers' *Blueprint for a Green Economy*, illustrates the argument in favour of precautionary environmental policies on major environmental issues such as global climate change and preserving biodiversity. Given scientific uncertainty about the future state of the world and a risk-averse decision maker, it is deemed better to take precautions now to avert possible future disaster, than to postpone such actions and risk disaster occurring.

		State of the world	
		Optimists right	Pessimists right
Type of policy	Reactive	HIGH GAINS	DISASTROUS LOSSES
	Precautionary	MODERATE GAINS	TOLERABLE GAINS

Depict the policies, future events and payoffs shown in the matrix in the form of a decision tree. What assumptions about the future consequences of reactive and precautionary policies are built into the pay-off matrix/decision tree? do you think the assumptions are reasonable? Explain why or why not.

17

The economics of global pollution: ozone depletion and climate change

The hardest environmental problems—where the costs of better co-ordination in use of the environment are highest—are those where the consequences of an action extend worldwide and far into the future. Global pollution problems such as stratospheric ozone depletion and possible global warming due to human emissions of greenhouse gases can affect billions of people, many not yet born. Future benefits and costs of current actions are uncertain and often unknowable. In contrast to domestic environmental problems, pollution control is only possible via voluntary agreements between countries. However individual countries have little incentive to control their emissions if they get few of the benefits of their actions but bear all of the costs, so international deals must incorporate carrots as well as sticks. This worldwide cooperation over environmental problems may be impossibly costly, yet successful international cooperation, exemplified by the Montreal agreement on ozone-depleting substances, suggests that this is not always the case. Success appears to depend heavily on scientific consensus about the problem and on the countries which bear most of the costs also getting most of the benefits.

17.1 The 'mother' of all pollution problems

Some pollutants are dispersed worldwide. Some persist in the environment for a long time, capable of harming our descendants as well as ourselves. Chlorofluorocarbons (CFCs), linked to thinning of stratospheric ozone, and greenhouse gases, predicted to cause warming of Earth's climate, have both of these characteristics. If I dump an old refrigerator containing CFCs, or burn garden refuse in my backyard, the CFCs and carbon dioxide thus released change the atmospheric concentrations of CFCs and carbon dioxide worldwide, not just over Australia, and not just for a few days or months, but for one hundred years or so. My action imposes non-rival and non-excludable costs (or possibly benefits) on all or most of the people alive in the world during the next one hundred years.

There is no possibility of a multilateral deal among all the people who benefit and suffer from CFC or greenhouse gas emissions. Even if I want to compensate all those I damage, there is no way they could all be identified (many are not born yet) or could agree among themselves about sharing the compensation. In fact, since I am just one of millions or billions of such polluters worldwide, I probably prefer to do nothing, free riding on the compensation payments or emissions reductions of other polluters. The story is the same as for acid rain in Section 5.7—an absence of agreed individual property rights, prohibitive costs of measuring transfers between individual pollution sources and sufferers, unwillingness of individual sources and sufferers to reveal their true willingness to pay or to accept compensation for more or less pollution—but on a grander scale.

When negotiation between polluters and pollution sufferers becomes impossibly costly, we usually expect planners to step in, as explained in Chapter 15. However, since almost all CFC and greenhouse gas emissions cross international frontiers, there is no possibility of balancing the interests of pollution sources and sufferers using the democratic planning processes of a single government; no government can coerce polluters outside its borders. On the contrary, in the case of CFCs and greenhouse gases, countries have incentives to free ride on the pollution-control efforts of other countries.

The ozone depletion and global climate change problems differ from the local or regional pollution problems discussed in Chapter 15 in ways that make social coordination to control pollution even more difficult. To better appreciate the differences, it helps to understand a little of the science of ozone depletion and greenhouse.

17.2 Stratospheric ozone depletion[1]

The stratosphere, that portion of Earth's atmosphere between 15 and 50 kilometres above the surface, contains small amounts of ozone gas, popularly known as the 'ozone layer'. This ozone absorbs most of the sun's ultraviolet (UV) radiation. Thus depletion of the ozone layer allows more UV radiation, specifically UV-B, to reach Earth's surface. Increased UV-B radiation is damaging to human health, increasing the incidence of skin cancers and cataracts. There is also evidence that it reduces the productivity of many crop species and of aquatic ecosystems.

CFCs are chemically inert, and hence non-toxic and non-flammable, compounds invented in the 1930s as cheap, safe refrigerants. They are used in refrigeration and air-conditioning, as blowing agents to manufacture plastic foams, as solvents for cleaning electronic components and clothing and, formerly, as aerosol propellants. Halons, which substitute bromine for the chlorine of CFCs, are used in fire extinguishers.

CFCs' and halons' chemical stability means that they cause no harm in the lower atmosphere, but also that they survive in the atmosphere for a very long time—50 or more years. Once released in use or from products such as refrigerators, they take 5 to 10 years to drift upwards to the stratosphere, where UV radiation slowly breaks them down, releasing chlorine or bromine atoms.

Once free in the upper atmosphere, single chlorine and bromine atoms are each capable of destroying many thousands of ozone molecules, converting them to oxygen gas. The result is that ozone destruction exceeds the natural creation of ozone, the concentration of ozone in the stratosphere falls, and there is an increase in UV-B radiation reaching Earth's surface. However, the long delays involved in CFC transport to the stratosphere and breakdown mean that current UV levels may be a poor guide to the future levels resulting from the stock of CFCs already released into the atmosphere.

The first unequivocal evidence of thinning of the ozone layer emerged from a British Antarctic station in the mid-1980s. Average stratospheric ozone levels in September–November were found to be 50 per cent lower than in the 1960s. Subsequent observations confirmed major thinning above the Antarctic continent each spring—the famous ozone 'hole'—and a similar, less pronounced, 'hole' over the Arctic. The 'holes' are believed to result from the interaction of seasonal climatic conditions with the reservoir of stratospheric chlorine; thus the severity of depletion varies from year to year, and ozone levels recover each summer.

17.3 Global climate change due to greenhouse gas emissions[2]

Earth receives shortwave radiation, including sunlight, from the sun. Some is reflected, but most is absorbed in the atmosphere and at the surface, warming Earth. For Earth to maintain a constant temperature, the incoming energy absorbed must be balanced by an equal amount of longwave infra-red radiation emitted back into space. If Earth had no atmosphere, this energy balance would be achieved at a temperature of –18 degrees Celsius, the temperature of the moon. However, some atmospheric gases (so-called greenhouse gases), including water vapour, carbon dioxide, methane, nitrous oxide and CFCs, which are transparent to solar radiation, absorb and re-emit infra-red radiation. Thus Earth

Table 17.1 Greenhouse gas emissions: characteristics

	Carbon dioxide	Methane	Nitrous oxide	CFC–11	CFC–12
Concentration:					
• pre-industrial	285	0.75	0.29	–	–
• 1989 (ppmv)[a]	350	1.70	0.31	0.00025	0.00045
Sources	Fossil fuels, biomass burning, soil tillage, volcanoes, oceans	Rice paddies, ruminants, gas fields, coal mines, landfills, wetlands, oceans	Biomass burning, soil tillage, fertilisers, oceans, soil bacteria	Human manufacture and use	
Removal	By plants, soil, oceans	Changed to formaldehyde	Changed to nitrogen	Broken down by UV radiation	
Atmospheric lifetime (years)	60	10	150	75	110

Notes: [a] ppmv= parts per million by volume.
Source: Industry Commission, *Costs and benefits of reducing greenhouse gas emissions*, vols 1 & 2, AGPS, Canberra, 1992.

receives warming radiation from its atmosphere as well as from the sun. Earth and its atmosphere then achieve the required balance between incoming solar energy and outgoing radiation at a much more congenial temperature, 15 degrees Celsius.

Human activity during the last two hundred years has increased the atmospheric concentrations of some greenhouse gases. Table 17.1, based on the Industry Commission report on reducing greenhouse gas emissions, shows the changes to 1989. The concentration of carbon dioxide has risen by about one-third since 1850, and that of methane has about doubled. CFCs, which are also important greenhouse gases, did not exist before the 1930s. The average global temperature has risen by about half a degree Celsius over the last one hundred years, but not necessarily due to the increase in greenhouse gas concentrations. In fact, most of the warming of the last century took place prior to the last fifty years, when greenhouse gas concentrations have risen fastest.[3]

Table 17.1 also shows the natural and human sources of major greenhouse gases and how they are removed from the atmosphere. The contribution of each gas to eventual global warming depends on its physical properties and atmospheric lifetime and concentration. (The lifetimes of the gases are also given in Table 17.1) Based on these data, the Industry Commission report stated that the carbon dioxide in the atmosphere in 1989 would account for about 68 per cent of the resultant warming of Earth over the next one hundred years, compared to about

17 per cent for methane, 5 per cent for nitrous oxide and 10 per cent for CFCs.

According to the 1995 assessment prepared by the United Nations-sponsored Intergovernmental Panel on Climate Change (IPCC), computer simulations of world climate change based on current theories indicate that increasing concentrations of greenhouse gases are expected to raise the average global temperature by between 1 and 3.5 degrees Celsius by the year 2100.[4] This assessment suggests less global warming than was expected earlier; the Panel's 1990 and 1992 reports suggested average temperature rises in the range 1.5 to 4.5 degrees Celsius. Any such rise will be superimposed on natural climate variability due to other factors. Thus, while increasing greenhouse gas concentrations due to human activity are a fact, their effect on Earth's average temperature is subject to substantial uncertainty.

Changes in the average global temperature are not what matters to people. The computer models predict that temperatures will rise more at the poles than at the equator, more in winter than in summer, and more at night than in the daytime. People's lives will be most affected by the related changes in sea-level and local and regional temperatures, precipitation, droughts, floods, freezes, cyclones and so on. Unfortunately, existing models of world climate change contain too little detail to yield reliable predictions of regional, let alone local, climate changes.

Some possible impacts of global warming could create major costs and benefits for humanity. Substantial sea-level rises, due to ocean warming and expansion and/or melting of icecaps, would require major expenditures on sea walls or on relocation of people from low-lying areas such as southern Bangladesh and Pacific atolls. Changes in temperature, evaporation and rainfall would affect the productivity of agriculture and forestry. The computer models of warming suggest the possibility of higher productivity in areas such as Siberia, due to longer growing seasons, and lower productivity in areas such as Australia's wheat belt, due to more erratic rainfall and increased evaporation. Increased levels of atmospheric carbon dioxide will speed the growth of most plant species, and reduce plant water requirements. The impacts on recreational activities are likely to be similarly diverse; in southern Australia, skiing may suffer, but water-based recreation could benefit. Climate changes will also affect land and ocean ecosystems, with unpredictable effects on species populations and survival, and the incidence of pests, weeds and diseases that harm people. If the changes are more rapid than those experienced in the previous history of evolution, the rate of extinction of species may accelerate. Finally, it is possible that changes in sea temperatures and currents (the El Niño phenomenon on a larger scale) could increase the frequency of extreme weather events, such as tropical cyclones, droughts and floods. However, all of these impacts of possible global warming are subject to even more uncertainty than the warming itself.

17.4 Delayed damages and information problems in pollution control

In the cases of CFCs and greenhouse gases, damages (and any benefits) to people are based on the long-lived stocks present in the atmosphere, so that current emissions of pollutants that add to future as well as present stocks cause damages far into the future. Where there are long delays in experiencing damages from pollutants, the information problems described in Section 15.2 are greatly exacerbated. Compare the problems of resolving conflicting interests over coal smoke pollution, discussed in Section 15.2, with those that arise in the case of CFCs. In each case, the marginal costs of reducing smoke and CFCs, which are the same as the marginal benefits of additional smoke and CFCs, are borne by existing households and businesses. However, in the coal smoke example, smoke sufferers were assumed to experience breathing difficulties, extra medical expenses, cleaning bills and so on shortly after smoke emissions. In the case of CFCs, emissions may persist in the atmosphere for as long as one hundred years, and so most of the damages are experienced far in the future.

The passage of time makes a big difference to our ability to identify the true marginal costs of pollution (*MCP*). Recall from the discussion of the *MCP* curve in Section 15.2 that sufferers have alternative ways of dealing with pollution: they may simply bear the costs of smoke inhalation or skin cancer; they may alter their activities to avoid damage, (e.g. staying inside out of the smoke and the sun) and they may adopt measures to insulate their activities from damage (for example wearing face masks and sunscreen). The *MCP* curve is made up of the least-cost combination of these alternatives for each sufferer. The difference between coal smoke and CFCs is that, for smoke, these choices are made by *existing* households and businesses whose exposure and adjustment alternatives are reasonably well known. This is not the case for CFCs, which persist long into the future. The identity and exposure of future sufferers from additional current emissions of CFCs are not clear. Current emissions affect people in many future time periods, depending on the processes of accumulation and decay of atmospheric stocks of CFCs over time, which are not perfectly understood. Also, with the damages resulting from extra emissions of CFCs spread through future time, not only the identities but also the adjustment costs of future sufferers (which depend on future technologies, resource costs and preferences) are highly uncertain. We cannot know what sunscreens or skin cancer treatments will exist in thirty years time, or what they will cost, or future preferences for outdoor versus indoor recreation.

Another problem due to the passage of time is that the costs of reducing CFC and greenhouse gas emissions are borne by people now, while most of the damage or avoidance or insulation costs are borne by people in the future. Current decision makers cannot avoid the practical and ethical problems involved in deciding the relative weights of present and future benefits and costs, discussed in Chapter 9.

17.5 Free riding in international pollution control

The essence of the international free-rider problem is that emissions reductions by any one country benefit all countries but all the costs are borne by the reducing country. Australia is responsible for 1–2 per cent of total global emissions of CFCs and greenhouse gases. Assume, for the moment, that there are no international conventions on global pollutants and that what Australia does has no effect on the emissions policies of other nations. (As a party to the Montreal Protocol On Substances that Deplete the Ozone Layer, Australia's CFC use will in fact elicit responses from other parties. See Section 17.6.) Now consider the benefits and costs to Australia, acting in isolation, of a decision to reduce CFC use, if there are no other technologies currently available to provide the same services at comparable cost. In the absence of the Montreal Protocol, the Australian government would estimate the benefits and costs of reduced CFC use as follows:

1 Because Australia contributes a very small fraction of global CFC emissions, the reduction in Australian emissions will have a very small impact on UV radiation in Australia. The marginal benefit to Australians of the consequent slight reduction in stratospheric ozone will therefore be very small.
2 The marginal cost to Australians of reduced CFC emissions will equal the incremental costs involved in switching to the least costly alternatives in areas such as refrigeration and electrical cleaning. These costs would be substantial, in the millions of dollars per year, until cheaper alternative technologies are available.
3 Since reducing CFC emissions will benefit Australians less than it costs them, Australia will be better off doing little or nothing to reduce CFC use.

Thus, *providing that other countries do not respond to Australia's emissions choices*, it will pay Australia to free ride on the CFC reductions made by other countries.

Figure 17.1, which is comparable to Figure 6.1, illustrates the logic of the decision to free ride. The marginal costs of Australian emission reductions rise as increasingly severe reductions require switching to increasingly costly and/or inferior substitutes for CFCs. Australian emission reductions are a worldwide non-rival and non-excludable good; therefore, the global benefit of each unit reduction by Australia is the vertical sum of the benefits enjoyed by citizens of all countries. Thus in Figure 17.1 the marginal benefits to Australians are only a small proportion of the marginal benefits to the world as a whole. Acting in isolation, the Australian government will reduce Australian CFC

Figure 17.1 Marginal benefits and costs of emissions reductions by Australia

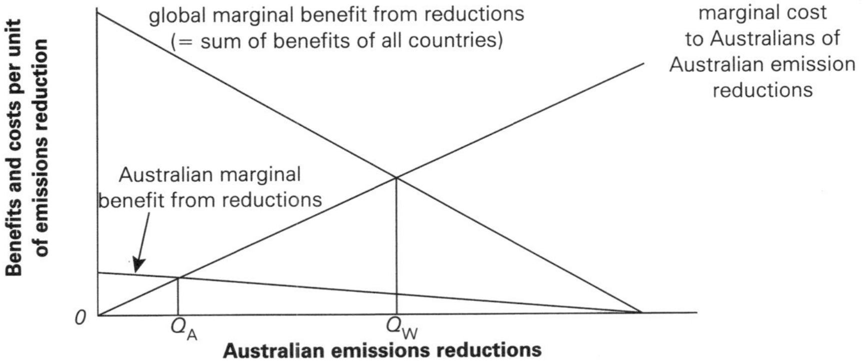

emissions to the point where the marginal cost to Australians equals the marginal benefit to Australians; that is, Q_A in Figure 17.1. In doing so, it will ignore the benefits of Australian reductions for people elsewhere in the world, the positive externality represented in Figure 17.1 by the vertical distance between the Australian and world marginal benefit curves. If Australia took these benefits into account, it would reduce emissions to Q_W, where the marginal cost to Australians equals the worldwide marginal benefits.

Countries with large areas and populations have stronger incentives to control global pollutants, because their emissions have a greater impact on the global environment and they obtain a larger share of total world benefits of emissions reductions. Assume that Figure 17.1 applies to the USA, rather than to Australia, and that their marginal cost curves are the same; that is, equiproportional reductions in CFC emissions have the same marginal costs in the two countries. Because of the larger size of the USA, equiproportional reductions in its emissions generate larger marginal benefits worldwide, and the USA gets a larger share of those marginal benefits. Therefore the marginal benefit curve for the USA will be much higher than that for Australia, for equiproportional reductions in emissions. Thus, acting in isolation, the USA will reduce its CFC emissions by a larger proportion than indicated by Q_A in Figure 17.1.

Non-cooperation among countries does not mean that global emissions reductions will be negligible. Nevertheless, if every country ignores the impact of its emissions on other countries, each will act like Australia in Figure 17.1, and reduce emissions less than is efficient for the world as a whole. Global emissions will exceed the level that would equalise global marginal benefits and costs. The result: lower ozone levels and higher UV damage than would be the case if countries acted cooperatively.

Where a group of countries adheres to a common emissions reduction policy, as is the case for the European Community under the

Montreal Protocol, they will act like a single large country, and the emissions reductions of each will exceed the non-cooperative level. The larger the share of total world emissions emanating from the cooperators, the less is the difference between their combined marginal benefit curve and the worldwide marginal benefit curve. If all countries in the world cooperated, each would reduce emissions to a level equivalent to Q_W in Figure 17.1.

Viewed from another perspective, countries acting in isolation are equivalent to the individual village herders of Section 7.4. Individual countries' emissions choices together determine the intensity of use of the common pool of stratospheric ozone. In choosing its level of CFC emissions, each country ignores its impact on UV radiation damage in other countries. The consequent overexpansion of emissions results in the marginal costs of emissions exceeding marginal benefits, as is the case at Q_A in Figure 17.1. Collectively, countries 'overgraze' the common ozone pool.

17.6 Successful cooperation to reduce ozone depletion[5]

International cooperation has to be voluntary; there is no world government with the power to force countries to reduce CFC or greenhouse gas emissions. Free riding aside, are there other reasons why countries might refuse to cooperate with others to solve a global environmental problem? How can the barriers to cooperation be overcome? One way to answer these questions is to examine a recent case of international cooperation. Control of ozone-depleting substances is, so far, a success story among international attempts to cooperate to protect the global environment. What explains this success? How were negotiators able to overcome the incentives to free ride and other impediments to cooperation?

The Montreal Protocol on Substances that Deplete the Ozone Layer was signed in September 1987. The Protocol, which requires a progressive reduction in production and consumption of CFCs and halons, has been in force since January 1989. In 1990 the parties to the Protocol agreed to an accelerated timetable, with CFCs and halons to be phased out by developed countries by the year 2000. Developing countries are allowed an additional ten years to comply, provided their consumption does not exceed 0.3 kilograms per capita.

The Montreal Protocol and the Vienna Framework Convention for the Protection of the Ozone Layer (which preceded it and provides its legal framework) have been adopted by all the major producers and consumers of ozone-depleting substances. By 1992 the Protocol covered almost 90 per cent of world production and consumption. In the year 2000, with the elimination of CFCs and halons by developed countries, world consumption is projected to be one-third of its peak level in 1988.

International action on ozone-depleting substances took time. Scientific predictions of chlorine-catalysed destruction of stratospheric ozone and consequent threats to human health were made in the early 1970s, and the USA banned the use of CFC propellants in aerosol sprays in 1978. However, there was little impetus for a binding international agreement until after the discovery of the Antarctic ozone 'hole' in 1984. The 'hole' was important for two reasons. First, together with other results of worldwide monitoring of ozone, it provided evidence in support of the scientific theories of the relationship between CFCs and serious ozone depletion, evidence strong enough to convince most scientists of the seriousness of the problem. Second, widespread publicity about the 'hole', and consequent risks to human health, lead the governments of most developed countries to put a high priority on negotiating an international agreement to control CFC production and use. In other words, by the mid-1980s the benefits of reducing CFCs were judged to be high, both by scientific experts and by citizen-voters in developed countries, who perceived increased risks of skin cancers if worldwide CFC use was not reduced. As a result, politicians in developed countries were receiving signals from voters that created strong incentives to act to negotiate international controls on CFCs.

The immediacy of the danger was an important force behind the Montreal agreement. At the time, CFC emissions were linked to the increasing incidence of skin cancers. People could personally identify substantial short-term benefits to weigh against the costs of phasing out CFCs.

On the cost side, the cost of phasing out CFCs and halons is only a small fraction of global income. Their share of income at the time of the Montreal agreement was small in developed countries, and smaller still in developing countries. The anticipated cost of reducing emissions was further reduced by the availability of CFC substitutes, hydrochlorofluorocarbons (HCFCs) and hydrofluorocarbons (HFCs), developed by the same multinational chemical companies that produce CFCs.

The location and structure of the CFC industry is probably the most important clue to international cooperation in reducing CFC production and consumption. In 1987 the USA and the European Community (EC) were both the dominant producers (30 and 45 per cent of world production, respectively) and the dominant consumers (30 per cent each) of CFCs. Each was big enough to have a major impact on the UV-B radiation received by the other; each therefore had an interest in trying to get the other to participate in coordinated emissions reductions. In addition, the CFC industries in the USA and the EC are highly concentrated, and industry cooperation is important for effective control of emissions. Once regulation became politically attractive in the mid-1980s, it was clear that companies' market shares of CFCs and the new CFC substitutes were best protected if the USA and EC jointly controlled CFC production, preferably along with Japan and the then USSR, the other important producers (each with about 10 per cent of world production). It is therefore no surprise that the Montreal meeting agreed to equal percentage reductions in countries' CFC production over the phasedown period. The interests of the existing multinational CFC producers were further served by the introduction of internationally tradeable permits in 1990, allowing producers to transfer production internationally between plants by agreement between the parties.

The effect of the Montreal Protocol, then, was to award production rights to the existing producers, who were also the main producers of the new substitute chemicals, and to create barriers to the entry of new producers. In other words, the Montreal Protocol created a cartel for the small number of CFC producers in the developed signatories, allowing these companies to charge higher prices than otherwise during the phasedown period.

Protection of existing producers, and emissions reductions, depend on the deterring of cheating by parties to the Montreal agreement and of free riding by non-parties. What deters cheating and free riding in the case of CFCs and halons? First, the small number of producers reduces monitoring costs. Second, the dominant producing countries and companies are also the countries and companies that stand to benefit most from rigorous enforcement of production limits. Third, international trade in CFCs and halons is relatively easy to monitor, and finally, the Protocol includes trade provisions that penalise non-cooperation. Countries that are parties to the Protocol agree to reduce production and consumption of CFCs by limiting their production and banning imports of CFCs and CFC-containing products from non-parties. Exports of CFCs to non-parties are also banned. A country that cheats on the Montreal agreement is liable to be converted to a non-party. Thus, if Australia were to fail to comply with the agreed phasedown of its CFC production, it could be subject to a ban on CFC exports to, and CFC imports from, parties such as the USA, the EC and Japan, plus a ban on exports of CFC-containing products to the same countries.

The incentive to join the Montreal agreement is weaker for poor countries, which would like continued access to the cheap refrigeration that the rich countries have enjoyed in past decades. The cartel that suits the wealthy producers of CFCs and CFC-substitutes threatens growth in living standards in poor countries. Because poor countries were seen as likely to experience higher costs relative to benefits than rich countries, the Montreal Protocol allowed them a ten year delay in CFC reductions, which must begin in 1999. The 1990 revision of the Protocol created a second incentive for poor countries to join: a multilateral fund financed by the developed parties, designed to meet all *agreed incremental* costs of compliance by developing countries. The fund can be used, for example, to finance the incremental costs of importing the more expensive CFC substitutes.

17.7 International cooperation: lessons from Montreal[6]

What does the experience of the Montreal Protocol teach us about the prospects for international cooperation on climate change and other perceived global environmental problems? Briefly, it suggests that effective international cooperation is easier if there is scientific and political consensus on the seriousness of the problem, if the costs of cooperating are relatively low, if a small number of producers and countries are responsible for most of the problem, if cheating and free riding can be effectively and economically deterred, if the same countries that bear most of the costs get most of the benefits, and if countries whose costs of cooperating exceed their benefits are compensated for their cooperation. The first three of these points will be clear from the preceding discussion of the Montreal agreement. The latter three require further elaboration.

Deterring international free riding The free riding described in Section 17.5 occurs because individual countries (Australia in Figure 17.1) are subject to signals and incentives about the costs of reducing emissions, but not about the benefits of their reductions for other countries. This is due to the assumption that other countries do not respond to Australia's emissions choices. In practice, they may respond in ways that reward Australia for emissions reductions, and penalise it for increases, reducing the incentive to free ride. In other words, the international community may be able to create feedback to national decision makers to deal with externalities, equivalent to the feedback mechanisms discussed in Chapter 5.

A variety of international feedback mechanisms are possible. As we have seen, the Montreal agreement incorporates both negative and positive feedback to discourage free riding: trade restrictions that penalise non-parties (which may include parties detected cheating) and a multilateral fund to compensate poor countries for their emissions reductions. Another possibility is that countries can penalise another country's non-cooperation by suspending or reversing reductions in their own emissions. The idea is to have the non-cooperator understand that it will suffer increased marginal costs of its actions as other countries match its emissions performance. A similar feedback mechanism was incorporated in the Montreal Protocol, which specified that it did not come into force until at least eleven countries accounting for at least two-thirds of 1986 CFC consumption had ratified the agreement. In other words, the initial signatories were prepared to forgo the benefits from a less comprehensive agreement in order to promise new signatories that their individual choices were likely to generate major, rather than minor, benefits.

Feedback mechanisms designed to encourage countries to reduce their emissions, whether stick or carrot, will only be effective under certain conditions. First, it must be possible to monitor accurately other countries' emissions. Second, promised penalties and rewards must be credible and swiftly applied. Punishments that will impose high costs on the country doing the punishing are less credible. This may be the case when countries respond to non-cooperation by threatening to reverse reductions in their own emissions.

Penalties for non-cooperation do not have to take the form of extra emissions by other countries. Countries benefit from international cooperation in many other areas, and penalties imposed in some of those areas may harm the non-cooperator more than they harm the countries administering the penalties. International aid, technology transfers and trade are obvious candidates, especially if the technology or trade relates to the pollutant of concern. Recall that the Montreal agreement bans imports of CFCs and CFC-containing products from non-parties. More wide-ranging threats are beneficial if they do not have to be implemented, but may undermine international cooperation in other important areas, such as the international trading agreements operating under the World Trade Organization.

The direct money payments from the Montreal agreement's multilateral fund are one way of rewarding emissions reductions. Other ways include issuing marketable permits that the recipient country can sell to other countries wishing to increase their total emissions; technical assistance to meet the costs of reducing emissions; and improved access to developed country markets and technology. However, in such circumstances there is likely to be less emphasis on the benefits a country gains as a free rider, and more on the benefits to other parties from its emissions reductions. In the extreme, a country cannot be classed as a free rider if it obtains insignificant benefits but bears a major share of the costs of providing benefits to other countries.

Uneven distribution of costs and benefits and compensation The free-rider problem aside, countries will not join an international pollution control agreement if they believe that cooperative action will cost them more than they will benefit. As explained earlier, this is so for poor countries under the Montreal agreement. In the case of greenhouse gas emissions, poorer countries that are large emitters, such as China and Brazil, produce much higher levels of emissions per unit of national income than rich countries such as Japan and the USA.[7] Thus China and Brazil will sacrifice a greater proportion of national income in achieving given percentage reductions in emissions. The people of China and Brazil would get few benefits and bear high costs as their emissions were reduced—in terms of Figure 17.1, Q_A for China and Brazil would be close to zero. In these circumstances, if developed countries that gain from an international agreement want poorer countries to join, they must compensate them for the net costs borne as a result; otherwise the poor countries will be better off not cooperating.

Countries' marginal benefits and costs of emissions reductions vary according to their environmental and technological circumstances, industry structure, income levels and social and cultural attitudes. For example, low-lying island states and countries heavily reliant on agriculture, fisheries and forests are more vulnerable to climate change. Rich countries with better transport and early warning systems are likely to suffer less from climate-based natural disasters; cyclones of the same strength are likely to kill and injure far more people in Papua New Guinea than in northern Australia. On the other hand, people in poor African countries living on the edge of starvation are likely to discount the future benefits of emissions reductions much more heavily than do Australians or Americans.

The essence of a successful international pollution control agreement is to design compensation mechanisms—'carrots'—to draw the countries that lose from emissions reductions into the coalition. The Montreal agreement includes two carrots to induce poor countries to join: the multilateral fund and the ten year delay in the phaseout of CFCs.

17.8 Bases for an international agreement[8]

An international agreement could be negotiated on any of a number of bases. Parties could agree to country emissions quotas involving specific percentage reductions in emissions, starting from emissions in some base year, as in the Montreal Protocol, with or without trading in quotas. Alternatively, parties could agree to implement pollution taxes, or to introduce new production and consumption technologies, or to eliminate subsidies that encourage global pollution.

Consistent with the previous discussion of pollution control instruments in Section 15.4, an international agreement can specify pollution standards—allowable levels of emissions and/or inputs and/or technology. Alternatively, it can put a price on emissions, either by taxing production or consumption of pollutants, or by allowing international trading of emissions quotas. As explained in Chapter 15, instruments that put a price on emissions are more cost-effective. Taxes or tradeable permits allow countries to make use of their private information about the costs of reducing emissions to achieve least-cost emissions reductions.

The choice of pollution control instruments is important for several reasons. It will affect the total emissions reduction achieved; for example, quotas involve a target for worldwide emissions, while a tax leaves the target unspecified because parties to an agreement cannot know the worldwide costs of reducing emissions. The choice will also affect the costs of achieving emissions reductions, and therefore the overall net benefits from an agreement. Most importantly, the choice will affect the costs incurred and net benefits obtained by individual countries—both parties and non-parties—thus influencing which countries choose to join the agreement. For example, if an international agreement on greenhouse gas emissions involves country-by-country emissions quotas, a quota allocation based on current emissions or national income would probably be acceptable to the USA, but not to China. On the other hand, China would almost certainly agree to emissions quotas based on population, but the USA almost certainly would not.

The problem with taxes on international pollutants, say a tax on carbon-based energy sources, is that the tax could be imposed by either consuming or producing countries. In the former case most of the revenue would accrue to large energy users such as the USA and the EC; in the latter much more revenue would accrue to countries such as Saudi Arabia and Australia. An international agreement that includes quotas makes countries' property rights in international pollution explicit at the outset. Thus internationally tradeable pollution permits, which combine quotas with incentives to minimise the costs of emissions reductions, appear more likely to be internationally acceptable than taxes.

17.9 Other impediments to international cooperation

Consistent with the importance of benefits-versus-costs calculations for individual countries, uncertainty about countries' benefits and costs will be a major barrier to international pollution control agreements. Since no country will join an agreement where it expects to lose, the acceptable allocations of emissions quotas and emissions reductions, and the acceptable limits to compensation of losers by gainers, are set by each country's benefits and costs of reducing emissions. Therefore disagreements over quotas, emissions reductions and shares of compensation will be less tractable when there is scientific and behavioural uncertainty about countries' benefits and costs of reducing emissions. Thus uncertainty about the future benefits of reductions in greenhouse gas emissions, discussed in Section 17.4, is likely to impede an international agreement on reducing greenhouse emissions.

Disagreements over the form of transfers between countries may also impede international cooperation. For example, recipients may prefer financial transfers, and donors transfers of pollution-control technology and expertise.

Another reason for refusing to cooperate is the belief that other countries' adherence to the agreement is impossible to enforce. This could be based on a judgment that human emissions cannot be satisfactorily monitored, as may be the case for methane from rice paddies and waste dumps, or that a global pollution agreement contains inadequate rewards or penalties to compel compliance.

Finally, international competition can undermine cooperation. If Australia agrees to tax fossil fuels in proportion to their carbon dioxide emissions and China and South Africa do not, the foreign buyers of coal, aluminium and steel may be able to shift from the Australian to the Chinese and South African products. Thus Australia is unlikely to cooperate if its major competitors do not. In adopting the National

Greenhouse Response Strategy, the Australian government stated that it is subject to:

> Australia not implementing response measures that would have net adverse economic impacts nationally or on Australia's trade competitiveness, in the absence of similar action by major greenhouse gas producing countries.[9]

17.10 Is international cooperation on climate change likely?

The United Nations Framework Convention on Climate Change was negotiated at the Rio Earth Summit in 1992. It sets out the principles to be followed and the commitments required of nations in pursuit of stabilisation of greenhouse gas emissions at levels that would avoid dangerous human interference with Earth's climate system. It establishes a legal framework for agreed international reductions in greenhouse gas emissions, which would be agreed to in subsequent protocols, as in the case of the Montreal Protocol. Like Montreal, it provides for separate treatment of developed and developing country parties, including financial and technical assistance to developing parties. The Convention came into force when ratified by fifty countries. By late 1996, 164 ratifications had occurred.

The prospects for international cooperation on greenhouse emissions reductions are less promising than for CFCs. There is less consensus among scientists about the long-term impacts of greenhouse gas emissions than about CFCs; a minority of scientists are sceptical that greenhouse-gas-induced warming is occurring. There is also less public concern about the possible dangers of climate change, no doubt partly because those dangers will not materialise until some decades hence. The costs of major reductions in carbon dioxide and methane emissions will be large in relation to the incomes of most countries. Fossil fuel consumption for power generation, heating and transport is a major economic activity in most countries, and many poorer countries rely heavily on rice and livestock industries that are major emitters of methane. Unlike CFCs, no small group of countries or companies is responsible for most human emissions of greenhouse gases. The rich countries of the OECD produce less than half of global carbon dioxide emissions, and a similar proportion of the global warming potential due to carbon dioxide, methane and CFCs combined.[10] Effective international action to reduce greenhouse gas emissions would require the cooperation of at least some other major emitters, of which Russia, China and Brazil are the most important.[11] The large number and diversity of signatories required for an effective agreement on reducing greenhouse gas emissions means a wide disparity in costs and benefits of any emissions reduction policy, and increased difficulty in obtaining agreement on both the allocation of responsibility for reductions and the compensation mechanisms necessary to draw 'loser' countries into an international agreement.

In the case of greenhouse gas emissions, the incentive to free ride is increased by the price effects that will follow emissions reductions by the cooperating countries. As the parties to an agreement reduce their fossil fuel use, world prices of fossil fuels will fall, encouraging greater use by non-parties. Non-parties will gain a cost advantage in international competition, and energy-intensive industries will relocate to those countries. For example, if Australia complies with an international agreement by raising coal and electricity prices, steel and aluminium producers will have strong incentives to shift production from Australia to non-signatory countries where these prices are relatively lower.

Should we be worried about the poor prospects for an agreement on greenhouse gas emissions? It depends on our subjective evaluation of the uncertain consequences of precautionary versus reactive environmental policies, discussed in the preceding chapter. An international agreement to reduce emissions is a precautionary policy. As explained in Section 16.8, precautionary policy makes sense if people are pessimistic about future climatic catastrophes and optimistic about the effectiveness of emissions reductions in averting severe damage in the future. Conversely, if people believe that we know little about global climate change at present, but are optimistic about our future knowledge and the ability of the world's economies to grow rapidly in the future, it makes sense to put off costly preventative actions, because a much richer future world will be better equipped to deal with unforeseen climatic changes.

Discussion questions

1. In order to negotiate effective international agreements on the control of global pollution, some countries will require positive incentives such as technical aid or compensation payments for reducing emissions. True, false or uncertain? Explain why.
2. Why is it much harder to implement international pollution controls dealing with problems such as acid rain and stratospheric ozone depletion, than it is to implement local pollution controls within Australia?
3. Why is a large rich country, such as the USA, less likely to free-ride on other countries' actions to control global pollutant emissions than a small rich country such as Australia?
4. Suppose that you are an international bureaucrat whose job is to assist countries in negotiating an international treaty for effective control of carbon dioxide emissions which contribute to global climate change. What sorts of scientific, technical and

economic information do you require to do your job effectively?

5. Australia is heavily dependent on carbon-based fuels, for transport, power generation and foreign exchange earnings. Suppose that Australia is allocated a carbon-based emissions quota under an International Climate Change Protocol. Explain why Australia is likely to be much better off if the quotas are internationally tradeable.

18

Management of common pool resources

Common pool resources such as ocean fish or groundwater are the subject of competition between users in extraction, but require user cooperation to maintain the productivity of the resource system. Thus successful management of common pool resources requires definition and enforcement of two types of rules: rules for sharing the harvest, and rules governing the total size of the harvest and other investments necessary to maintain the productivity of the common pool. Such rules may be defined and enforced by the user community, by government planners or by users and planners working together. However collective agreement on and enforcement of harvesting and investment rules typically involves large amounts of information and high costs, so that communities must decide whether the gains from protecting common pool resources will justify the sacrifices involved.

18.1 Rushes for Australia's marine resources

Most Australians know of the gold rush of the 1850s and the squatters' rush for grazing land in the 1830s and 1840s. These were not Australia's first rushes to appropriate natural resources. The continent's first commercial industries were sealing and whaling, conducted initially by English and American ships, but then increasingly by Australians, both in ships and from the shore. In *The Tyranny of Distance*, Geoffrey Blainey reports that the Bass Strait fur seals were decimated by the 1820s, and the bay-breeding whales by the 1840s. The sperm whales, which bred in the deep ocean, were granted a temporary reprieve by the discovery of gold in California and Australia, because this made the opportunity costs of whaling prohibitive for shipowners and crews alike.[1]

In the early 1800s Australia's gold, pastoral land, seals and whales were all open access resources, subject to the overharvesting incentives described in Chapter 7. 'Finders keepers' applied in the absence of enforceable exclusive rights. However, the high value and immobility of mining claims and land ensured that exclusive rights to those resources were soon defined and enforced. On the other hand, neither the colonial authorities nor the industries themselves were willing to police rights to seal and whale populations dispersed along thousands of kilometres of mainland and island coasts. Consequently seal and whale populations were hunted as long as the individual's return from hunting exceeded its cost, ignoring the impacts of hunting on the populations, and hence on the net returns of other hunters, present and future. The result was extinction or near-extinction of most local populations of seals and bay whales.

These days most of us recoil at the killing of marine mammals for commercial gain, but excessive harvests of marine resources continue. Graeme O'Neill, writing in the Melbourne *Age*, lamented the overfishing of school and gummy shark off south-eastern Australia, which threatens the future supply of flake (shark), the most popular choice of Victorians buying fish and chips.[2] The story is similar for many other Australian commercial fisheries, both inshore (such as prawns and scallops) and deep water (such as orange roughy and tuna).

18.2 Overharvesting of a common pool resource: the southern bluefin tuna

The flesh of the southern bluefin tuna is prized for sashimi in Japan. Adult fish (9 to 20 years of age and weighing over 50 kilograms) caught by the Japanese longline fleet in the Southern Ocean are worth thousands of dollars landed in Japan. Younger, lighter tuna are far less valuable, but are also accessible to Australian fishers because schools of immature fish migrate relatively close to the Australian coast. In the past, access to the tuna was uncontrolled. The result was not just overfishing of the adult tuna stock, but also premature harvesting of tuna. Immature fish weighing 5 to 20 kilograms were being caught for canning at around one dollar per kilogram; these fish would have been worth a thousand dollars or more each, and could have spawned to renew the tuna stock, if allowed to survive 5 or 6 more years at sea.

Because it is the subject of regular negotiations between Australia and Japan, the southern bluefin tuna fishery provides a well-researched example of the consequences of overharvesting a common pool resource. Adult tuna spawn south of Java and young fish migrating around the

Table 18.1 Catches of southern bluefin tuna

	Australia		Japan		Indonesia, Korea N.Z., Taiwan, others	Total
Year	Weight (t)	No. ('000s)	Weight (t)	No. ('000s)	Weight (t)	Weight (t)
1952	264	17	556	6		820
1953	509	35	3 809	49		4 318
1954	424	35	2 183	27		2 607
1955	322	28	2 915	36		3 237
1956	964	65	14 948	186		15 912
1957	1 264	94	21 878	400		23 142
1958	2 322	161	12 417	225		14 739
1959	2 486	189	63 896	1 032		66 382
1960	3 545	259	75 672	1 188		79 217
1961	3 678	282	77 491	1 209		81 169
1962	4 636	335	40 852	675		45 488
1963	6 199	427	59 200	1 009		65 399
1964	6 832	693	42 718	743		49 550
1965	6 876	448	40 627	721		47 503
1966	8 008	588	39 607	683		47 615
1967	6 357	546	59 086	931		65 443
1968	8 737	917	49 482	828		58 219
1969	8 679	1 151	49 644	844		58 323
1970	7 097	956	40 622	699		47 701
1971	6 969	846	38 120	697	600	45 689
1972	12 397	1 010	39 604	806	117	52 118
1973	9 890	847	31 205	651	112	41 207
1974	12 672	1 193	33 924	672	183	46 504
1975	8 833	1 132	24 118	441	108	33 059
1976	8 383	996	33 714	634	42	42 139
1977	12 569	1 352	29 595	536	12	42 176
1978	12 190	1 293	22 974	451	120	35 284
1979	10 783	1 384	27 715	520	62	38 560
1980	11 195	1 619	33 364	586	206	42 765
1981	16 843	1 482	28 056	477	367	45 266
1982	21 501	2 368	20 809	331	480	42 790
1983	17 695	2 063	24 735	424	262	42 692
1984	13 411	1 447	23 323	365	351	37 085
1985	12 589	973	20 393	304	214	33 186
1986	12 531	999	15 522	213	162	28 215
1987	10 821	817	13 964	na	248	25 033
1988	10 591	na	11 422	na	556	22 569
1989	6 118	na	9 222	na	2 449	17 789
1990	4 586	na	7 056	na	2 117	13 759
1991	4 489	na	6 474	na	2 438	13 401
1992	5 248	na	6 137	na	2 590	13 975
1993	5 373	na	6 320	na	2 355	14 048
1994	4 724	na	6 064	na	2 522	13 310
1995	4 413	na	na	na	na	na

Sources: ABARE, *Individual Transferable Quotas and the Southern Bluefin Tuna Fishery*, AGPS, Canberra, 1989; A.E. Caton and K.F. Williams, 'The Australian 1994–95 and 1995–96 southern bluefin tuna seasons' Working Paper, Bureau of Resource Sciences, Canberra, August 1996.

western and southern coasts of Australia are in turn caught by Western Australian, South Australian and New South Wales fishers. The adult fish move to deeper water, where they are caught by the Japanese fleet.

The initial overharvesting of southern bluefin tuna in the 1950s and 1960s was the responsibility of the Japanese. Table 18.1, listing catches from 1952 to 1987, shows that Australians were responsible for less than half the number of fish caught until 1968. The numbers and weight of the Japanese catch peaked at around 77 000 tonnes in 1961 and declined to less than 21 000 tonnes in 1982. The numbers of young fish caught by Australians increased eight-fold over the same period, reducing subsequent additions to the adult stock fished by the Japanese; in 1982 Australians were taking 88 per cent of the fish caught. The Australian catch peaked at 21 000 tonnes in 1982. In 1983 Australia and Japan agreed to limit their tuna catches to 21 000 and 29 000 tonnes respectively, by which time the New South Wales fishery, based on larger 4 to 5 year-old fish, was sustaining heavy losses due to the decline in numbers of tuna off New South Wales. Despite successive reductions in subsequent catch limits, tuna catches continued to fall. The stock of adult southern bluefin tuna is estimated to have fallen from 600 000 tonnes in the early 1960s to just over 100 000 tonnes in the late 1980s.[3] In 1989 the global catch quota was reduced to 11 750 tonnes, with national allocations of 5625 tonnes for Australia. 6065 tonnes for Japan and 420 tonnes for New Zealand. The quotas for the three countries have been maintained at this level since that time.[4] However the restraint observed by Australia, Japan and New Zealand has been undermined by the expansion of catches by countries outside the trilateral agreement, in particular Indonesia and Taiwan, who increased their tuna catches from 1989.[5]

How can tuna fishers escape the pressures of the rush to harvest? Since tuna are a migratory species, it would do no good to restrict each fisher to a private patch of ocean, even if it were technically feasible. Fishing in a particular location still affects the tuna stock available to other fishers. There is no technology that permits individual tuna to be monitored and collected by individual owners, like cattle in a common grazing herd. So if tuna fishers want to preserve the tuna stock and maximise the net return from the fishery over time, they must agree among themselves about, or be forced to adopt, harvest rates and shares and any other measures designed to preserve the stock for the future.

18.3 Harvesting and investment problems

As explained in Chapters 6 and 7, individual units of common pool resources are rival and excludable in use (i.e. when they are withdrawn from the pool). On the other hand, actions that affect the total stock of a common pool resource, such as harvesting tuna or restricting fishing gear used or the size of fish caught, have non-rival and non-excludable impacts on all other users of the resource. Tuna fishers, or the users of any other common pool resource, have interests in both the current harvest (excludable) and in the future productivity of the resource stock (non-excludable). Successful collective management of any common pool resource therefore requires resolution of two types of problems: first, how to divide a given harvest while avoiding wasteful methods of competition among harvesters, thereby maximising harvesters' combined net returns; and second, how to create incentives for investments in the future productivity of the resource stock, given the incentive to free ride on the investments of others.[6]

Harvesting involves withdrawals of resource units from a resource system. *Investment* involves the restoration, maintenance or enhancement of the resource or its supporting environment or both.

Where the common pool resource is a living, self-renewing resource, harvesting and investment are closely related. Restraint in fishing, which augments the future breeding stock, is the dominant form of investment in the future stock of southern bluefin tuna. Harvesting techniques that alter the supporting environment, such as seabed trawling for prawns or scallops, may also affect the future productivity of the resource system. However, the future stock of a common pool resource is usually also affected by actions other than harvesting. The stocks of some species of fish, animals and birds can be increased by releasing captive-bred individuals, or by protecting or enhancing the environments on which those species depend. For example, protecting old-growth vegetation may improve nesting opportunities for particular species of parrots and small mammals. In the case of oil pools, the oil recovery rate can be increased by injection of other fluids into oil-bearing strata.

Recall from Section 7.2 that the harvesting and investment problems would disappear if the resource system was under the exclusive control of a single decision maker. If a single individual controlled the entire tuna fishery, the harvest would not have to be divided between competing fishers. Unitary ownership would also eliminate the incentive to disinvest by premature harvesting; to do so would sacrifice the increased profits obtainable from waiting for the tuna to increase in size and value per kilogram. Alternatively, the harvesting and investment problems could be eliminated if the resource system could be subdivided into separate sections, each capable of functioning independently and exclusive to a particular resource user. However, neither of these options is available in the case of the southern bluefin tuna, whose life cycle involves migration over thousands of kilometres of open ocean, unsubdividable and accessible to fishers from several nations.

18.4 Management of southern bluefin tuna

Investment measures Table 18.2 summarises the past management of the southern bluefin tuna fishery. Various forms of restraint on fishing,

Table 18.2 Management in the southern bluefin tuna fishery

Year	Event
1952	Targeted commercial exploitation of southern bluefin tuna by New South Wales and South Australian boats and Japanese longline fleet.
1961	Japanese fleet takes record catch of 77 kt.
1968	First commercial catches by Western Australian boats.
1971	Japanese industry introduces voluntary area closures to protect spawning fish and juveniles.
1974	Successful introduction of five purse seiners to the southern bluefin tuna fleet.
1975	Entry of additional purse seiners in the fishery prohibited. Purse seiners banned from fishing off Western Australia.
1976	'Freeze' on further entry of pole boats into the south-eastern sector of the fishery. Number of boats limited to 76.
1979	Australian biologists warn that fishery is fully exploited.
1981	Lifting of 'freeze' on entry of additional pole boats. Restriction on number of purse seiners maintained.
1982	Record Australian catch of 21 kt.
1983	Biologists warn that the global catch of southern bluefin tuna should be urgently reduced to arrest the decline in size of the spawning stock.
	Australia, Japan and New Zealand agree to prevent any further growth in catches. Catch limits are 21 kt, 29 kt and 1 kt, respectively. Australia introduces a minimum landing size for southern bluefin tuna and area closures for purse seiners.
1984	Individual transferable quota system introduced into the Australian fishery. Australian quota reduced to 14.5 kt. Japanese refuse to reduce their catch. Japanese longliners prohibited from fishing for southern bluefin tuna in Australian waters (south of 34°S).
1985	Japanese agree to limit their global catch to 23.15 kt and are readmitted to the Australian fishing zone around Tasmania.
1986	Biologists warn of the risks of recruitment failure if catches are not further reduced. Voluntary 3 year agreement reached between Australian and Japanese industries to limit the global catch to 31 kt. Catch limits for Australia and Japan are 11.5 kt and 19.5 kt. respectively, but the Japanese fleet catches only 15.5 kt.

designed to protect the tuna stock, date from 1971 in the case of the Japanese and from 1973 for Australia. The Japanese industry has closed areas known to contain spawning fish and small fish. Australia has restricted the catching of tuna using nets, the number of boats, the size of fish caught and the total Australian tuna catch. However, the catch data in Table 18.1 indicate that these measures failed to slow the decline in the tuna stock up to the early 1990s. A recent innovation is tuna 'farming'. Young tuna are captured live by purse seine netting (where boats catch tuna by encircling a school of fish with a large net). The small tuna are placed in cages moored offshore and grown to a size suitable for the Japanese sashimi market.

Harvesting measures Table 18.2 records that the first measures to divide the southern bluefin tuna harvest were adopted in 1983, when Australia, Japan and New Zealand agreed to national catch quotas.

Table 18.2 (*continued*)

1988	Joint venture agreement between Australian and Japanese industries, allowing Japanese longline boats to take Australian quota fish in the Australian Fishing Zone.
1989	Biologists recommend major catch reductions. Catch limits are reduced to 5.3 kt for Australia, 6.1 kt for Japan and 0.42 kt for New Zealand.
	Major expansion in southern bluefin tuna catches by vessels from Taiwan and Indonesia, who are not parties to the voluntary trilateral quota arrangement between Australia, Japan and New Zealand. In subsequent years Taiwanese and Indonesian vessels have been responsible for about 15 per cent of the global catch.
1991	A joint venture agreement between Australia and the Japanese industry allows up to 2900 tonnes of Australian quota to be leased and fished in the Australian Fishing Zone by Japanese longliners. In subsequent years up to half of the Australian catch is taken in these joint venture operations.
	First sales of tuna fattened in farm cages at Port Lincoln in South Australia. In 1995 approximately 2000 tonnes of cage-reared tuna were produced, but mortality in cage-reared fish can be high.
1993	Australia, Japan and New Zealand sign a convention to formalise the three countries' former voluntary management arrangements. The non-participation of other countries, in particular Taiwan and Indonesia, continues to be a serious obstacle to southern bluefin tuna monitoring and management.

Sources: ABARE, *Individual Transferable Quotas and the Southern Bluefin Tuna Fishery*, AGPS, Canberra, 1989; P. Neave (ed.) *The Southern Bluefin Tuna Fishery 1993*, Fisheries Assessment Report, Australian Fisheries Management Authority, Canberra 1995; A.E. Caton and K.F. Williams 'The Australian 1994–95 and 1995–96 southern bluefin tuna seasons', Working Paper, Bureau of Rural Sciences, Canberra, August 1996, Tables 1 and 2.

However, because individual fishers' rights to tuna continued undefined prior to harvest, the incentive to rush to harvest tuna remained. This deficiency was partially rectified in 1984, when individual transferable catch quotas (ITQs)—entitlements to a percentage share of the total allowable catch determined annually by the Australian government—were allocated to Australian bluefin tuna fishers. However, the value of ITQs to Australian fishers was in part dependent on the behaviour of the Japanese fishing the same tuna stock. As indicated in Table 18.2, after initially refusing to reduce their catch, since 1985 the Japanese have agreed to catch limits in consultation with Australia. This has increased the specificity and value of Australian ITQs, since marine biologists are better able to predict future catches if they know the size of the Japanese catch, and limits on the Japanese catch augment the tuna breeding stock. New Zealand joined in the voluntary quota arrangement in 1989; the three countries have since adhered to a total allowable catch of 11 750 tonnes annually. However this investment in the future tuna stock has been at least partly undermined by increased catches by countries not party to the tripartite agreement, in particular, Indonesia and Taiwan.

With reasonably well defined ITQs, Australian southern bluefin tuna fishers who have purchased quota have invested in the longline equipment required to catch the larger, more valuable tuna suited to the Japanese sashimi market. Equally important, the introduction of ITQs has enabled Australian quota holders to enter joint venture agreements with Japanese longliners.[7]

18.5 Information requirements for managing the tuna fishery

If tuna fishers want to preserve the tuna stock and maximise the net return from the fishery over time, either the fishers themselves or fisheries planners must specify and enforce rules governing tuna harvesting and investment—in effect, these are property rights to shares of the tuna harvest and corresponding responsibilities to maintain the tuna stock. The aim of the rules should be to eliminate incentives to overfish and underinvest in the tuna stock; in terms of Figure 7.1, this means restricting the total fishing effort to the equivalent of Q*. What information is required to achieve these goals?[8]

Identity of users Coordination to maximise the net return from the tuna fishery requires the cooperation of all significant fishers. Fishers with access to the tuna stock must therefore be identified and encouraged or forced to cooperate.

Physical and biological characteristics of the resource system Several types of knowledge of the tuna stock and its supporting ocean environment are needed:

1 *The extent of the resource system.* This is needed to identify and monitor fishers and the state of the tuna fishery. If the physical or biological nature of a common pool resource system is poorly understood, for example if resource managers have a poor understanding of subsurface geology or animal movements or life cycles, identification of users and monitoring of resource use may be difficult or impossible.
2 *System responses to natural environmental changes, causing variations in resource availability in time and space.* Regular and irregular changes in the marine environment, such as seasonal ocean temperature variations and disease infestations, will cause the tuna biomass to vary across time and space. Information about such natural changes is required in order for fishers to agree on variations in the allowed tuna catch. Land-based resource systems will also be subject to changes affecting resource productivity (e.g. weather variations).
3 *System responses to changes in the rate of harvest.* This information is necessary to identify the marginal returns from additional fishing effort.

4 *System responses to other actions*. In the case of tuna, the only other actions likely to significantly affect the stock are changes in fishing techniques. For example, in terms of the size of fish caught, fishing for tuna with nets is a less discriminating fishing technique than fishing with poles and lines, hence it is banned in the Western Australian fishery, where the fish are youngest. In other resource pools, the response of the stock to artificial breeding or habitat preservation may be important.

Characteristics of harvesting and investment technologies Tuna fishing technologies must be known in order to understand their impacts on current and future catches of tuna, and also to estimate fishing costs. Where investment in the resource system and hence in future stocks may take other forms besides harvesting restraint, knowledge of both harvesting and investment technologies is necessary.

Input and product prices and discount rates These are required to estimate the present value of returns from different rates of harvesting and levels of investment in the tuna fishery.

Possible monitoring and enforcement techniques and their costs Harvesting and investment rules have to be policed and enforced, either by the resource users themselves, as is largely the case for the separate Australian and Japanese tuna catch quotas, or by planners, as is the case for Australian tuna fishers' ITQs. Rules should not be adopted if the total costs of policing and enforcing the rules exceed the benefits resource users derive from collective management of the resource pool. Thus rule makers require information about the inputs required for policing and enforcement and their costs, and the likely collective losses suffered due to imperfect enforcement of rules. In the case of the quota regime for the southern bluefin tuna, policing and enforcement of domestic and foreign fishing involves inspection of fishers' logbooks, surface and aerial surveillance by state fishing agencies and the Australian Fisheries Management Authority, placing observers on Japanese vessels and administration of penalties for violations.[9]

Likely responses to possible harvesting and investment rules and rule enforcement Tuna harvesting and investment rules are of no use if not generally observed, and so the choice of rules, whether collectively agreed by fishers or imposed by fisheries planners, depends on judgments about tuna fishers' responses to the rules and associated monitoring and enforcement regimes. This requires information about fishers' alternatives to observing the rules, and the net benefits of those alternatives. For example, when Australia first introduced ITQs for southern bluefin tuna in 1984, Japanese fishers refused to cooperate by reducing their catch, undermining the tuna stock assumptions upon which the total Australian quota was based. As a result, Australia prohibited Japanese longliners from fishing in southern Australian waters, reducing the benefits of non-cooperation. The Japanese agreed to limit their catch in 1985, and were readmitted to Australian waters in that year.

The information requirements to achieve well-specified property rights in an extensive common pool resource such as the southern bluefin tuna fishery are formidable. Compare the tuna fishery with an excludable fish stock under the control of a single manager, such as a fish or oyster farm. In the latter case the boundaries of the resource system are clearly defined, and its physical and biological characteristics and environmental responses correspondingly better understood. Management and harvesting of a farm resource system are in the hands of a single individual, whereas many people's actions affect the tuna fishery. Thus coordinated fishery management requires either multilateral bargains and mutual enforcement among fishers, or management by planners, who must know fishers' alternatives and the enforcement options, or some combination of the two.

It is apparent that the overharvesting model of Figure 7.1 oversimplifies the task of common pool rule makers in several important respects. It assumes that the returns to harvesting effort are known, which requires precise knowledge of the characteristics of the resource system and prices. It ignores spatial and temporal variation within the resource system. Ignoring differences in resource harvesting skill and equipment, it assumes that all resource users have identical harvesting costs, and that resource managers know those costs. Finally, it assumes that the optimum level of harvesting can be achieved costlessly. Figure 7.1 makes no allowance for managers' costs of specifying and enforcing appropriation and provision rules, or for resource users' costs of compliance.

18.6 Common pool user numbers and management costs

The geographic extent of a common pool resource system, discussed in Section 7.3, is one important determinant of the number of users and potential users of the pool. The greater the area, the more people are likely to have low-cost access to units of the resource. For example, more people can readily harvest migratory species such as salmon, tuna and muttonbirds. The second important factor affecting numbers is the likely reward from the harvest, which depends on the value of the resource versus the costs of harvesting. Our ancestors were prepared to bear very high costs to harvest whales; the same is true of Japanese tuna fishers today. However, when the gold rush raised the opportunity costs of whaling and the development of the oil industry lowered the value of whale oil, the number of people with an interest in the resource system fell dramatically.

Numbers of users and potential users make a big difference to the costs of managing common pool resource systems. As the number of users increases, the costs of negotiating and enforcing harvest sharing and investment rules increase. The situation is equivalent to the problem of sharing the costs of preserving cranes, discussed in Section 6.4. As the number of users rises, the amount and costs of the multilateral information exchange required rises even faster. For this reason, agreement on rules is more likely if all users are similarly placed—if each has similar attitudes, skills and technical options and anticipates similar net benefits of cooperation and non-cooperation. This lowers users' or planners' costs of identifying users' alternatives. Suppose that all tuna fishers had similar attitudes to fishing, similar fishing and non-fishing skills and education, similar fishing equipment, and fished at the same time in the same location. Then a proposed tuna harvesting rule would create symmetric benefits and costs for all fishers, and consensus on the rule would be relatively likely.

Increased user numbers also increase the attraction of free riding. The more users, the less is the perceived impact of an individual's harvesting restraint and other contributions to collective action upon his or her future harvests. Hence there is greater individual incentive to forgo negotiation and free ride on the contributions of others.

18.7 User self-management: common property

When common pool resources are managed as common property, the rules governing resource use are negotiated and monitored and enforced by the user community. Characteristics of the resource system and the community that affect the success of such collective actions were discussed in Section 7.6.[10] The common pool must be local to enable policing by a defined user community, and the community must be able to exclude outsiders at low cost. Resource users' attitudes and skills and production alternatives must be sufficiently similar so that they anticipate similar benefits and costs of harvesting and investment rules. The user community must be sufficiently close and stable that monitoring and enforcement are cheap, and all users must have a long-term stake in adherence to the rules. Finally, for local community rules to be effective, external planners must refrain from interfering. Nothing undermines carefully crafted local understandings about rights and responsibilities for natural resources more than the knowledge that external agents with coercive powers are likely to intervene.[11]

18.8 Management by planners

Government's coercive power can facilitate management of a non-excludable resource system such as the southern bluefin tuna fishery in at least two ways. First, it can enforce harvesting and investment rules, either its own rules or rules agreed by resource users. Its coercive power allows penalties beyond those available to the users themselves, and thus can increase the level of cooperation. The Australian Fisheries Management Authority can fine Australian and foreign fishers, or confiscate their equipment, or deny them access to the Australian fishing zone, if they violate catch quotas or other fisheries regulations.

Second, government coercion can overcome free-rider problems that bedevil resource users' own attempts to negotiate and enforce rules. Remember that rules applying to numbers of people are non-rival and non-excludable goods from the point of view of those affected. Tuna harvesting and investment rules are costly to negotiate and enforce; since no single Australian or Japanese fisher takes more than a small proportion of the total tuna catch, none may be willing to voluntarily contribute to the research and surveillance activities necessary for effective rules. The Australian government levies all fishers to fund fisheries management.

The benefits of government coercion will be greater when resource systems are large and complex, and the users numerous and diverse, as in the case of the southern bluefin tuna. Free riding will deter voluntary negotiations among common pool users as the number of users becomes large. However, political and bureaucratic planners not directly involved in the use of common pool resources also suffer some disadvantages compared to the users themselves. For example, planners may lack local knowledge of the tuna fishery known to fishers. Planners also lack information about individuals' fishing skills and alternatives to current fishing activities, and hence about the net costs or benefits of adjustment to new fishing rules. In each case they are likely to rely on fishers for additional information, with the attendant risks of deliberate distortions of information referred to in Sections 8.2 and 8.3. Also, as pointed out in Section 8.4, imperfect political feedback to politicians and bureaucrats may allow them opportunities to advance their own interests to the disadvantage of common pool users in general.

Despite the risks attending government intervention, if user communities are directly involved in public management of common pool resources, the combination of users' specific knowledge and government's coercive power offers the best prospect of overcoming the major information and incentive problems involved in managing extensive common pool resources. Johnson and Libecap, discussing government regulation of fisheries, point out that, given the different impacts of uniform government regulation on fishers who differ in fishing skills, equipment etc., many fishers are likely to strongly oppose regulation unless developed in close consultation with the industry. On the other hand, fishers will lobby government for enactment of regulation that takes such differences into account.[12] Ostrom, discussing the conditions for successful management of southern California's groundwater basins, points to the crucial role of government in providing technical and legislative assistance to groundwater user communities.[13]

18.9 Planning for the common pool: alternative signalling and incentive systems

The decisions facing planners responsible for managing common pool resources parallel those faced by their colleagues dealing with pollution control. Pollution control planners have to determine total allowable emissions, common pool planners total allowable harvests. Each lacks accurate information about the true marginal benefits and costs of changes, in emissions or harvests. Pollution control planners have to coordinate emissions reductions across many polluters; common pool planners have to coordinate harvest reductions and other investments in the resource system across many resource users. Each has to commit resources to monitor and enforce adherence to their plans. Finally, both sets of planners have similar options for signalling plans to resource users and providing incentives for compliance: direct intervention in production decisions, backed by legal penalties for violations, or specifying individual users' rights to shares of a scarce natural resource, and allowing markets to price access to the resource. We now review these signalling and incentives options as they apply to the southern bluefin tuna fishery. It is assumed that a total harvest target has been decided on the basis of physical, biological and economic research. (In practice, harvest targets may reflect political pressures as well as the true marginal benefits and costs of reduced harvests. Politicians will be subject to pressures to preserve people's livelihoods in the short term by allowing unsustainably high harvests. This appears to have been the case in the setting of tuna catch limits, summarised in Table 18.2.)

Direct controls on inputs Historically, regulation of harvesting and investment in fisheries has usually involved controls on inputs. As shown in Table 18.2, input controls in the southern bluefin tuna fishery have included area closures, limits on the number of boats and limits on the fishing gear used. Time-based closures are another common method of restricting input use.

The tuna catch data in Table 18.1 suggest that the input controls adopted up until the early 1980s had no appreciable impact on the number of fish caught, the most important determinant of the future productivity of the tuna fishery. Nor did they prevent the growth of fishing capacity. As a result, the profitability of Australian tuna fishing fell in the early 1980s.[14] Why did these outcomes occur in response to policies designed to reduce the catch and increase returns?

Imagine yourself to be a tuna fisher restricted to a single boat and particular fishing gear. These input controls provide no information about the benefits of restricting your fishing to preserve the tuna stock for yourself and other fishers. More importantly, they give you no incentive to do so. Input controls do not deal with the division of the tuna catch between fishers, and so they do not directly affect your access to tuna.

You still own any tuna that you catch. You and your competitors are in an analogous position to the children sharing the milkshake in Section 7.1. He (or she) who hesitates loses out. Those who fish early and fast will end up with a greater share of the total catch. So you have a strong incentive to rush to harvest tuna.

In order to harvest fish quickly and maximise your share of the tuna catch, you also have an incentive to invest in more or better equipment. However, the input controls restrict your fishing methods; you can replace or refit your present boat, but you cannot invest in a second boat. You cannot switch from pole and line fishing to netting of tuna, despite your judgment that this would lower the per unit costs of your catch. So input controls encourage you to commit more resources to fishing and ensure that your costs of fishing are higher than they would be in the absence of the controls.

Assume that all tuna fishers respond to input controls in the manner described above. We then get the result depicted in Figure 18.1, a modified version of Figure 7.1(a). The total costs of harvesting tuna are raised by the input controls, lowering the open access level of harvesting from Q_m to Q^*. Harvesting pressure on the tuna stock is reduced, but the fish are no longer caught in the least-cost manner, and the total revenue from the fishery is completely offset by unnecessarily high fishing costs.

Total harvest limits These target the right variable, the total tuna catch, and therefore can protect the resource stock if effectively policed. However, there is still no incentive to restrict fishing. So long as individuals' rights to tuna remain undefined, so that you can only own tuna by catching them, total harvest limits are likely to produce the same rush to harvest as input controls. The fishery is again over-capitalised and harvesting costs greater than the situation where each fisher has a right to a specific share of the total tuna catch.

Individual transferable quotas ITQs based upon a total allowable tuna harvest protect the tuna stock and define individual fishers' rights to catch tuna. If properly enforced, ITQs eliminate most of the pressure to rush to harvest. With enforced penalties for overfishing, there are incentives for restraint. Fishers can be confident that they can catch their quota of tuna when and how it suits them. As a result, harvesting costs will be lower; fishers no longer need to invest in the bigger and better boats and equipment necessary to maximise their share of the total catch.

ITQs can still lead to waste. Compliance with quotas may cause fishers to discard above-quota fish. Also, in attempting to catch their quota of the target fish at the lowest cost, fishers may adopt methods that increse by-catch (the incidental destruction of non-target species, e.g. dolphins). Stricter policing of ITQs can reduce these problems, but it increases the cost of enforcing fishers' property rights.

Figure 18.1 Effect of input controls on fishing

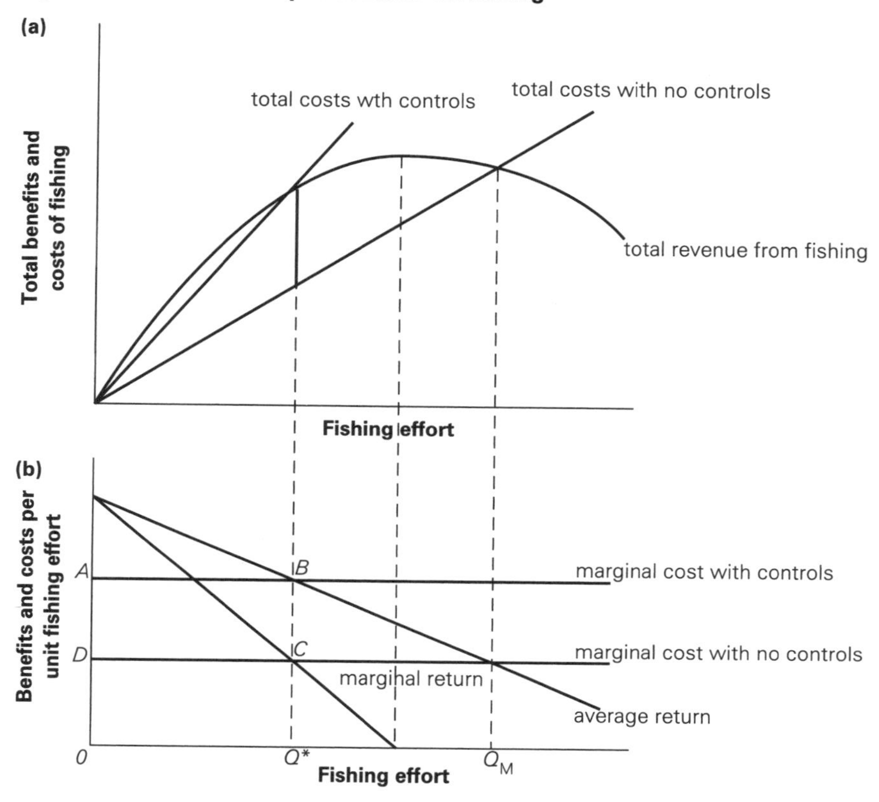

As transferable rights, ITQs permit fishers who anticipate little or no profit from fishing to sell their quotas to others who anticipate substantial profits. Thus fishers with good alternatives outside the industry are compensated for giving up fishing, and fishers with low opportunity costs of fishing are able to expand their allowed catch. The result is a lower total cost of the total allowed tuna catch. Figure 18.2, which is analogous to Figure 15.2, illustrates this consequence of ITQs. It assumes just two fishers, *A* and *B*, whose marginal profit gains from increasing catches are shown in panels (b) and (c) respectively. Fisher *A* derives modest profits from additions to his catch, while catch increases are highly profitable for *B*, because she has poor earning opportunities outside fishing, and/or because her superior fishing skills result in a low unit cost of fish caught, or catching of more valuable fish. Suppose that the fishery planner sets the total allowed catch at Q_t in panel (a). Recall, from Section 18.8, that the planner lacks information about *A*'s and *B*'s fishing skills and alternatives to current fishing activities, in other words, about their marginal profit curves. Assume that they receive equal ITQs, $\frac{1}{2}Q_t$ each.

Figure 18.2 Individual transferable quotas

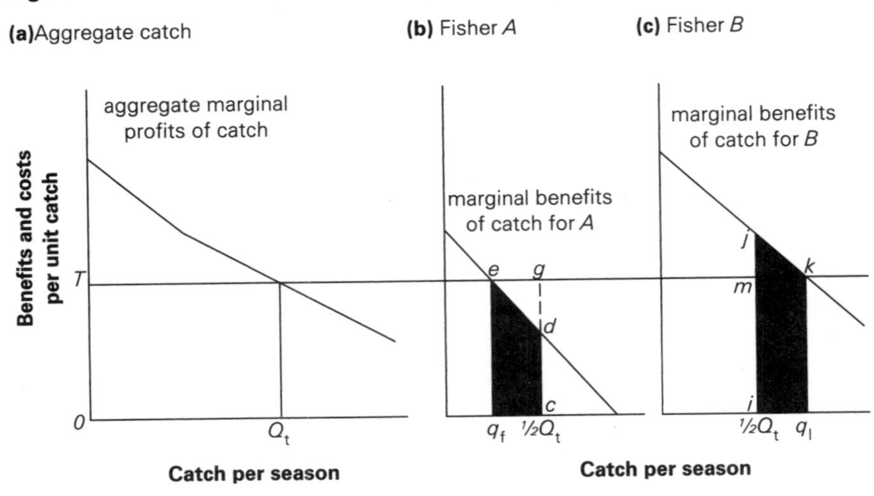

If ITQs are strictly enforced, *A* and *B* can both be better off if *A* sells some quota units to *B*. If both honestly reveal their marginal profit curves, which determine what *A* is willing to accept to relinquish quota and what *B* is willing to pay to acquire quota, they will trade $q_f c$ ($= iq_l$) units of quota. *A* will gain an amount equal to *deg* and *B*'s gain will be *jkm*. The sum of these gains represents the increase in fishery profits due to shifting fishing to the individual (*B*) who can fish more profitably.

The gains from successful implementation of ITQs can be large. In 1992, six years after the introduction of ITQs for most New Zealand fisheries, the New Zealand Ministry of Agriculture and Fisheries estimated that the total market value of quota was close to $NZ 1 billion, far more than the direct costs to the government of the quota system.[15] New Zealand ITQs also exemplify the cooperation between resource users and planners mentioned in Section 18.8; by legislation, the total allowable catch for each species is determined annually by the Minister of Fisheries in consultation with industry representatives.[16]

Note that ITQs not only create incentives for harvesting restraint; they also generate additional information for planners and fishers. When tuna ITQs are traded in the market, quota buyers find it in their own interests to reveal their benefits from increased fishing, and quota sellers to reveal their losses from reducing fishing. The revelation of this private information enables the total catch target to be achieved at least cost.

The results of trade in ITQs in the southern bluefin tuna fishery seem consistent with the logic of Figure 18.2. Trading after Australian ITQs were created in 1984 resulted in transfers of quota holdings from New South Wales and Western Australia to South Australia; by 1987, the South Australian fleet held entitlements to 91 per cent of Australia's total allowable catch, compared to 66 per cent in 1984. (A likely problem with an ITQ scheme that facilitates adjustment out of a particular fishery, such as the southern bluefin tuna fishery, is that the excess fishing capacity may simply shift to other fisheries.) Some of this quota was leased to South Australians by New South Wales fishers who wished to retain the option to catch tuna in the future. At the same time the cost of catching a tonne of tuna is estimated to have fallen about 25 per cent.[17]

ITQs have also enabled trades between Australian and Japanese tuna fishers. Under a joint venture agreement, Australian quota holders hired Japanese fishing companies to harvest up to 3000 tonnes of Australian quota between 1991/92 and 1993/94. This increased the Australian fishers' profits because the Japanese are more skilled in the longline fishing used to catch larger and more valuable tuna.[18]

An ITQ is an property right to a share of the total allowable tuna catch. It is a relatively vague property right, because enforcement of catch quotas, based on logbooks and fishery surveillance, is costly and imperfect, and because of variations in the total allowable catch set by Australian fisheries' planners. In the 1980s the planners apparently misjudged the resilience of the southern bluefin tuna parent stock, leading to the reductions in Australian and Japanese catch quotas recorded in Table 18.2.

18.10 Collective management of common pools can be too costly

With excludable resources, specification of rights and management are separable and sequential; the owner of a private oyster farm knows his or her rights prior to harvesting or investing in oysters, and reaps the full benefits and costs of those decisions because of the ability to exclude others. The oyster farmer does not need to know the status of the oyster fishery beyond the specified boundaries, or other farmers' reactions to his or her decisions. Coordination with others is achieved by mutual respect for private property and by market prices. In this way, effective exclusion and monitoring technologies and effective administration of penalties for illegal use of private property allow great economies in information costs. Information about individually managed resource systems and individual resource users' alternatives can remain private, contrary to collective management of the common pool.

The crucial difficulty in managing a common pool resource system is that specification of rights and resource use and management are not separable. The only way to obtain legal rights to tuna is to catch tuna which alters the future productivity of the tuna fishery. Individual resource users' benefits and costs of harvesting are dependent on the actions of other resource users. Multilateral coordination of harvesting and investment, by resource users or planners, is essential to maximise collective benefits from resource use. As we saw in Section 18.5, the information requirements for effective multilateral coordination of harvesting and investment in tuna are very large. In particular, the collective managers need to know physical and biological conditions throughout the fishery, and the technologies and economic alternatives available to all fishers. Thus collective management of common pool resources is very costly. The New Zealand experience with ITQs suggests that the information and enforcement problems are most likely to be overcome by cooperation between resource users, who possess detailed information about the resource and their own alternatives, and government planners, who are able to enforce management rules and fund enforcement by levying all resource users.

As previously explained, if natural or created resources are valuable enough, technological or administrative changes may result in effective definition and enforcement of private or national property rights. Barbed wire, cable television and the Montreal Protocol on Substances that Deplete the Ozone Layer are examples of this. If such solutions are not forthcoming—if exclusion remains prohibitively costly—will it always pay to negotiate and enforce rules designed to overcome the rush to harvest common pools?

The answer is no. Sometimes it may cost too much to put the matter right. Depending on the value of the resource concerned, the extent and complexity of the resource system, the monitoring and enforcement technologies available, and the number and diversity of resource users, it may be better to put up with the losses due to overharvesting and underinvestment than to incur the very large costs involved in negotiation and enforcement of management rules for a common pool. No doubt the early twentieth-century whalers knew that overharvesting threatened their industry's future. Would we have acted any differently, given contemporary attitudes to and scientific knowledge of whales, and the costs of collecting information about whalers' alternatives and of monitoring and enforcing catch limits in a world without swift and reliable transport and communication links? Are we not relying on 20–20 hindsight when we say that past whalers should have managed the resource better?

Discussion questions

1. If the nineteenth-century sealers who harvested fur seals along the southern coast of Australia had been equipped with today's harvesting, communication and monitoring technologies, the seals would have disappeared much faster. True, false or uncertain? Explain why.
2. If a fishery is open access, a large improvement in fish har-

vesting technology makes the extinction of the target species more likely. True, false or uncertain? Explain why.

(Hint: How does the change in technology affect the curve(s) in Figure 18.1?)

3. Allocation and enforcement of ITQs in a marine fishery eliminates overfishing and damage to the associated marine environment. True, false or uncertain? Explain why.
4. Suppose that a small Pacific island country asks you to negotiate with Asian fishers to preserve valuable fish stocks inside the country's 320 km exclusive fishing zone. What objectives would you aim at, and what scientific, technical and economic information would you require to have a chance of reaching a satisfactory agreement on fishing rights? Explain the significance of your chosen objectives, and the importance of each type of information.

19

The economic significance of biodiversity

Biodiversity—the variety of life on the Earth—benefits people in several ways. We value the products we get from animals and plants and microorganisms. We recognise the potential for discoveries of future uses for living organisms, and derive satisfaction from their very existence. Most of us are less aware of the importance of biological diversity for the continued functioning of the world's ecosystems, which perform the photosysnthesis, soil formation, waste decomposition, etc. necessary to sustain human life. Unfortunately, while it is clear that some levels of biodiversity loss would threaten human living standards and survival, incomplete scientific knowledge of our evolutionary inheritance of genes and species, and of ecosystem functioning, make it impossible to measure biodiversity losses precisely, let alone measure their costs.

19.1 Mangroves: wastelands or treasure troves?

The Esplanade running north along the shore of Trinity Bay in Cairns may remind the casual tourist of Surfer's Paradise or Waikiki; ocean and beach on one side, high-rise hotels on the other. On closer inspection, however, the 'beach' turns out to be a mudflat, punctuated by occasional forlorn sticks of mangrove. A productive mudflat, too. Each morning when the tide is in there are people catching prawns with handcast nets. If you follow the waterfront north, towards Cairns International Airport, the mudflat is abruptly replaced by a 4 to 5 metre wall of mangroves—salt-tolerant shrubs and trees that grow in the intertidal zone. The 'beach' is not a work of nature, but the result of clearing of the mangrove forest on the shore of Trinity Bay.

Why clear the mangroves? Presumably the tourist industry believes that visitors like unobstructed sea views and open beachfronts, even if swimming is out of the question. Muddy tangled mangrove forests have little aesthetic appeal and, pragmatically, provide cover for saltwater crocodiles.

On the other hand, the traditional view of mangroves as unappealing wasteland, best cleared and converted to other, more valuable, land uses, is giving way to recognition of their ecological, environmental and economic value.[1] Ecologically, they provide a unique habitat for a wide range of fauna, from microscopic invertebrates to crocodiles, and support the food chains that link these species. Eighty-six Australian bird species are either confined to mangroves or regularly use them as foraging or nesting sites.[2] The mangroves around Cairns also enhance the coastal and reef environments enjoyed by tourists in a number of ways. They mitigate storms and reduce coastal erosion; they trap silt and filter harmful effluents, improving coastal and reef water quality; and they provide food sources for birds and marine organisms, such as barramundi and prawns, valued by tourists and the fishing industry.

Actual and potential benefits of mangrove ecosystems may extend beyond the areas where they are found. Mangroves are a potentially rich source of material for genetic engineering of more salt-tolerant crop species. The adaptation of mangrove species to their unique environment involves the synthesis of unique organic chemicals, some of which are likely to be of value in medicine and in industry. Some mangrove species have been used as traditional medicines by indigenous peoples around the world. Finally, mangrove ecosystems, having evolved at the land–sea interface, are likely to be well adapted to any future changes in sea-level associated with global climate change, and thus can assist humans in adapting to such changes.

Clearing along the Cairns waterfront alone is unlikely to reduce the diversity of genes and species associated with mangroves, but this will not be true for clearing throughout north Queensland. Widespread clearing will also reduce diversity of mangrove ecosystems themselves. As a result, in a changing world, mangrove ecosystems and human societies that depend on them may have less options to deal with change without themselves having to undergo major or catastrophic changes.

19.2 Biodiversity and human dependence on other living things

Biological diversity or *biodiversity* is the variety of life on Earth. It includes diversity at three levels of biological organisation: genes, species and ecosystems. Ecosystem diversity refers to the variety of habitats, interdependent communities of organisms and ecological processes within nature.

Table 19.1 Countries with greatest species richness

Mammals	Birds	Reptiles
Indonesia (515)	Columbia (1721)	Mexico (717)
Mexico (449)	Peru (1701)	Australia (686)
Brazil (428)	Brazil (1622)	Indonesia (600)
Zaire (409)	Indonesia (1519)	India (383)
China (394)	Ecuador (1447)	Colombia (383)
Peru (361)	Venezuela (1275)	Ecuador (345)
Colombia (359)	Bolivia (1250)	Peru (297)
India (350)	India (1200)	Malaysia (294)
Uganda (311)	Malaysia (1200)	Thailand (282)
Tanzania (310)	China (1195)	Papua NG (282)

Source: J. McNeely et al., *Conserving the World's Biological Diversity*, International Union for the Conservation of Nature, Gland, Switzerland, 1990.

Living organisms satisfy human needs in two ways.[3] First, the individual organisms in the natural environment have specific properties that satisfy people's consumption needs (e.g. food and firewood) and production needs (e.g construction timber). Second, organisms interacting in ecosystems are an essential component of humanity's natural life support system, described in Section 1.3. As indicated in Figure 1.1, living organisms play a key role in maintaining and modifying the global environment. Changes in the composition of life on Earth can cause changes in the chemical and water cycles and in Earth's climate. Given time, natural selection and the evolution of ecosystems enable life to adapt to natural stresses, such as El Niño climatic fluctuations, or human-created stresses, such as major oil spills or acidification due to acid rain. Thus, when combined in ecosystems, individual organisms are also of indirect value to people because of their role in supporting life, now and far into the future.

Species diversity is greatest in the tropics, particularly in the tropical forests, and decreases towards the poles. It is estimated that the majority of the world's terrestrial species are found in tropical forests. Table 19.1 lists the countries with the greatest numbers of mammal, bird and reptile species. Almost all of these countries contain large areas of tropical forests. With the exception of Australia, all of these biodiversity-rich countries are low-to-medium income countries experiencing rapid population growth. The implications of these facts for biodiversity loss and conservation are discussed in Chapters 20 and 21.

Separate levels of biological organisation—genes, species and ecosystems—require separate measures of biodiversity. In fact, there is no consensus about measurement even at a single level of biodiversity, in part because of incomplete knowledge of Earth's total array of genes and species, and of ecological functions and processes. While changes in biodiversity are most commonly measured by changes in numbers of species, these measures are handicapped by the fact that the majority of Earth's species are unidentified. Published estimates of species loss are based on empirical findings about the relationship between the area of a given habitat and the number of species present. Many biologists argue that species diversity measures should take account of the genetic distance between species, so that a region containing many closely related species would rank lower in terms of its biodiversity than one containing the same number of distantly related species. At the ecosystem level, possible measures of biodiversity include the relative abundance of organisms fulfilling different ecological functions and the diversity or patchiness of biological communities within a region.

19.3 The value of biodiversity

How does biodiversity lead to direct and indirect benefits to people? First, the more genes and species, the greater is the possible variety of useful living organisms. For example, in the 1970s scientists discovered a 6 hectare stand of a new perennial virus-resistant species of the genus *Zea*, which includes maize (*Zea mays*), in the Sierra de Manantlan in Mexico. Maize is an annual, the world's third most important crop, worth more than US$50 billion per year worldwide, thus the potential gain from incorporating genes from the new species in commercial maize varieties is very large.

Biodiversity is important for human life support because a variety of life forms are necessary to ecosystem functioning, and because it contributes to ecosystem resilience. Resilience is the capacity of an ecosystem to maintain its patterns and rates of photosynthesis, chemical and water cycling and so on, in response to variable environmental conditions. Humanity benefits from the genetic and functional diversity within ecosystems because it helps to maintain life support functions in the face of environmental stresses and shocks.

The many benefits that people derive from biodiversity fall in all of the categories described in Section 11.2. Direct-use values of biodiversity, both rival and non-rival, arise from the use of organisms or ecosystems in consumption and production activities (e.g. genetic material for plant and animal breeding, opportunities for birdwatching and fishing).

Indirect-use values of biodiversity arise from the dependence of humanity on the functioning of ecosystems that capture the sun's energy and cycle essential chemical elements and water. A multitude of interacting micro-organisms, plants and animals combine to provide the life support functions that enable human societies to exist.[4] For example, plants and soil bacteria cycle carbon, oxygen and water, regulating the gaseous composition of the atmosphere, precipitation and runoff, and thereby the global and local climate. Other life support services of ecosystems include soil formation, revegetation and waste decomposition and assimilation. Wetlands, such as mangroves, filter effluents, removing nutrients, heavy metals and suspended solids. Natural ecosystems also provide pollination and pest control services to commercially valuable plants and animals, the outcome of coevolution of these and other species. Finally, ecosystems and their constituent genes and species have the capacity to evolve, thus maintaining their life support functions despite changes due to either nature or people.

As explained in Section 11.2, the categories of non-use values are not clearly defined. Economists writing on biodiversity have paid most attention to its existence and option values. The existence value of biodiversity is a value placed on organisms, and complexes of organisms integrated in ecosystems, unrelated to their actual or potential use; it is based on the satisfaction that individuals experience simply from knowing that an organism or ecosystem exists. For example, many people value the remaining wild population of blue whales or the tropical rainforests of north Queensland, regardless of their usefulness to those individuals or anyone else.

Many benefits of biodiversity stretch into the future, and so the corresponding values should, at least in principle, include discounted future benefits as well as present benefits. However, our knowledge of the identity of many species, of the functioning of ecosystems and of future technological and environmental changes is at best imperfect. Therefore we are uncertain of both the future usefulness of particular components of today's portfolio of genes, species and ecosystems (and hence of our future biodiversity requirements) and of their future availability. In these circumstances, risk-averse individuals may be prepared to pay today to preserve biodiversity as a form of insurance against an uncertain future. The option value of biodiversity is the value attached to maintaining future options to use, learn more about, and simply enjoy the existence of, genes or species or ecosystems, where we are presently uncertain of their nature and future usefulness.

Part of the option value of biodiversity is based upon its contribution to ecosystem resilience. We know that our ecological life support systems will be subject to stresses and shocks, both natural and human-created. We do not know what form these will take. Existing ecosystems and their constituent genes and species represent the results of hundreds of millions of years of evolutionary experimentation, presently freely available to humanity. We can take out more insurance against future losses of ecosystem resilience and life support services by choosing to preserve a larger portfolio of biodiversity.

Option values of biodiversity are likely to be significant and positive if people are risk-averse, if the future values of genes or species or ecosystems are expected to be high although currently unknown, and if biodiversity losses are irreversible.[5] For example, indigenous Australian flora and fauna, the sources of traditional Aboriginal medicines and some current medicinal products, undoubtedly contain medicinal resources yet to be discovered and used in modern medicine.[6] Many people are risk-averse in respect of major diseases such as heart disease and cancer, and the values attached to biological knowledge and resources that assist in treatment of such diseases are likely to be high. Yet destruction of rare Australian ecosystems such as tropical rainforests, not duplicated elsewhere in the world, may lead to irreversible losses of presently unknown or imperfectly understood medicinal resources. Thus the biodiversity contained in rare Australian ecosystems is likely to have significant positive option values on medical grounds. For example, a substance called Prostaglandin E_2, which could be of importance in the treatment of gastric ulcers, was originally discovered in two species of gastric brooding frogs found only in the rainforests of Queensland. It is possible that both species have since become extinct. If the frogs cannot be located, the opportunity to learn more about the natural operation of this biological agent will have been lost.

19.4 Distinguishing species values from ecosystem values

Recognition of the separate contributions of biodiversity to specific human wants and to our life support system is critical in understanding the relationship between the values of particular species and the value of biodiversity. The total value of the biological resources in a mangrove ecosystem includes the values attached to its constituent species, but it exceeds the sum of those values. This is partly because some interactions between mangrove species are of direct value to people, for example for scientific research and recreational enjoyment of nature. However, the main reason is that the mangrove ecosystem provides life support services in addition to individual species' contributions to human consumption and production. It is the complex of interacting organisms, not individuals or species, that captures the sun's energy by photosynthesis and conveys that energy to people via food chains, absorbs carbon dioxide and emits oxygen, traps silt and filters effluents, and mitigates tropical storms.

The additional value of natural ecosystems above the value of their constituent species is the major source of scientific concern about biodiversity loss. Policies to preserve identified species alone will not maintain the life support services of the environment; we need distinct policies to maintain ecosystems. Unfortunately, popular discussion of biological conservation commonly identifies threats to ecosystems (such as forests, mangroves and coastal heathlands) with threats to particular species (such as owls and orchids). Many of the benefits of such species are readily identifiable, while the life support functions of ecosystems are generally obscure. However, the loss of an ecosystem is potentially far more costly than the loss of many individual species dispersed across ecosystems. In the hierarchy of nature, functioning ecosystems are necessary to the survival of their constituent species, including ourselves, but the reverse is rarely true. Also, substitutes are more readily available for the direct services of species than for the indirect life support services of ecosystems. In particular, species can be conserved *ex situ*, in laboratories or gardens or zoos, but the scale and complexity of natural ecosystems means that they have to be preserved *in situ*, in protected natural areas. *Ex situ* conservation also eliminates environmental changes, which are the driving force of the evolution required to maintain life support into the future.

19.5 The significance of biodiversity loss

Worldwide and at the level of local ecosystems, the current level of biodiversity is the product of hundreds of millions of years of evolutionary change on Earth and, during the last 100 000 years or so, the conscious actions of modern humans (such as hunting, vegetation clearing, migration and coal burning) that have, deliberately or inadvertently, changed the natural environment.[7] The current concern about biodiversity loss centres on the large discrepancy between the high rates of extinction of populations and species in human times and the far lower average rate of creation of new species over geological time. This would not matter much if genetic engineering had progressed to the point that extinct genes and species could be identified and recreated at an acceptable cost. Most biologists judge that this will not occur in the foreseeable future, since they perceive Earth becoming biotically impoverished.[8] Accepting the biologists' judgment, is this likely to matter much to people, in economic terms?[9] After all, human beings have reached their present state of technical and cultural development while contributing to the extinction of many other species. Is there reason to believe that current reductions in biodiversity will prevent humanity progressing in the future as we have in the past?

How can we assess the impact of recent and current reductions in biodiversity on human consumption and life support systems? First, we need to be able to measure biodiversity changes, to know what is being lost. Second, we need to understand what biodiversity losses mean for the direct and indirect contributions of species and ecosystems to human consumption and life support. Third, since people can devise and create substitutes for many of the services of species and ecosystems, we need to know the availability and costs of substitutes. In fact, none of these steps is easy.

Measurement Recall, from Section 19.3, that incomplete scientific knowledge prevents precise measurement of genetic, species and ecosystem biodiversity. However, measurement is necessary in order to assess the implementation and benefits of conservation policies. In practice, the appropriate measure of biodiversity is likely to depend on policy goals, scientific knowledge and the costs and accuracy of available measures. For example, if the prime reason for mangrove preservation was to learn more about salt-tolerant species, the appropriate biodiversity measure might be the number of identified species that were either highly salt tolerant or closely biologically linked to such species.

Impacts of biodiversity loss The greatest impediment to understanding the consequences of biodiversity losses is inadequate scientific understanding of the relationships between the biodiversity of ecosystems and their life support functions and resilience. Where species and ecosystems are directly useful to people, for example where they provide food or timber or drugs or recreational opportunities, the consequences of losing those resources can usually be estimated. However, this is not the case for the ecological processes that provide humanity's life support. A recent United Nations review of the conservation of global biodiversity highlights the complexity of the relationships involved:

> No simple relationship exists between the diversity of an ecosystem and such ecological processes as productivity, hydrology and soil generation. Nor does diversity neatly correlate with ecosystem stability, its resistance to disturbance, or its speed of recovery. There is also no simple relationship between a change in its diversity and the resulting change in its component processes. On the one hand, the loss of a species from a particular area or region . . . may have little or no effect on net primary productivity if competitors take its place in the community. On the other hand, there can be cases where the converse is true. For example, if zebra and wildebeest are removed from the African savannah, net primary productivity of the ecosystem would decrease.[10]

Thus, with present scientific knowledge, it is impossible or very costly to obtain reliable estimates of the indirect life support benefits of biodiversity conservation. All that can be said for sure is that there are levels of biodiversity loss that will undermine the resilience of individual ecosystems and, beyond that, reduce the life support services of ecosystems sufficiently to force catastrophic changes or fundamental reorganisations on human societies.[11]

Availability of substitutes It is easier to estimate the availability and costs of substitutes for the services of species and ecosystems. Organisms that directly satisfy people's consumption and production needs often have good substitutes. In the case of food and fibre production, humanity has gained greatly from substituting artificial managed ecosystems for natural ones. If tuna are scarce, we can readily shift to other fish or other meats. Ecotourists visiting north Queensland will lament the loss of a particular mangrove ecosystem but may not suffer too much if they can visit equivalent areas of mangroves. On the other hand, the complexity and scale of the ecological life support mechanisms that maintain the gaseous composition of the atmosphere, regulate climate, recycle nutrient elements, create soils and assimilate wastes is such that large-scale substitution would be impossibly costly in terms of other natural resources and human time and effort. Thus it is only possible to substitute for some of these ecological services at the local or regional level, for example by irrigation, revegetation, soil protection and fertilisation, and waste treatment. Finally, if ecosystems are lost, we cannot replace the creative evolutionary potential inherent in those ecosystems, the capacity for genetic modifications that will maintain human life support functions into the future.

What can we conclude about the seriousness of recent and current global biodiversity losses? Biodiversity loss will only be a problem for species or ecosystems that are directly useful to people if exclusive rights to those resources are not enforced, so that no one has an incentive to conserve the resource, as discussed in Section 7.2. Where exclusion is not too costly, the costs of losses of directly valued organisms (such as high-value timber species in mangrove forests) are readily recognised, and those who benefit have incentives to act to conserve the resource.

The seriousness of losses is much less clear where many of the benefits of biodiversity are non-excludable, such as the educational and existence and ecological benefits derived from the mangroves around Cairns. First, as explained in Chapter 6, beneficiaries of non-excludable goods have little incentive to reveal true values. Second, in the case of the indirect life support services of biodiversity, losses can only be measured when we have an agreed measure of biodiversity linked to the relevant ecological processes. Third, both the choice of the measure and the relationship between biodiversity and life support functions are clouded by inadequate scientific understanding of the functioning of ecosystems. However, significant reductions in the life support functions of major ecosystems would be serious because the costs of replacing these ecological services would be either very large or prohibitive.

Discussion questions

1. The value of an ecosystem equals the value of that ecosystem's constituent species. True, false or uncertain? Explain why.
2. Is the world's stock of genes an exhaustible or a renewable biological resource? Explain.
3. How is the option value of biodiversity (measured as today's portfolio of genes, species and ecosystems) related to Earth's long evolutionary history?
4. In the World Conservation Strategy published in 1980, the priority for saving a species is determined by its rarity and its uniqueness in relation to the biological system of classification (i.e. does loss of the species mean loss of a family or a genus, or just the species itself with other species of the genus surviving). Is this an economically sensible way to decide on priorities for saving species? Explain your answer.

20

The economics of biodiversity loss

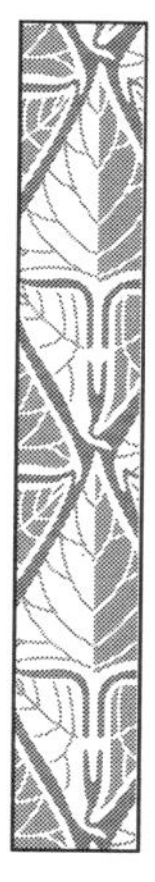

Most losses of biodiversity are the result of human appropriation of other species' habitats and increasing specialisation in production, especially agricultural production. Historically, human progress has involved losses of local biodiversity; otherwise, human populations and living standards would have remained at the levels achievable by hunter-gatherers. The somewhat ironic result is that most of the world's land-based and coastal biodiversity is located in poorer countries, while the major demands for its preservation come from the affluent countries which have been most successful at converting their local ecosystems. Assessment of the prospects for biodiversity preservation must begin with an appraisal of the information and incentives of the private and public decision makers who control land use in regions of high biodiversity.

20.1 What causes biodiversity loss?

Despite well-publicised threats to the existence of some commercially valuable species, such as whales, elephants and some parrots, direct human exploitation of living organisms is not the major threat to biodiversity. Only a tiny fraction of naturally occurring species are directly used by humans, and of these, extinction of species or local populations is against the interests of users if they enjoy exclusive rights to the resource. For example, the owners of commercial aviaries and deer farms undertake major expenditures to protect their valuable livestock against diseases and predators.

There are three reasons why people might eliminate genes, species and ecosystems. First, if valuable organisms are non-excludable like whales or tuna, there is a rush to harvest the resource, and users have no incentive to invest in the future productivity of the resource stock. Second, humans may try to eliminate organisms that endanger human lives or valuable resources, for example wolves and tigers in the Old World, the Tasmanian tiger in Australia and the smallpox virus worldwide. Third, as has happened with the mangroves at Cairns, biological resources may be lost as a result of habitat alteration or destruction if decision makers value the naturally occurring resources less than alternative land uses. Habitat alteration and destruction is by far the most important of these, and expansion of agriculture is the most important cause of habitat alteration, as becomes clear when we consider the three reasons in turn.

20.2 Non-excludable resources and dangerous organisms

As explained in Chapters 6 and 18, resources are non-excludable whenever the net returns from resource use are too low to justify the costs of fencing or monitoring resource use and penalising illegitimate users. For example, the tuna found in Australia's territorial waters are effectively non-exclusive if the Australian government does not allocate sufficient resources to police and penalise overseas fishers who illegally harvest tuna. Similarly, suppose that the African politicians and bureaucrats who have the legal rights to control the use of elephants in national parks find that *their* financial and/or political costs of protecting elephants exceed *their* benefits, for example because of the high costs of patrolling the parks and the political clout of ivory poachers and farmers who regard the wild animals as pests. Then the politicians and bureaucrats are better off leaving property rights in elephants unenforced, making elephants an open access resource.

Non-excludability makes extinction a possibility, not a certainty. While users have no incentive to invest in the future productivity of the resource stock, the intensity of harvesting, and hence the likelihood of extinction, depends on the value of the harvested resource and the costs of harvesting. Figure 7.1(b) demonstrates that the intensity of harvesting is lower, the higher the harvesting cost (the marginal cost line) and the lower the value of the harvest output (lower values lower the average revenue line). Even when a biological resource is non-exclusive, extinction in the wild is only likely when the species is very valuable, is cheap to harvest and reproduces slowly. This appears to be the case for the elephant and rhinoceros today, in countries where governments make only modest efforts to protect their wildlife. Both animals produce high-value products (ivory and rhinoceros horn), are cheap to hunt with modern rifles and reproduce too slowly to withstand high rates of harvesting.[1]

What about deliberate destruction of living things? Given modern scientific understanding, technology and culture, there are relatively few biological organisms today whose complete elimination would be an unambiguous benefit to humanity. The smallpox and AIDS viruses and some pathogens causing serious animal and plant diseases may qualify. Where such organisms do exist, their identification and elimination is a non-rival and non-excludable good. The problem is too little investment in their elimination, not too much. However, we cannot entirely rule out future positive values for even the most dangerous organism; remember that species that our ancestors regarded as dangerous vermin have positive values today.

20.3 Habitat alteration and destruction

Habitat alteration and destruction resulting from the expansion of human populations and activities are the main threats to the world's biodiversity. People change habitats directly, by changing land use, and indirectly, by introducing exotic species such as rats and goats, by taking resources from the environment and by discharging wastes. For example, in Papua New Guinea, some tropical forests are cleared for farming, others are logged, and offshore coral reefs suffer as a result of silt and nutrient runoff from the cleared areas.

In trying to understand the causes and significance of habitat loss, it is important to remember that appropriation of other species' habitats and increasing specialisation in production are facts of human history. Over time, some habitat has been lost to provide living and working space and transport links, but most has been converted to feed, clothe and house growing human societies. Increases in food and fibre production, which have underpinned human population expansion and growth in living standards, have involved the conversion of ecosystems and the sacrifice of biological diversity in favour of the cost reductions possible through specialisation, standardisation of production and exchanges of commodities between specialised producers.[2] Our hunter-gatherer ancestors, including Aboriginal firestick farmers, were capable of producing major changes in ecosystems, but still utilised a wide variety of naturally occurring organisms.[3] For human populations and living standards to grow beyond those achievable by hunter-gatherers, it was necessary for societies to appropriate an increasing share of the local biological product, the plant and animal materials that result from plant use of solar energy. This was achieved by replacing pre-existing ecosystems with new managed ecosystems, where 'domesticated' plant and animal species are favoured and competing species are reduced or removed. In this way, so long as the life support functions of ecosystems are not seriously damaged (e.g. where societies based on irrigated agriculture have declined due to salinity build-ups), human progress has historically involved losses of local biodiversity.

Local argicultural specialisation may help to preserve biodiversity at the regional and global level. Specialised technologies can reduce land use by raising yields, as is argued in the case of Asia's Green Revolution. On the other hand, modern industrially based agriculture can cause pollution of neighbouring ecosystems.

The process of agricultural specialisation has proceeded furthest during the last century, with the globalisation of agricultural technologies and exchange.[4] Because there are economies of scale in the production of large quantities of homogeneous agricultural inputs, such as chemical fertilisers and farm machinery, and because inputs and products can be transported more economically than ever before, agricultural technologies are now less diverse worldwide than ever before. Thus the number of species protected and managed for human sustenance is probably at an historic low.

We now begin to understand the geographic mismatch between the supply and the demand for land-based biodiversity, with most of the supply located in poorer countries (recall Table 19.1), and most of the demand for preservation coming from rich countries. Where poorer countries are not densely populated, their past relative lack of success in ecosystem conversion leaves long-established ecosystems relatively undisturbed (Brazil's Amazon region, for example). At the same time, rapid population growth, the adoption of modern farm and forest technologies, and the desire for higher incomes create very strong demands for access to additional land and for ecosystem conversion. On the other hand, in achieving affluence, rich countries have greatly reduced their domestic supply of undisturbed ecosystems (sparsely populated rich countries such as Australia being the exception), including many natural resources highly desired by the rich and leisured. (According to the World Resources Institute, if wilderness is defined as areas of at least 4000 square kilometres showing no evidence of settlement, then 3 per cent of the land area of Europe is wilderness, 5 per cent of the USA, and 30 per cent of Australia.)[5]

The historical perspective on the connection between biodiversity and human progress also serves to remind us of connections between biodiversity and culture. Recall the discussion in Section 1.5 and the economy–environment interactions depicted in Figure 4.1. In the very long run, say the thousands of years of Aboriginal occupation of Australia discussed by Flannery in *The Future Eaters*, human culture and ecosystems may be viewed as having coevolved. In other words, biodiversity in an area is a joint product of natural evolutionary change and past and present human activities, driven by the knowledge, values and institutions of that society.[6] Thus hunter-gatherer societies that continue to gain their sustenance from a wide variety of species in a relatively stable ecological setting have developed knowledge, values and rules governing resource use designed to maintain the diverse ecosystems on which they depend. In such circumstances, where biodiversity and culture have coevolved, for example in Arnhem Land and in the Amazon Basin, losses of biodiversity are likely to be linked to losses of indigenous cultures.

Despite the historical record, it is too simplistic to think that growth in human populations and living standards is always associated with increasing loss of habitats and biodiversity. For one thing, severe poverty, as observed in the drought-stricken Sahel in the 1980s, is a major cause of habitat destruction. Desperately poor people care more about today's meal than preservation of the trees or pastures that could feed them in the future. For another, living standards and population growth are interrelated. Increasing living standards generally reduce family size and population growth rates, partly due to increases in the opportunities and status of women, and partly to the reduced need to rely on children for security in old age in affluent societies. Finally, we observe that environmental consciousness and political concern about the environment are phenomenons of high-income societies.

The fact is, we cannot adequately understand the causes of biodiversity loss by focusing on changes in aggregate variables such as human populations, incomes and agricultural technology. Biodiversity loss results from the decisions of billions of individual natural resource users worldwide. Thus, if it is a problem, the ultimate causes must lie in inadequate incentives for individual resource users to preserve biodiversity, and/or faulty signals about biodiversity values.

20.4 Biodiversity incentives and signals: the case of Sarawak's forests

Some Australians feel strongly that the tropical rainforests of South-East Asia—in Indonesia, the East Malaysian states of Sarawak and Sabah, the Philippines and Papua New Guinea—should not be logged, for a variety of reasons, including to protect their biodiversity. Most of these forests are state-owned, under the legal control of political and bureaucratic decision makers, whether or not inhabited by indigenous peoples.

Put yourself in the place of a government land-use planner in Sarawak with rights to determine and monitor the use of forest resources. Broadly speaking, you have a choice between two land-use options. There are land uses that will destroy the present forest ecosystems, for example harvesting of all marketable trees followed by natural regeneration, or clear-cutting followed by agriculture or plantation forestry (the development option). Alternatively, there are land uses that will maintain ecosystems more-or-less unchanged, for example continued use of the forest for production of non-timber products by forest dwellers (the preservation option). What information is available to help you choose between these alternatives, and what are your incentives to choose one way or the other?

Your aim will be to choose the option that you believe will benefit Sarawak most over some chosen planning period. In line with the choice procedure described in Chapter 9, you will choose development if you believe that the present value of the expected stream of benefits and costs to Sarawak resulting from initial logging and subsequent forestry or agriculture exceeds the present value of the alternative stream of benefits and costs associated with uses consistent with ecosystem preservation, including the costs of monitoring and enforcing forest protection. Formally, you choose development if:

$$PV(B^d - C^d)(1 \ldots t) > PV(B^p - C^p)(1 \ldots t) \qquad (1)$$

where the *B*'s and *C*'s are the values to be discounted to present equivalents as outlined in Section 9.1, *d* and *p* refer to the development and preservation options, and your planning horizon extends over *t* time periods. Note that, unlike the present value equation in Section 9.1, the benefits and costs you may consider are not restricted to those that can be represented by money values. Thus the *B*'s and *C*'s could include values attached to winning the next state election, protecting the way of life of forest tribes, asserting Malaysia's rights to control its own resources and so on.

Equation (1) shows that your choice of development or preservation will depend on the benefits and costs you consider, the values you attach to them, your time horizon and (implicitly) the rate at which you discount future benefits and costs to arrive at their equivalent present values. Consider each of these in turn.

Which benefits and costs matter to a Sarawak land-use planner? We argued in Chapters 2, 3 and 8 that private and public decision makers' incentives to use scarce resources in a certain way depend on the personal rewards and penalties that the individual expects to derive from that resource use. In the most straightforward case, a private owner of a forest can maximise his or her commercial return by choosing the time sequence of husbandry and uses that yield the greatest possible present value of market profits. If you are a land-use planner in Sarawak, forestry revenues and profits are very important. The timber industry accounts for about one-quarter of Sarawak's GDP and employs about 10 per cent of the workforce. Your incentives to give substantial weight to the benefits of development options are reinforced by the Sarawak government's commitment to land development schemes designed to replace forests and traditional shifting cultivation with permanent cash cropping and plantation agriculture. Also, in Sarawak, leading government and opposition politicians have interests in major timber concessions and in logging contracts.[7]

On the other hand, what are your incentives to give weight to the benefits of forest preservation? You will worry about the non-market costs of development when they affect your political or bureaucratic fortunes. For example, in the 1980s logging damage to traditional native forest lands led to a series of native blockades of logging operations, creating considerable expense and adverse publicity for the Sarawak government. Also, you have incentives to consider preservation benefits where they translate into revenue and profits for Sarawak, as in the cases of rattan, a climbing palm used in cane furniture, and tourist revenues obtained from national parks. Nevertheless, there are few incentives for you to consider *all* the benefits of the preservation option. To understand why, consider Table 20.1.

The first column of Table 20.1 lists possible direct-use, indirect-use, option and existence benefits from land uses that preserve Sarawak's forests and their biodiversity. As a Sarawak politician or bureaucrat, whether you will care about particular benefits depends heavily on their impact on your political contributions and voter support, or your career prospects. The rest of Table 20.1 summarises the major factors that determine impacts on you personally. Column 2 indicates whether the main beneficiaries are local rural residents, or the wider communities of Sarawak and Malaysia, including politicians, bureaucrats and commercial interests, or people outside Malaysia. Column 3 indicates the excludability of the benefit, which determines the probability of beneficiary free riding, and hence of either compensation of the Sarawak government for benefits provided, or substantial economic or political pressure on the government by domestic or overseas beneficiaries, including overseas governments. Column 4 indicates likely sources of value information and pressure to preserve forests.

Table 20.1 suggests that it will be rational for you to ignore or pay little heed to many of the benefits of preserving Sarawak's forests. With the exception of limited areas of communally owned forest, indigenous forest people have no legal rights to forest resources. Therefore you will give little weight to the concerns of forest dwellers unless they are able effectively to pressure Sarawak politicians. As a Sarawak politician or public servant, you have little or no stake in benefits received by overseas beneficiaries, unless Sarawak is compensated for forest preservation, or punished for development. However, where the benefits of preservation are global, they are usually also non-excludable, so that little or no international compensation or penalties will be forthcoming. Similarly, although some of the non-excludable ecological benefits of preservation, such as watershed protection, are confined to Sarawak, local political pressure will be reduced by the fact that most beneficiaries reside in distant rural communities and some will choose to free ride on the lobbying efforts of others.

What about the costs of development and preservation, C^d and C^p in equation (1)? The private costs of the commercial activities, such as forestry, agriculture and tourism, matter because they affect private profits. You will also be concerned about, because your decisions partly determine, the Sarawak government's costs of regulating land use. Both development and preservation options will involve the preparation of land-use plans and administration and enforcement of the resulting

Table 20.1 Benefits of forest preservation and their characteristics

Types of benefit	Spatial incidence of benefits	Excludability of benefits	Planner's sources of information
Direct use–present and future:			
Sustainable timber products	Local, regional	Rival, excludable	Timber markets
Non-timber products (fruits, nuts, cane etc.)	Local, regional	Rival, excludable	Markets, except subsistence products; Political pressure from natives
Recreation and tourism	Local, regional, global	Non-rival at low intensities; excludability depends on fencing/monitoring costs	Markets, local and overseas tourist industry
Medicines/genetic material	Global	Rival, but no recognised property rights to natural materials and exclusion very costly	International scientific community, drug and biotechnology industries
Research and education	Global	Non-rival, non-excludable	International scientific community
Human habitat	Local	Non-rival at low intensities; excludable, but communal native rights rarely recognised	Political pressure from native communities
Indirect use–ecological functions:			
Watershed protection	Local, regional	Non-rival, non-excludable	Political pressure from downstream residents, natives
Nutrient cycling	Local, regional	Non-rival, non-excludable	Political pressure from downstream residents, natives
Waste decomposition	Local, regional	Non-rival, non-excludable	Political pressure from downstream residents, natives
Climate regulation	Local, regional, global	Non-rival, non-excludable	Political pressure from overseas environmentalists, governments
Carbon storage	Global	Non-rival, non-excludable	Political pressure from overseas environmentalists, governments

Options for future use:			
Future direct and indirect uses	Local, regional, global	Non-rival, non-excludable	International scientific community and environmentalists
Existence:			
Biodiversity—species, ecosystems	Global	Non-rival, non-excludable	Political pressure from overseas environmentalists, governments
Culture/heritage	Local, regional, global	Non-rival, non-excludable	Political pressure from natives, overseas cultural organisations

Sources: David Pearce, 'An economic approach to saving the tropical forests' Ch. 8 in *Economic Policy Towards the Environment*, ed. Dieter Helm, Blackwells, Oxford, 1991; Michael Wells, 'Biodiversity conservation, affluence and poverty: Mismatched costs and benefits and efforts to remedy them' *Ambio*, vol. 21, no. 5, May 1992, pp. 237–43

regulations. The development option involves the costs of regulating logging, transport of timber, reforestation, cultivation, agricultural input use and so on. Forest preservation involves the costs of regulating forest access and legitimate uses such as harvests of non-timber products, tourism and scientific research.

Which benefits and costs are likely to matter most? You live in Sarawak's capital, Kuching. In deciding the use of Sarawak forests, you have to trade off the support of the mainly rural and overseas beneficiaries of preservation against that of the Sarawak and national commercial interests who are the main beneficiaries of development. Who will you fear offending most, a combination of relatively poor and remote native communities who get to vote for single representatives in the state assembly, plus overseas conservation groups and governments who have no votes and little influence, or wealthy urban commercial interests and job seekers who benefit from development options in the forest hinterland?

Where does a Sarawak planner get value information? Politician or bureaucrat, you will be much better informed about the possible net benefits of logging and agriculture ($B^d - C^d$), than about the net benefits of preservation ($B^p - C^p$). The outputs of logging and agriculture, such as logs, foods and rubber, are valued in markets, and would-be loggers and farmers and the Sarawak government's forestry and agricultural staff can estimate likely yields, so that it is relatively easy for you to obtain estimates of the near-term benefits of development options. Similarly, the private costs of development options, in the form of labour, fuel, machinery time, fertilisers etc., can be estimated, to which you must add the government's costs of regulating logging and agriculture to calculate C^d. So the likely net benefits of development, excluding environmental damage resulting from logging or agricultural development (here included as a benefit of preservation), can be calculated by you and your political and bureaucratic colleagues.

What credible information will you have about the benefits and costs of preservation? As a land-use manager, you can estimate the government's likely costs of regulating forest access and use. Preservation benefits are another matter. Column 4 of Table 20.1 indicates that only the values of timber and non-timber forest products, and possibly of tourism, can be estimated from production and market sources. The medicinal and genetic resources in Sarawak's forests have commercial value, but market bids are unlikely. There are no internationally recognised property rights to naturally occurring organisms, although new pharmaceuticals developed from naturally occuurring organisms can be patented in the USA. Costa Rica's Instituto Nacional de Biodiversidad has a contract with Merck and Company which returns part of the profits derived from naturally occurring organisms to Costa Rica, but such agreements to sell 'nature prospecting rights' to drug companies have yet to withstand the tests of disputes between the parties and legal challenges. So it will be difficult for Sarawak's planners to value medicinal and genetic resources of the forests.[8] On the other hand, they will have little difficulty in valuing Sarawak's timber and mineral resources by offering forest timber concessions and mineral prospecting rights.

Other preservation benefits are even more difficult to value. Many are likely to be speculative, often based on the assertions of interested parties. The indigenous forest communities have no means of compensating the government for preservation; in any case, they believe that the forests are more theirs than the developers. The scientific and educational, ecosystem maintenance, option and existence benefits of preservation are all non-excludable; beneficiaries who know that they cannot be excluded if the forests are preserved have little incentive to reveal their true willingness to pay. On the contrary, if groups such as international environmental organisations do not have to put their money where their mouth is, by supplying funds for preservation, they have incentives to exaggerate the benefits of preservation.

Perhaps your biggest difficulty in coming up with credible preservation benefits is the lack of understanding of the functioning and ecological contributions of forest ecosystems. Your political and bureaucratic colleagues can argue: 'If scientists do not properly understand how preserving certain areas of the forests will affect Sarawak's and the world's environment and biodiversity, now and in the future, how can we sacrifice certain timber revenues, jobs and income, which can make our people's lives better now?'

What is the planner's time horizon and discount rate? Your planning horizon and discount rate together determine how you weight present versus future benefits and costs of alternative land uses. According to equation (1), you give zero weight to benefits and costs arising beyond your planning horizon. In an extreme case, if you are a Sarawak politician risking the loss of your assembly seat at a close election, you may favour immediate sale of timber concessions for cash that can be spent in your campaign, since you temporarily care little for benefits and costs arising after election day. More realistically, as a land-use planner, you may care about benefits and costs arising fifty or more years hence. However, your job is to choose between land uses for the people of Sarawak, and your political or public service career prospects depend on your satisfying community desires. Therefore, assuming that most people in Sarawak, rich and poor, distinctly prefer goods today to goods in the future, and that income earned today can be productively invested in Sarawak's future, say in education and communication facilities, you are likely to discount distant benefits and costs to small present values.

What if the planner cannot enforce the rules? Regardless of whether you choose development or preservation, if you cannot enforce your land-use plans, forest resources are non-excludable despite your wishes.

For example, you may award forest logging concessions subject to strict conditions about selective harvesting of certain types of trees and minimisation of collateral damage to other trees when trees are felled and moved. However, these rules may be impossible to enforce, due to scarcity of funds and field staff, or corruption or inefficiency of field staff, or political interference and so on. Similar problems could lead to illegal logging in protected forests. In these circumstances valuable forest resources are to some extent non-excludable, and the resulting rush to harvest will accelerate the destruction of forest ecosystems.

Summarising the development versus preservation choice Considering the likely incentives and information of Sarawak decision makers, it is not surprising that deforestation continues in Sarawak. Native forest dwellers who benefit from preservation have few rights to forest resources. The net benefits of development are heavily concentrated in Sarawak and Malaysia, especially among the urban and commercial elites. Sarawak bears almost all the private and public costs of forest preservation, while a large proportion of the benefits are enjoyed overseas. Most of the benefits of development are tangible and commercial, while most of the benefits of preservation are subject to scientific uncertainty and cannot be commercialised. Development promises extra income to Sarawak in the near future, while the ecological costs of that development, forgone by not preserving Sarawak's forests, are more distant. Finally, incentives and information aside, if enforcement of logging and forest protection regulations is imperfect, forest ecosystem destruction will exceed that planned.

20.5 Economic analysis of ecosystem destruction

Figure 20.1 illustrates the logic of continued destruction of Sarawak's forest ecosystems. Assume that our planner has a long time horizon—say fifty years. Sarawak's marginal cost of preserving extra forest ($PVMC$) equals the estimated present value of extra timber and agricultural profits sacrificed over the fifty years, less the costs of regulating development. While these profits are very uncertain more than a few years hence, they are depicted as definite, partly because discounting reduces the relative importance of the more distant returns and costs. The marginal cost of preservation is likely to rise as the area preserved is increased, because the more forest preserved, the higher is the likely average productivity of the land lost to timber and agriculture.

The global marginal benefit of preserving extra forest in Sarawak ($PVMB_W$) equals the present value of the direct, indirect, option and existence benefits to people worldwide over the next fifty years, less Sarawak's costs of forest protection. Again, while depicted as definite in Figure 20.1, the global marginal benefit curve is even less certain than

Figure 20.1 The economics of tropical deforestation

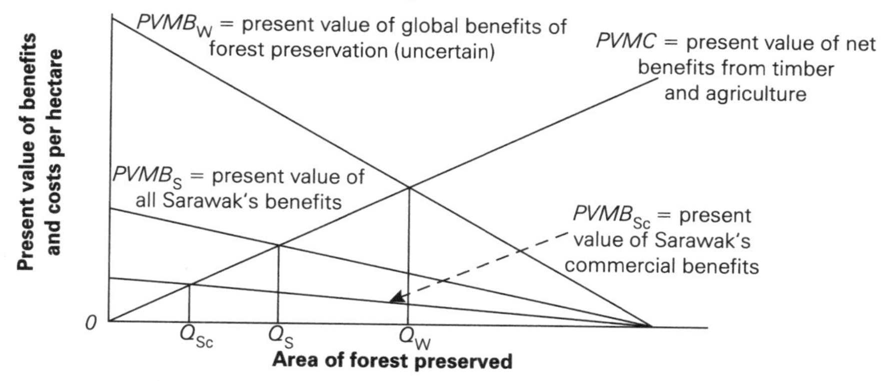

the marginal cost curve, due to a lack of market valuations and scientific uncertainties about the ecological contributions of tropical forests. Because many of the preservation benefits are global non-excludable goods, the present value of the marginal benefit of forest preservation for people in Sarawak ($PVMB_S$) is only a portion of the global benefit. Further, since some of Sarawak's benefits are also non-excludable, and some of the excludable benefits, such as forest fruits, are non-commercial, the present value of the commercial marginal benefits of forest preservation to Sarawak ($PVMB_{Sc}$) is smaller still.

If our planner bases his or her decision on commercial benefits to Sarawak, the choice will be the level of forest preservation where the expected marginal commercial benefit to Sarawak equals the expected marginal commercial cost to Sarawak that is, Q_{Sc}. The planner might, in response to political pressure from rural areas, choose a level of preservation based on all benefits to Sarawak, Q_S. In any case, the planner is almost certain to ignore the benefits of Sarawak forest preservation to people elsewhere in the world. If the planner took those benefits into account, he or she would preserve an area of forest equal to Q_W, where the expected marginal cost of preservation to Sarawak equals the expected worldwide marginal benefits.

We get a similar low level of forest preservation if we assume that our Sarawak planner has a very high discount rate and/or a very short time horizon. This greatly diminishes the global and local ecological benefits of forest preservation, which are likely to increase into the future, relative to development benefits, which are mostly realised in the short term.

The model in Figure 20.1 can also be used to indicate the impacts of specific government and international policies on ecosystem preservation. For example, some countries with tropical forests have effectively subsidised deforestation by government-assisted agricultural land settlement schemes (e.g. Brazil, Indonesia and Malaysia, including Sarawak), subsidising agriculture (e.g. cattle ranching in Brazil), tax concessions for capital investments (e.g. wood processing in Brazil) and road building (e.g. Brazil). How do such policies affect our Sarawak planner? As long as the Malaysian government is footing the bill, the result is to make development more profitable for Sarawak, relative to forest preservation. In terms of Figure 20.1, the *PVMC* is raised, shifting Q_{Sc} and Q_S to the left.

20.6 Biodiversity signals and incentives in Australia

Sarawak is representative of developing countries suffering losses of ecosystems and biodiversity. How will signals and incentives differ in the case of ecosystems in rich countries, such as undisturbed parts of Australia's arid lands, forests, mangroves and coral reefs? The types of ecosystem benefits and the benefit characteristics remain as described in Table 20.1. As in the case of Sarawak, the net benefits of the development option are concentrated in Australia, as are the costs of preservation, while a substantial portion of the benefits of preserving Australian ecosystems is enjoyed overseas. As before, the benefits of development are more tangible and commercial than those of ecosystem preservation. So how do the Australian land use planner's incentives and information differ from those of the planner counterpart in Sarawak?

The affluence of Australians makes a big difference to the planner's incentives. Economists and others studying behaviour of people in affluent countries observe that people's demands for environmental protection rise rapidly with income once annual income per head passes about US$5000 (in early 1990s dollars).[9] Because most important environmental goods are non-excludable, this increase in demand for environmental protection is expressed mainly by interest group monitoring of environmental changes and by pressure on politicians to adopt 'Green' policies, and on bureaucrats to implement them. In Australia, environmental monitoring and the mobilisation and communication of political pressure is undertaken by specialist lobby groups such as the Australian Conservation Foundation, the Wilderness Society and Greenpeace, by minor parties such as the Australian Democrats and the Greens, and by an environmentally sensitive media. Thus an Australian land use planner, typically a state politician or bureaucrat, generally has much stronger incentives to attach substantial weight to preservation options than does a Sarawak counterpart. In terms of Figure 20.1, in the Australian case state or national preservation benefits will be a larger proportion of global preservation benefits (compared to the Sarawak case), moving $PVMB_S$ upwards and the amount of preservation represented by Q_S to the right.

Incentives aside, we emphasised the Sarawak planner's lack of accurate information about the location of the marginal curves, especially the global marginal benefits curve. Despite the greater amount of information likely to be available from scientists and interested parties in Australia, an Australian land-use planner may be little better off. Inadequate scientific understanding of ecosystem functioning allows substantial uncertainty about the environmental consequences of development and ecosystem preservation, for example the impact of Great Barrier Reef tourist development and accompanying marine pollution on future coral growth. The non-excludable nature of many benefits of preservation allows interested parties, both developers and preservationists, ample opportunities to overstate or understate people's willingness to pay for preservation. For example, conservation groups may overstate people's willingness to pay to enjoy the continued existence of old-growth forests.

While Australian politicians and bureaucrats have strong incentives to give weight to the benefits of ecosystem preservation, political events of the late 1980s and 1990s suggest that their planning horizons are not much longer, and their discount rates not much lower, than those of Sarawak planners. Both the Hawke and Keating Labor governments moved to adopt more preservationist stances in the run-up to elections (state and federal), a recent example being the February 1995 decision to release forest coups for logging in all states except New South Wales, prior to that state's March election. At the same time, industry pressure, such as the loggers' blockade of Federal Parliament in February 1995, which immediately preceded the release of logging coups, also appeared to achieve short-term results. Thus, despite the long time perspective involved in many development-versus-preservation issues, most Australian politicians act as though voters, or at least swinging voters, have short memories.

Discussion questions

1. Enforcing limits on the commercial harvesting of valuable flora and fauna will do little to slow biodiversity losses worldwide. True, false or uncertain? Explain why.
2. Economic growth is both the major cause of past biodiversity losses, and the stimulus for biodiversity protection in the future. True, false or uncertain? Explain why.
3. Unlike Sarawak, a large proportion of Costa Rica's remaining tropical rainforest is in national parks where no logging is permitted. The Costa Rican economy is dependent on tourism and agriculture, not forestry and agriculture and the government is considered a world leader in conservation of natural areas. Yet deforestation continues in Costa Rica. Suggest signalling and incentive-based reasons why, even in countries where government planners are committed to conservation, the

remaining areas of natural ecosystems are likely to continue to decline.

4. Bans on the import of tropical timber are likely to impact on both the economy and the environment in both Australia and the South-East Asian countries which export such timber. The complex relationships between and within the economy and the environment in exporting (e.g. Malaysia) and importing (e.g. Australia) countries make it very difficult to predict the consequences of such bans. What do you see as the major types of complexities of the combined economic–environmental systems which make prediction difficult? Based on your answer, suggest reasons why an Australian ban on imports of rainforest timbers might increase, rather than decrease, environmental damage in Malaysia and/or Australia.

21

Measures to preserve biodiversity

Improved worldwide cooperation to preserve biodiversity depends on either changing the information and incentives of those who currently control land use or shifting control to others with stronger incentives for preservation. Commercialisation of the products of undisturbed ecosystems, foreign acquisitions of rights to biodiversity-rich ecosystems and vesting rights in indigenous communities may all assist preservation. Since poor biodiversity-rich countries bear most of the costs of preservation in the form of reduced food, timber, etc. production but get limited benefits from preservation, international cooperation to preserve biodiversity will require transfers from rich to poor countries. However, as in the case of proposed controls on greenhouse emissions, international cooperation to preserve biodiversity is likely to be impeded by scientific and behavioural uncertainty about the long-term impacts of international action.

21.1 Improving incentives to preserve biodiversity[1]

Consider the government land-use planners in Sarawak discussed in Section 20.4. How can distortions and gaps in the planners' incentives and information be reduced? Alternatively, can control of some forest resources be shifted to other decision makers whose incentives are likely to be less subject to distortion?

Incentives for forest preservation can be increased by creating institutions that give people overseas and Sarawak's forest communities more economic and political influence over the use of Sarawak's forests. There are several ways in which this might be achieved. One is to create property rights and markets for the benefits of forest preservation, thereby creating both incentives and signals in favour of additional preservation. A second institutional change is that envisaged in the United Nations Convention on Biological Diversity negotiated at the Rio Earth Summit in 1992, with biodiversity-rich countries such as Malaysia being rewarded for undertaking preservation actions. A third change involves the implicit purchase, by overseas interests, of rights to exercise some control over areas of forests or other ecosystems, exemplified by recent debt-for-nature swaps. A fourth possible change is to vest property rights in forest resources in indigenous forest dwellers, who have a major stake in continuing sustainable forest use. We consider these in turn.

21.2 Commercialisation of preservation benefits

There are a number of changes in property rights that would increase the commercial value of sustainable uses of Sarawak's forests. Sedjo has proposed international legal recognition of property rights in naturally occurring genetic materials.[2] This would enable countries or communities or individuals to profit from the discovery and preservation of organisms found on territory under their control. In terms of Figure 20.1, the effect of such a change would be to raise $PVMB_{Sc}$, thereby shifting Q_{Sc} to the right, and increasing forest preservation. Legalisation of international trade in individuals or products of valuable forest species, such as butterflies, crocodiles and parrots, will have similar effects.

The creation of property rights and markets will only assist in the preservation of Sarawak's forests if property rights in legally acquired forest organisms and products can be clearly defined and enforced at an acceptable cost. The existence of a lucrative illegal international trade in the products of rare and endangered species, for example parrot eggs and rhinoceros horn, suggests the difficulties of policing legal rights. Legal markets will provide few rewards for forest preservation if buyers, international monitoring agencies and forest and wildlife managers cannot economically distinguish between (say) birds produced by sustainable harvesting or captive breeding, and birds acquired by unsustainable raids on wild stocks. Similar enforcement problems could arise if Sarawak or individual Sarawakians were able to sell 'nature prospecting rights' in their forests, or were entitled to royalties for every use of genetic material derived from Sarawak's forests. Potential overseas buyers of genetic material, such as international pharmaceutical companies, will only be willing to pay substantial sums for rights if they believe that the property rights thus acquired are secure. Yet an internationally recognised 'patent enforcement agency' may find it very costly to identify genetic material originating in Sarawak, and to distinguish legally from illegally acquired material. Consider the costs of defining and enforcing property rights to Sarawak's timber resources versus its genetic resources. Timber products are bulky and excludable, making it cheap to document both their origin and mode of acquisition. Genetic material is more akin to computer software; concealment and transfers to other parties are cheap and very costly to monitor.

21.3 International cooperation to protect biodiversity

The Rio Convention on Biological Diversity provides a legal framework for protection of global biodiversity and for international transfers of biotechnology. Implementation of the Convention depends on international ratification of subsequent protocols setting out agreed protection and transfer procedures. Once implemented, the Convention will provide biodiversity-rich countries such as Malaysia with significant incentives, including financial assistance and biotechnology transfers from developed countries, to consider the international benefits of biodiversity preservation.

What are the prospects for successful international cooperation to preserve biodiversity? Based on the criteria listed at the beginning of Section 17.7, effective implementation of the Convention will be difficult. First, international agreement will be impeded by less-than-complete scientific and political consensus on the extent and seriousness of biodiversity losses. Scientific uncertainty about species and ecosystems in turn creates uncertainty about individual countries' benefits from preservation, leading to disagreements over the extent of protection measures in the biodiversity-rich countries and the amount of compensation due from the developed countries that are the major proponents of preservation.

Second, biodiversity preservation is neither low-cost nor restricted to just a few countries. Recall that most of the world's undisturbed ecosystems and biodiversity are located in countries like Sarawak: low-to-medium income countries experiencing rapid population growth. As for Sarawak, these countries' costs of cooperating by forgoing agricultural or forestry or mining developments are high, especially relative to their citizens' evaluation of the benefits of ecosystem preservation. Also, a large number of countries contribute significantly to global biodiversity losses (see Table 19.1), raising the costs of international negotiations.

What about the distribution of costs and benefits from a successful biodiversity agreement? As explained in Section 20.3, developed countries have a high demand for biodiversity preservation but are mostly biodiversity-poor (Australia is a fortunate exception), while poorer countries like Malaysia have a lower demand for biodiversity and are often biodiversity-rich. Thus most biodiversity preservation must take place in poorer countries, with the developed countries' main contribution being compensation, including biotechnology transfers.

The asymmetry between the positions of most developed countries and most poorer countries has advantages. It may facilitate agreements on biodiversity policy within each group, and negotiations between the two. It also facilitates recognition by the rich that they have to pay, directly or indirectly, to acquire any rights over biodiversity controlled by poorer countries.

What about cheating once the Convention on Biodiversity is implemented? Gross cheating by biodiversity-rich signatories (e.g., extensive logging in Sarawak's forests) can be detected by satellites or local environmentalists. On the other hand, international enforcement of parts of the Convention may be undermined in the same way that legal markets in forest organisms and products may be undermined; it may be too costly to identify and penalise illegal use of species and other genetic material.

Finally, what about free riding by countries that expect to gain from an international agreement on biodiversity, but choose not to cooperate? The countries that control the world's biodiversity cannot, in general, free ride on one anothers' preservation efforts, because one country's contribution is not a good substitute for another's. A 1000 tonne reduction in Australia's CFC emissions has exactly the same global impact as a 1000 tonne reduction in Malaysia's emissions, but differences in ecosystems mean that biodiversity protection in Malaysia is not a good substitute for biodiversity protection in Australia. Thus the rationale for free riding presented in Section 17.5 has limited application to biodiversity. International demands for Australia to preserve its unique species and ecosystems will be little affected by the progress of preservation in Sarawak or Brazil or Tanzania.

Free riding on other countries' compensation contributions is possible; one country's transfers of funds or technical assistance are about as good as another's. Like the crane lovers of Section 6.4, rich but biodiversity-poor countries have strong incentives to avoid international contributions to biodiversity preservation, as long as they believe that other rich countries will not follow suit. However, their free riding may be deterred by feedback mechanisms such as those discussed in Section 17.7. On the other hand, Malaysians are unlikely to contribute much to protecting Brazil's biodiversity, not because they wish to free ride on Brazil's efforts, but because they expect few benefits from biodiversity preservation outside their own borders. Even if people in poorer countries such as Malaysia anticipate higher benefits in the future, high discount rates may reduce those benefits to small present values.

21.4 Acquisition of rights over ecosystems

The Convention on Biodiversity recognises that states have sovereign rights over their own biological resources, but appears to duck the issue of direct purchase of such rights by overseas interests. If people overseas wish an owner-state like Sarawak to forgo the commercial gains stemming from development in favour of ecosystem preservation, one option is the purchase of development rights, assuming that Sarawak is willing to cede the necessary control of resource use in return for financial compensation. This could involve overseas interests, countries or conservation organisations or individuals, acquiring direct control over parks or conservation reserves, or international payments to Sarawak in return for its agreement to restrict the uses of specified parcels of land. International biodiversity preservation contracts could include provisions to contract out the management of parks or reserves to agreed bodies, such as Malaysian conservation organisations.

Recent debt-for-nature swaps involve de facto acquisition of rights over other countries' biological resources. In a debt-for-nature swap, an overseas country or conservation organisation purchases the discounted secondary debt of a debtor country, and offers to give it back to the debtor in return for guarantees of continuing protection for specific natural areas.[3]

All acquisitions of rights over resources in other countries, whether oilwells or buildings or conservation reserves, are subject to sovereign risk, based on the fact that the government of the host country cannot be legally compelled to respect and enforce the previously agreed rights. In the case of biodiversity, sovereign risk is compounded by difficulties in defining and monitoring land use rights. Suppose that an Australian-based conservation organisation agrees to finance ecosystem preservation in a designated area of Sarawak forest in return for land-use guarantees. Subsequently Sarawak conservationists report evidence of logging or shifting cultivation or flora and fauna removal in the area, possibly with official complicity. Suppose also that Sarawak officials deny the latter, but admit the possibility of illegal incursions that cannot be policed with their current budget. They say that the integrity of the forest ecosystem is not compromised, and ask for more aid for enforcement, but resist requests to involve conservationists in the policing process. It is very difficult for the overseas conservationists to know whether their contracted-for rights are being enforced in good faith, yet they may fear that a cessation of their payments will see much of the forest lost to development. Presumably, in the long run, experiences of this kind would adversely affect Sarawak's and Malaysia's access to international environmental assistance.

21.5 Vesting property rights in indigenous peoples

Swanson and Barbier remind us that, until a few decades ago, almost all of the developing world's habitat was used by indigenous communities for subsistence and trade.[4] Indigenous peoples such as Australian Aborigines and Amazonian Indians were the repositories of large amounts of specialised knowledge about ecosystem products and management. Colonial regimes and new national governments shifted most biodiverse ecosystems from local community to government control. As a result, local communities, usually those best placed to control access and use, were excluded from and/or lost interest in the use and preservation of ecosystems, and their specialised knowledge declined with their involvement. Shifting rights back to indigenous communities would partly reverse these changes, increasing incentives for sustainable use of biological resources, and for community policing of unsustainable uses.

In the case of Sarawak, reinstatement of the former communal and individual rights to forest resources enjoyed by indigenous peoples would not halt all development on indigenous land. It would almost certainly lead to more careful and longer-term comparisons between the values of (say) different types of logging and the direct and indirect uses of the forests recognised by local people, such as non-timber products, cultural sites, watershed protection and nutrient recycling. It should also lead to careful local monitoring of logging activity, to see that such non-development benefits were maintained. On the other hand, indigenous communities have no stake in non-excludable benefits received by beneficiaries overseas. Thus they would need to be compensated to preserve additional areas on behalf of people in the developed countries.

Reliance on indigenous communities to preserve biodiversity depends on two important assumptions. First, owner-governments must be prepared to cede substantial control, of both biological resources and the resulting income flows, to locals. Second, consistent with the discussion in Chapter 2 and Sections 7.6 and 18.7, effective indigenous control of resources will only be possible if communities maintain, or can create, the property rights and other rules required to allocate amicably scarce resources, and the technical knowledge of ecosystems required to determine and police such rules. Recall, from Section 20.3, that indigenous culture is highly complementary to indigenous resource management; the latter is unlikely to succeed without preservation of the former.[5] Thus we might be reasonably confident that recognition of Aboriginal Australians' rights to control land use in Arnhem Land, where indigenous culture and lore remain well entrenched, will assist in the preservation of pre-existing biodiversity. At the same time, it is unlikely that such recognition will be much help in the more densely populated parts of Australia, where indigenous communities' culture and biological knowledge are either lost or fragmented.

21.6 Uncertainty about biodiversity

Creating institutions that strengthen decision makers' incentives to consider all the benefits of preserving biodiversity does not overcome all deficiencies in information about those benefits. As explained in Section

19.4, incomplete scientific knowledge of genes, species and ecosystem functioning rules out reliable estimates of how the preservation of Sarawak's forests will affect the world's environment and biodiversity, now and in the future. In particular, the present and future ecological benefits listed in Table 20.1, such as nutrient cycling and climate regulation, are impossible to estimate reliably.

The consequences of current losses of biodiversity extend far into the future—indefinitely, where genetic blueprints and evolutionary processes are lost and cannot be recreated. Thus the information problems involved in calculating the costs of biodiversity losses are similar to those for global pollution, discussed in Section 17.4: scientific uncertainty about future environmental consequences of present actions; ignorance of the identity and personal preferences of those who suffer from biodiversity loss in the future; ignorance of future technologies and resource costs that, together with future generations' preferences, will determine their adjustment costs and disappointment at living in a biologically impoverished world; and finally, the difficulty of deciding on the relative weights of present and future benefits and costs.

Uncertainty affects outcomes at two levels—individual decisions by private or public planners, and international negotiations over rules to apply across countries. As explained in Section 21.3, uncertainty about individual countries' benefits and costs will impede international agreement to reward poorer countries for preserving biodiversity. Within a single political unit like Sarawak, the effect of uncertainty on biodiversity preservation depends on how the responsible decision maker deals with an uncertain future. Recalling the exposition in Section 16.2, one approach involves identifying objectives and possible action event scenarios, and making judgments about outcomes, values and probabilities for each scenario.

Our earlier discussion of the Sarawak planner's choice between development and preservation of the forest did not specify the possible alternative events and outcomes resulting from each choice. The decision tree in Figure 21.1 provides an illustrative, highly simplistic, picture of possible actions and events as seen by the planner. The planner's choice is between logging followed by forest regeneration and forest preservation. Logging results in timber income for Sarawak and political protests by forest communities, with uncertainty about consequences restricted to the magnitude of the resulting ecological damage (such as increased flooding, stream pollution and loss of indigenous species). Logging may lead to either major or minor ecological damage costs for Sarawak. Forest preservation results in income from non-timber products, such as rattan, and from tourists, and political action against the government by urban commercial interests. The government's costs of regulating land use are the same for both logging and forest preservation.

According to the decision tree, the planner's choice depends on his or her judgments about the probabilities of major or minor ecological damage following logging, and on the relative values he or she attaches

Figure 21.1 Forest-use decision tree

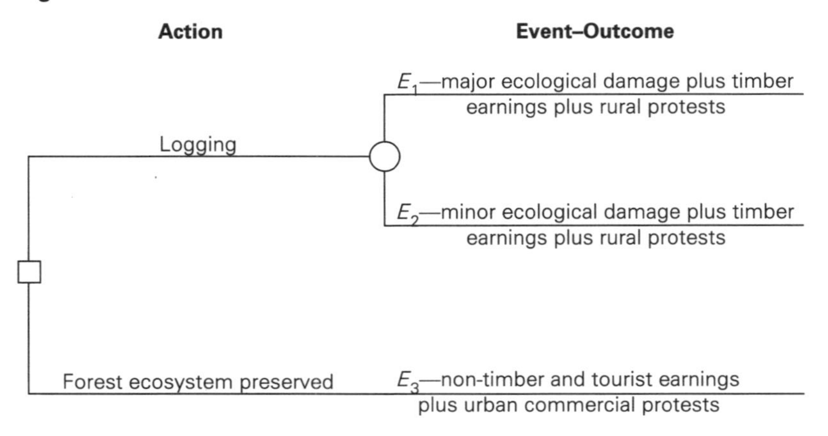

to the various possible consequences of a choice: timber and non-timber commodities, tourism, ecological damage and political opposition by different groups. In Chapter 20 we argued that the planner would choose development; in terms of Figure 21.1, the probability-weighted average of the values of E_1 and E_2 would exceed the value of E_3. If the planner is sceptical about serious ecological damage due to logging, the probability of E_1 is judged much lower than that of E_2, increasing the likelihood of choosing logging. Figure 21.1 also formalises our reasoning in Chapter 20, that logging is more favoured the more that timber values exceed non-timber and tourism values, the more that political costs of offending commercial interests exceed those of offending forest communities, and the more that the planner discounts costs of ecological damage falling well into the future.

In the case of Sarawak, we argued that the planner would choose logging, but different judgments about probabilities and values could lead to preservation. Suppose that the planner is deciding the use of a forest ecosystem in north Queensland, rather than Sarawak. In the Australian decision context, preservation could be preferred to logging because the probabilities of major ecological damage resulting from logging are judged to be high, or because Australians are believed to be very concerned about ecological damage imposed on their descendants, with the consequent political costs of logging judged to be higher than those of preservation.

21.7 [illegible]e precautionary approach to biodiversity preservation

[illegible] people would argue that the decision model depicted in Figure [illegible] not appropriate for regional or national or international decisions [illegible]odiversity preservation. If it is impossible to obtain reliable, scientifically based estimates of the consequences of widespread biodiversity loss for humanity, a decision maker has no firm basis to postulate action–event scenarios and their probabilities. Yet, as explained in Section 19.5, some level of biodiversity loss must force fundamental changes on human societies.

Advocates of *precautionary* and *safe minimum standard* approaches to biodiversity preservation reject the expected benefits versus expected costs approach outlined in Section 21.6.[6] They do so on at least three grounds. One is the spurious precision of such calculations in the face of great uncertainties about both the biological consequences of current actions and the future values attached to those consequences. Another is the inappropriateness of applying normal decision processes to possible losses when damage to human life support systems may endanger the institutions and organisation of society itself, in other words when it is believed possible that human actions will create major and irreversible changes in the economic–environmental system depicted in Figure 4.1. The third is ethical. Benefit–cost calculations are unfair to future generations because most of the benefits of resource use are received by this generation, while many of the costs will fall on future generations who have no say and are not necessarily compensated for reductions in their biological resource endowment.

Application of the precautionary principle to biodiversity is based on the premises that people are risk-averse and that some level of biodiversity loss may impose serious and irreversible losses on humanity, either now or in the future. Advocates of a precautionary approach suggest that, in the absence of scientific certainty about the consequences of using biological resources, the burden of proof should be on the would-be resource user. A logger or miner or resort developer should be required to provide appropriate proof that the proposed development will not result in some defined level of species loss or damage to ecosystems.[7] Otherwise, development should not proceed.

Strict application of the precautionary principle to the decision problem in Figure 21.1 could require that forest preservation be chosen unless it could be demonstrated that the probability of major environmental damage from logging (event E_1) is zero. Given uncertainty about the biological consequences of logging and how future generations will deal with ecological change, it is impossible for the proponents of logging to meet this condition. Proponents have no way to estimate reliably the probability and value of E_1. So strict application of the precautionary principle would bring a great many development projects to a halt, because of the impossibility of proving conclusively that projects will or will not result in particular biological and economic outcomes.

The history of the 1973 Endangered Species Act (ESA) in the USA illustrates the difficulties of a strict precautionary approach to biodiversity protection. As interpreted by the US Supreme Court, Congress intended the ESA to protect listed species regardless of cost. This has had a number of important consequences:

1 Government agencies and the courts were originally supposed to take no account of the value of alternative resource uses sacrificed to preserve listed species. The ESA was subsequently amended to allow a high-level government committee (nicknamed the 'God Squad') to grant exemptions in cases where overriding economic or national-interest concerns were deemed to outweigh species preservation.
2 Given the limited budget appropriated for the ESA, all available public funds are used up in actions to protect a few high-profile endangered species, such as the bald eagle and the grizzly bear. The only means left to protect other listed species is to prohibit all development that might damage the species or their habitats.
3 With limited funds and scientific uncertainty about many species' true viability status, there are long delays in listing species as endangered. By the time some species are listed, their wild populations may be past the point of no return.
4 In the absence of sufficient funds to rigorously monitor all actual and potential habitats of listed species, land owners and users have strong incentives to protect their commercial interests by destroying species and/or their habitat.
5 Antidevelopment interests have strong incentives to use the ESA in pursuit of objectives other than species preservation.[8]

It appears that the precautionary stance adopted in the ESA, combined with its limited funding and creation of perverse user incentives, has resulted in failure to achieve its intended objective. During the twenty years the ESA has operated, only four species have been removed from the endangered list, several listed species are believed to have become extinct, and the act has resulted in major economic costs to resource-using industries and major political controversy (e.g. over the halt to logging in millions of hectares of forest in the Pacific Northwest, to protect the northern spotted owl).

The apparent wastefulness of the ESA is not conclusive evidence against a precautionary approach, but rather against precautionary policies that take no account of the value of the alternatives sacrificed. Remember that humanity has gained in the past and will continue to gain from ecosystem conversions. On the other hand, the loss of ecosystems, with their ecological functions and evolutionary potential, is ultimately much more serious for humanity than loss of individual species. We need to balance, however crudely, the expected benefits against the expected costs. This is the function of the 'God Squad' set up under the ESA to deal with cases where species preservation might impose intolerable costs on people in the USA. Such incorporation of crude judgments about relative benefits and costs in the choice procedure is what differentiates the safe minimum standard approach to biodiversity preservation from the strict precautionary approach.

21.8 The safe minimum standard approach to biodiversity preservation[9]

The safe minimum standard (SMS) approach follows the precautionary approach in assuming that the benefits of preservation are positive and large; it follows the benefits-versus-costs approach outlined in Section 21.6 by estimating the costs of preservation, which equal the benefits sacrificed by not choosing development. The planner balances (assumed) large benefits of biodiversity preservation against its (estimated) costs. Applying the SMS approach to the decision problem depicted in Figure 21.1, the planner simply assumes, in the absence of reliable evidence, that forest preservation results in large benefits from non-timber products, tourism, avoidance of ecological damage and avoidance of political opposition from forest communities. The costs of preservation, sacrificed timber earnings and the political opposition of Sarawak commercial interests, are estimated as accurately as possible, in financial terms where possible. The planner chooses to preserve the forest unless the estimated costs of preservation are judged to be intolerably high.

Under SMS, the barriers to projects that threaten biodiversity are lower than when the precautionary approach is adopted. Extinction of species or destruction of ecosystems is permissible if their preservation is intolerably costly for society. Consider the difference between the two approaches in the case of Figure 21.1. A strict precautionary approach rules out logging unless it can be proved that event E_1, involving major ecological damage, will not occur. An SMS approach would allow logging if the estimated benefits of logging were so large that preservation would impose intolerable costs on Sarawak society, for example if large up-front payments by loggers were available to fund major improvements to health, education and other human services throughout Sarawak. The reasoning is that if the benefits of development are very large, society is no worse off as a result of logging followed by a worst-case outcome (in Figure 21.1, major environmental damage, E_1) than following the worst possible outcome of preservation (in Figure 21.1 there is only one, E_3).

SMS still places the burden of proof on the proponents of development, but it changes the question to be answered from 'Can the community be guaranteed that particular ecological outcomes will not occur?' to 'Can the community afford to sacrifice the benefits of development?' As illustrated in the change in the criteria applied to species preservation under the ESA, the nature of the proof required changes from scientific calculations of the biological consequences of development to calculations of its financial and social benefits.

The virtue of the SMS approach in circumstances of great uncertainty is that, contrary to the strict precautionary approach, it prods the decision maker towards balancing benefits and costs, using the information available about the benefits of development, but does not treat the loss of biodiversity as a routine resource sacrifice. Randall summarises the idea this way:

> The SMS rule places biodiversity beyond the reach of routine trade-offs, where to give up ninety cents worth of biodiversity to gain a dollars worth of ground beef is to make a net gain. It also avoids claiming trump status for biodiversity, permitting some sacrifice of biodiversity in the face of intolerable costs . . . The idea of intolerable costs invokes an extraordinary decision process that takes biodiversity seriously by trying to distinguish costs that are intolerable from those that are merely substantial.[10]

Despite the different treatment of preservation benefits and costs under SMS, changing the decision process from benefits-versus-costs to SMS may make little difference to biodiversity preservation if the responsible decision makers and their incentives remain the same. Consider again decisions about the use of Sarawak's forests, discussed in Section 20.4. Would the adoption of SMS by the government of Sarawak be likely to make much difference to the choice of development or preservation?

The Sarawak politician or bureaucrat with the right to decide land use makes the judgment about what represents an intolerable cost; he or she also determines the discount rate, the relative weighting of present and future costs. The political and social and professional pressures on the planner, and hence his or her incentives, remain the same. Assuming that the ecological benefits of preserving forest ecosystems are highly uncertain, SMS eliminates the requirement for spurious precision in documenting such benefits. However, assumed large ecological benefits of preservation can be offset by defining the sacrificed earnings from logging as 'intolerable costs'. Yet what is politically or commercially intolerable to the decision maker may be perfectly tolerable for the people of Sarawak.

Rolfe argues that in real world decision making, SMS is in fact likely to lead to similar decisions to cost–benefit analysis.[11] This is because SMS and cost–benefit analysis ultimately rely on the same societal (or interest-group) preferences. Rolfe's argument may also apply to real world implementation of the precautionary principle. Recall the IGAE definition of the precautionary principle in Section 16.7 above. Someone has to determine what is 'serious environmental damage', and how to identify and assess the 'risk weighted consequences' of alternatives. If this is done by the same people as before—scientists, economists, politicians, bureaucrats—it is unclear that adoption of a SMS or precautionary approach to environmental policy will make a great deal of difference to environmental management in practice.

Perhaps the most serious limitation of SMS is its disregard of available information about biodiversity and its functions. SMS in effect treats one gene, species or ecosystem as being just as good as another, so that the priorities for preserving biodiversity depend solely on the costs of preservation. This ignores the fact that there may be some scientific and value information available about benefits. Also, the SMS assumption of very large ecological and other benefits from extra species or areas of forest preserved becomes less plausible as the total amount of preservation becomes very large.

21.9 Don't expect too much of biodiversity policy

We need not worry about genes or species or ecosystems that are directly useful to people if exclusive rights to those resources exist, so that someone has an incentive to conserve the resource. As explained in Section 19.5, it is the indirect usefulness of biodiversity, its contribution to ecosystem resilience and thus to human life support, that is of most concern. Unfortunately, we know so little about ecosystem functioning that the appropriate measure of biodiversity, measuring its contribution to resilience, is unclear.

Once we recognise the formidable incentive and informational barriers to national and international cooperation over biodiversity, two conclusions follow. First, the costs of better cooperation over biodiversity will sometimes be too high. The anticipated benefits of preservation may not justify the costs of acquiring additional information about ecosystems and values, and of creating institutions that strengthen resource users' incentives to respond to other people's concerns about biodiversity. In the extreme case, there is no way we can acquire reliable information about the value of additional biodiversity to future generations. Second, errors in biodiversity preservation are unavoidable. Imperfect knowledge of how preservation affects the environment and how people value those changes mean that decisions about the appropriate level of preservation will frequently turn out to be incorrect with the benefit of hindsight.

We can demonstrate the error problem using Figure 21.1. Forest preservation, involving major sacrifices of logging income, may subsequently be shown to have little or no impact on the welfare of the people of Sarawak, now or in the future. On the other hand, if logging proceeds, destruction of forest species and ecosystems may impose major costs on people, present and future. Suppose that uncertainty is too great for credible analysis of benefits versus costs, and that the people of Sarawak are risk-averse; that is, they put more weight on avoiding large future losses than on achieving large future gains. Which type of error is more serious for Sarawak?

With major uncertainty about the impacts of biodiversity loss, there is no right answer. As discussed in the case of greenhouse policy in Section 17.10, it depends on the responsible decision maker's subjective evaluation of the uncertain consequences of precautionary policies (preservation) versus reactive policies (development). If a decision maker is pessimistic about the ecological consequences of biodiversity loss, and about the real returns to development, so that he or she judges that the worst possible outcome of development will be more costly than the worst possible outcome of preservation, he or she will choose preservation. If optimistic on both counts, the decision maker will choose development. In each case, the chosen course of action is consistent with the SMS criterion: choose the policy that minimises costs if the choice turns out to be wrong.

The preceding paragraph reminds us of the importance of individual planners' knowledge and perceptions in determining biodiversity policy. As Wildavsky points out, people may choose anticipatory policies such as biodiversity preservation not because they are risk-averse but because of their perceptions of the alternative outcomes—they more readily perceive irreversible and catastrophic consequences of development than of preservation.[12] Remember also that a lack of local and international cooperation to preserve biodiversity is as likely to result from planners' pursuit of their perceived self-interests as from their imperfect knowledge of environmental and economic outcomes.

Discussion questions

1. Indigenous ownership and control of former public lands throughout Australia, based on the *Mabo* judgment and the Native Title legislation, will greatly assist biodiversity preservation in Australia. True, false or uncertain? Explain why.
2. Government regulations which require private and public actions to preserve designated species and ecosystems regardless of the cost of doing so may reduce the amount of effective preservation. True, false or uncertain? Explain why.
3. In what respects will negotiation and enforcement of an effective international convention to protect the world's biodiversity be easier, and in what respects will it be more difficult, than negotiation and enforcement of the Montreal Protocol on ozone-depleting substances?
4. The advocates of private property rights and free markets suggest creation of private rights and commercialisation of flora and fauna as a way of preserving biodiversity. Discuss the advantages and limitations of this approach in providing signals and incentives to protect biodiversity.
5. How does the safe minimum standard approach to biodiversity preservation differ from a cost–benefit analysis of preservation versus 'development', similar to the analysis of preservation of the South-East Forests in Chapter 10? What are the advantages and disadvantages of the two techniques as alternative methods of assessing preservation policies?

22

Economics and environmental problems: important lessons

22.1 Environmental problems will always be with us

Users of the natural environment will sometimes impose uncompensated harms on others. Why? In a world of self-interested people and dispersed and costly private information, coordination of natural resource use can never be perfect because it sometimes costs too much relative to its benefits. Defining and enforcing people's property rights, signalling, and rewarding and penalising private and public decision makers all require resources which have other valued uses. The gains from better social coordination (e.g. reduced air pollution improving people's health and views, people's satisfaction at knowing that an endangered species or ecosystem is preserved) are finite. So it does not always pay societies to strive for perfect coordination between resource users and those they harm. It is sometimes better to leave environmental problems 'unsolved' until improved technologies or better information become available.

To illustrate the point, consider the problem of unsightly roadside litter. It imposes harms on most road users and adjacent property holders. However, the problem persists because the costs of better coordination of solid waste disposal are almost certainly much greater. There is presently no technology for identifying mobile litterers of public or private land at an acceptable cost, and even if this was possible, the costs of penalising offenders are likely to be prohibitive. Also, since many people are harmed by the same litter, it is difficult for either private parties or government to establish the true value of the harms caused; individual sufferers have no particular reason to reveal their true costs where they do not expect that their individual responses will make any difference to litter volumes. Given the very high costs of establishing precise property rights and of market or planning coordination, it is not surprising that governments put major effort into campaigns such as 'Clean up Australia', that are designed to reinforce traditional societal norms or to create new norms.

22.2 Costs and benefits of coordination change, so the problems do not necessarily remain the same

Today's environmental problems may be resolved by improved social coordination in the future. Factors determining coordination costs and benefits, discussed in Chapters 5 to 8, include:

1. Shared cultural norms, which reduce the need for costly market or political exchanges and associated enforcement procedures when one person's resource use harms others.
2. The state of science and technology, which determines:
   - the costs of communicating about property rights, and in markets and politics;
   - knowledge of environmental cause and effect, and hence ability to identify both causal and affected parties;
   - the costs of measuring transfers between people. Technological innovations such as isotope tracers added to emissions may lower measurement costs;
   - the costs of excluding potential users. Innovations such as electronic signal scramblers and decoders can lower exclusion costs.
3. The state of the social sciences, which can lead to innovations in institutions, organisations and decision-making, which may reduce costs of identification, measurement, exclusion, enforcement and valuation (e.g. freedom of information legislation, the Productivity Commission, cost–benefit analysis).
4. The costs of scarce resources used in identification, measurement, exclusion, enforcement and valuation (e.g. fencing and electronic monitoring equipment, police and courts).
5. The values of the assets or services preserved or enhanced due to improved social coordination (e.g. cleaner air, more live whales).

Changes in these factors—in people's attitudes, technology, social knowledge, resource costs and asset values—alter the costs and benefits of the coordination activities listed in Table 8.1. Thus the efficient scope of social coordination by property rights, markets and government regulation of resource use will change over time. For example, it may soon be possible to economically measure exhaust emissions from identified vehicles in urban traffic, permitting pollution costs to be signalled back to individual vehicle owners.

22.3 Major environmental problems involve non-rivalry and non-excludability

Major environmental problems, such as air and water pollution, ozone depletion and biodiversity loss, almost always involve many contributors and many sufferers. Sometimes, as in the case of the oil spill from BHP company's ship *Iron Baron* off north-east Tasmania in 1995, there is a single contributor and many sufferers. Environmental problems where an action affects only one or very few people, for example, the costs imposed by a neighbour's barking dog or by a new building blocking a property owner's view or solar access, are much less serious. Where the benefits of coordination between two parties exceed its costs (e.g. where a new building will shade the solar collectors of an existing building) differences can usually be resolved by bilateral negotiation or private legal action.

Society cannot deal with multilateral environmental problems by splitting them into a large number of separate bilateral deals between the parties who are acting and those who are suffering the resulting costs. As explained when discussing acid rain in Section 5.9, bilateral deals between multiple contributors and sufferers are impossible if goods are non-rival and non-excludable. It is the combination of non-rivalry and non-excludability which leads to multilateral, as opposed to bilateral, bargaining, to free riding, and to serious breakdowns in market coordination in resource use. For example, it would be impossibly costly to define and enforce individual property rights in the sea birds harmed by the *Iron Baron* oil spill, therefore few individual bird lovers are likely to act to penalise BHP. Free riding by individual bird lovers undermines social coordination; by not acting, individuals reduce incentives to control marine pollution and conceal information about the values they attach to the birds.

22.4 Environmental and economic complexities make errors in environmental policies inevitable

Major environmental problems commonly involve major scientific and behavioural uncertainty. Scientific uncertainty is due to the complexity of interactions within the environment, discussed in Chapters 1, 4 and 12. Behavioural uncertainty is largely due to the non-rival nature of major environmental problems. When the impacts of a single action are widely dispersed across space and time, the number of affected parties is very large. This greatly increases the costs of identifying impacts, the people affected and their preferences and adjustment options. In cases such as ozone depletion, global climate change and biodiversity loss, the impacts extend far into the future. It is then impossible to identify many of the people affected by current resource use (many not yet born) and the future preferences, technologies and resource costs which together determine the future costs and benefits of current actions.

Major uncertainty about the environmental and economic consequences of actions, discussed in Chapters 16 and 21, makes errors in environmental policies inevitable. Further, the greater the uncertainty, the smaller the scientific and economic basis for policy choices, and the greater must be the reliance on the preferences and subjective judgment of the person responsible for the choice. This is not a concern when the chooser also bears all or most of the important consequences of the choice; an individual contemplating risky heart surgery has strong incentives to consider all the possible outcomes of his or her choice. But political and bureaucratic decision makers do not personally bear all the consequences of their decisions. Faced with major scientific and behavioural uncertainty, current citizens will have great difficulty in establishing the possible long-term consequences of environmental policy choices. Thus planners will have more than their usual scope for pursuing personal goals inconsistent with democratic preferences. This reasoning suggests two interrelated conclusions about environmental policy making in situations of major scientific and behavioural uncertainty, as in the case of greenhouse gas emissions policy. First, decision making on such policies needs to be less, rather than more, centralised. Choices based on imperfect information will be at least partly subjective, based on the decision maker's attitude to risk and beliefs about the effectiveness of environmental policies. Planners cannot know people's preferences and perceptions on these matters as well as do the individuals themselves. Thus such subjective choices are better made by well-informed individuals choosing for themselves, rather than having planners choose for them. Second, for people to be well informed about the possible outcomes of environmental policy choices (including the possibility of outcomes that no one foresees), there is a need for public support for and dissemination of the results of scientific and economic research into the possible consequences of policy choices (remember that most of the research results are non-rival and non-excludable).

22.5 It is more important to understand processes than to design 'solutions'

We live in an age of 'positive thinking'. As a rule, it is unfashionable in social interchanges, and not least in politics, administration and business, to admit that we do not understand problems, and do not have a plan to solve them. True scientists know, of course, that all disciplinary knowledge about how the world works is partial and provisional and therefore sensible decision making must allow for the possibility and costs of failure. But television viewers, and even company boards of directors, rarely want to hear all the ifs and buts. Generally, to 'sell' a policy in the competitive worlds of politics, administration or business, it is necessary to provide asymmetric information about proposals—to emphasise one's positive 'vision' and de-emphasise the 'downside'. This attitude conflicts with a recurring theme in the latter chapters of this book: faced with major environmental problems involving many people, environmental decision makers rarely have enough reliable technical and value information, or the altruistic incentives, to justify confidence that their planning on behalf of others is likely to improve the welfare of people in general.

Whatever the virtues of 'positive thinking' in situations where the decision maker responsible has substantial control of relevant variables and a good understanding of possible outcomes (compare former Prime Minister Bob Hawke's ability to deliver industrial peace via the Accord with the unions, with his ability to deliver on his promise to reduce child poverty in Australia), it may be very costly when considering changes to complex and poorly understood economic–environmental systems. In a world of self-interested people and dispersed and private information, even the most altruistic and well-informed planner may know too little to avoid costly policy errors. Remember that in the early 1970s, the prevailing wisdom among climate scientists was that Earth was cooling; any policy actions based on those scientific judgments would probably be perceived as very costly in the 1990s.

This book rejects the view held by some scientists and economists that 'scientific' and 'objective' planning based on computer modelling of the environment and the economy can do much to 'solve' major environmental problems. This is not so much due to uncertainties about the functioning of the environment and the economy; in that respect, individual resource users and computer modellers are not so differently placed. Environmental and economic planners suffer much more from their ignorance of what Hayek termed 'the particular circumstances of time and place'—individual resource users', beneficiaries' and sufferers' private knowledge of the resources they use and enjoy and the harms they suffer, and of their personal preferences and adjustment alternatives.[1] Rather, the economist's main task is to help people to understand how society's signalling and incentive systems lead well-intentioned people to use the environment in ways which harm their fellows; in other words, how existing social coordination processes create environmental problems. Then people can begin to think about ways of modifying coordination processes to generate better information and to create stronger incentives for individuals to respond to the interests of others.

In practice, it is a sad fact that much media and public discussion of environmental problems, real and supposed, ignores information and incentives and characterises environmental problems in terms of 'good' and 'bad' actors. While doubtless good media copy and emotionally satisfying to some, such comments divert attention from the central problem of coordination between ordinary people. Recognition of the importance of institutions, information and incentives in environmental decision making would lead to fewer demands on public decision makers for quick solutions, and hence less temptation to promise what cannot realistically be delivered.

Discussion questions

1. It is sometimes more efficient for a society to have no specified private or public property rights for valuable environmental assets. True, false or uncertain? Explain why.
2. Growth in the real incomes of people in the developed world is one of the major factors contributing to the resolution of environmental problems. True, false or uncertain? Explain why.
3. Advances in science and technology, leading to much more precise information about the future consequences of today's resource use, must ultimately lead to resolution of present international environmental conflicts over issues such as high seas fishing, greenhouse gas emissions and use of tropical rainforests. True, false or uncertain? Explain why.
4. Most environmental problems are caused by the actions of people who are either thoughtless or malicious. True, false or uncertain? Explain why.

Notes

1 Introduction

1 Tim Flannery, *The Future Eaters*, Reed Books, Chatswood, 1994.
2 Flannery, *The Future Eaters*, Chapters 21, 25 and 33.
3 Terry L. Anderson and Donald R. Leal, *Free Market Environmentalism*, Pacific Research Institute for Public Policy, San Francisco, 1991, Chapters 1 and 2.
4 Daniel B. Botkin, *Discordant Harmonies: A New Ecology for the Twenty-First Century*, Oxford University Press, New York, 1990, pp. 6–8.
5 Botkin, *Discordant Harmonies*, pp.133–51.
6 Industry Commission, *Costs and benefits of reducing greenhouse gas emissions*, AGPS, Canberra, 1992, Ch. 2.
7 W.C. Clark, 'Managing Planet Earth' *Scientific American*, Sept. 1989, p. 22.
8 J.H. Gibbons, P.D. Blair and H.L. Gwin, 'Strategies for Energy Use' *Scientific American*, Sept. 1989.
9 Charles Perrings, *Economy and Environment: A theoretical essay on the interdependence of economic and environmental systems*, Cambridge University Press, London, 1987, Section 1.3.
10 Flannery, *The Future Eaters*, Ch. 23.
11 David W. Pearce and R. Kerry Turner, *Economics of Natural Resources and the Environment*, Harvester Wheatsheaf, Hemel Hempstead, 1990, Ch. 1, reviews the historical development of economic thinking about the natural environment.
12 Rachel Carson, *Silent Spring*, Penguin, Harmonsworth, UK, 1965.
13 For example, see Dieter Helm (ed.), *Economic Policy towards the Environment*, Blackwell, Oxford, 1991, Introduction.
14 Robert Costanza, Herman E. Daly and Joy A. Bartholomew, 'Goals, agenda and policy

recommendations for ecological economics' in *Ecological Economics: The Science and Management of Sustainability*, ed. Robert Costanza, Columbia University Press, New York, 1991.

15 Herman E. Daly, 'Elements of environmental macroeconomics' in *Ecological Economics*, ed. Costanza, Ch.3.

16 Richard Dawkins, *The Selfish Gene*, new edn, Oxford University Press, New York, 1989.

17 Flannery, *The Future Eaters*, pp. 288–91.

18 Pearce and Turner, *Economics of Natural Resources and the Environment*, Ch. 15, reviews major issues in environmental ethics. Valuation of environmental assets is discussed in Ch. 11 of this text.

19 Roderick F. Nash, *The Rights of Nature*, University of Wisconsin Press, Madison. 1989.

20 Aldo Leopold, *A Sand County Almanac*, Oxford University Press, New York, 1949, p. 203.

21 Quoted in Nash, *The Rights of Nature*, p. 64.

22 Leopold, *A Sand County Almanac*, pp. 205, 226.

23 ibid. pp. 211, 224–5.

24 Christopher D. Stone, *Should Trees Have Standing? Toward Legal Rights for Natural Objects*, Los Altos, California, 1974.

25 Flannery, *The Future Eaters*, p. 389.

26 ibid. Ch. 25.

27 ibid. Chs. 19–21.

28 Comment of Mr Justice Blackburn in the 1971 Northern Territory Land Rights case *Milirrpum v. Nabalco Pty Ltd*; see *Federal Law Reports* (1971), pp. 270–1.

29 Flannery, op. cit., Chs. 23, 30 and 31.

2 Scarcity and systems of social coordination

1 But not always, as shown by the example of the former Yugoslavia, 1991–94. For a simple explanation of the emergence of property rights, see Richard B. McKenzie and Gordon Tulloch, *Modern Political Economy*, McGraw-Hill, New York, 1977, Ch. 5.

2 For an introduction to the nature and role of institutions, see Douglass North, *Institutions, Institutional Change and Economic Performance*, Cambridge University Press, Cambridge, 1991, Ch. 1.

3 Mr Justice Brennan, in 'Mabo and Others v State of Queensland', *Australian Law Reports*, vol. 107, no. 1, 1992, p. 42.

4 Randy T. Simmons and Urs P. Kreuter, 'Herd mentality: banning ivory sales is no way to save the elephant' *Policy Review*, Fall 1989, pp. 46–9.

3 Social coordination in market and planned economies

1 On costs as sacrificed opportunities, see Paul Heyne, *The Economic Way of Thinking*, 7th edn, Macmillan, New York, 1994, Ch. 3.

2 ibid. Ch. 1 and pp. 389–92 provides an excellent introduction to the economic way of thinking about human behaviour.

3 ibid. Chs. 2 and 3.

4 For a brief exposition of how an ideal market yields efficient outcomes, see Heyne, pp. 208–11.

5 For a discussion of how a profit maximising monopolist sets prices, see Heyne, Ch. 9.

6 The reasons why this is so are briefly explained in Ch. 9. A simple explanation is given in Heyne, pp. 279–82.
7 For an explanation of discounting, see Heyne, pp. 303–8.
8 These allowances are discussed in Heyne, pp. 282–4.

4 The economy and the environment

1 John Passmore, *Man's Responsibility for Nature*, 2nd edn, Duckworth, London, 1980, pp. 108–10.
2 Charles C. Mann and Mark C. Plummer, 'The Butterfly Problem' *The Atlantic Monthly*, Jan. 1992, pp. 47–70, 62–6.
3 For more detail, see Michael Common, *Sustainability and Policy*, Cambridge University Press, Cambridge, 1995, Chs 1–3 and Ian Wills, 'The ecologically sustainable development process: An interim assessment' *Policy*, Spring 1992, pp. 8–12.
4 Common, pp.46–9 outlines some economists' thinking about sustainability.
5 The ecological conceptualisation of sustainability, and ecological knowledge about ecosystem resilience, are discussed in Common, pp. 49–54.
6 This example is inspired by Robert Solow, *An Almost Practical Step Toward Sustainability*, Resources for the Future, Washington, 1992.
7 For example, see the Annex in David Pearce, Anil Markandya and Edward Barbier, *Blueprint for a Green Economy*, Earthscan Publications, London, 1989.
8 Steven Jay Gould, *Bully for Brontosaurus*, Norton and Co., New York, 1991, pp. 16–18; Daniel B. Botkin, *Discordant Harmonies: A New Ecology for the Twenty-First Century*, Oxford University Press, New York, 1990, pp. 6–13 and 154–7.
9 Botkin, *Discordant Harmonies*, pp. 158–60.
10 Common, *Sustainability and Policy*, p. 6.

5 Limitations of market signalling and incentives: high costs of markets

1 Based on Edwin G. Dolan, 'Controlling acid rain' in *Reconciling Economics and the Environment*, eds J. Bennett and W. Block, Australian Institute of Public Policy, 1991.
2 In the case of acid rain damage in the eastern USA, there is evidence that natural factors in the environment, such as acid soils and certain mosses that thrive in acidic conditions, interact with acid rain to produce observed damage. In these circumstances, it may be almost impossible to determine how much damage is due to particular types of pollutants. See Volker A. Mohnen, 'The challenge of acid rain' *Scientific American*, 259, Aug. 1988, pp. 33–4.
3 Mark Crawford, 'Scientists battle over Grand Canyon pollution', *Science*, 247, 23 Feb. 1990, pp. 911–12.

6 Limitations of market signalling and incentives: non-excludable goods

1 See Terry L. Anderson and Peter J. Hill, 'From free grass to fences: Transforming the commons of the American West' in *Managing the Commons*, eds Garrett Hardin and John Baden, W.H. Freeman, San Francisco, 1977, pp. 200–16.
2 See Elinor Ostrom, *Governing the Commons: The Evolution of Institutions for Collective Action*, Cambridge University Press, New York, 1990, Ch. 2.

7 Limitations of market signalling and incentives: common pool resources

1 For example, see P. Hurst, *Rainforest Politics: Ecological Destruction in South-East Asia*, Zed Books, London, 1990.
2 Don Coursey, 'The Demand for Environmental Quality', Working Paper, School of Business, Washington University, St Louis, 1992.

8 Limitations of government signalling and incentives

1 Anderson and Leal, *Free Market Environmentalism*, Chs 2 and 5.
2 For further discussion, see Heyne, *The Economic Way of Thinking*, pp. 389–92.
3 Resource Assessment Commission, *Kakadu Conservation Zone Inquiry—Final Report*, AGPS, Canberra 1991.
4 ibid. Ch. 5.
5 ibid. Ch. 7.
6 Resource Assessment Commission, *Kakadu Conservation Zone Inquiry*, p. 254.

9 Decision making over time

1 See Heyne, *The Economic Way of Thinking*, pp. 274–80.
2 For diagrams explaining how this is done, see Tom Tietenberg, *Environmental and Natural Resource Economics*, 3rd edn, Harper Collins, New York, 1992, pp. 29–35.
3 These allowances are discussed in Heyne, *The Economic Way of Thinking*, pp. 258–60.
4 For a fuller discussion of issues in choosing a discount rate, see David Pearce, Anil Markandya and Edward Barbier, *Blueprint for a Green Economy*, Earthscan Publications, London, 1989, Chs 1 and 6.
5 ibid. Chs 2 and 5.

10 Cost–benefit analysis of environmental changes

1 R. Routley and V. Routley, *The Fight for the Forests*, ANU Research School of Social Sciences, Canberra 1975.
2 Resource Assessment Commission, *Forest and Timber Inquiry—Final Report*, AGPS, Canberra 1992.
3 Mark Streeting and Clive Hamilton, *An Economic Analysis of the Forests of South-Eastern Australia*, Resource Assessment Commission, RAC Research Paper no. 5, AGPS, Canberra, 1991; Resource Assessment Commission, *Forest and Timber Inquiry—Final Report*, vol. 2B, Appendix U.
4 Streeting and Hamilton, *An Economic Analysis of the Forests of South-Eastern Australia*, Chs 3 and 5.
5 Resource Assessment Commission, *Forest and Timber Inquiry—Final Report*, vol. 2B, Appendix U.
6 ibid. vol. 2B, Appendix U.
7 ibid. vol. 2B, Appendix U, pp. U16–17.
8 D.W. Pearce, *Cost–Benefit Analysis*, 2nd edn, Macmillan, London, 1983.
9 For example, see Pearce, *Cost–Benefit Analysis*, Chs 1 and 3, and Alan Randall, *Resource Economics*, 2nd edn, Wiley, New York, 1987, Ch. 13.
10 For discussion of the ethics of cost–benefit analysis, see Stephen Kelman, 'Cost-benefit analysis: An ethical critique' *Regulation*, Jan.–Feb. 1981, pp. 33–40 and

'Defending cost–benefit analysis: Replies to Steven Kelman' *Regulation*, March–April 1981, pp. 39–43.

11 Valuing the environment

1 For an economics-based classification of values, see R. Kerry Turner, David Pearce and Ian Bateman, *Environmental Economics: An Elementary Introduction*, Harvester Wheatsheaf, Hemel Hempstead, 1994, Ch. 8.
2 Turner, Pearce and Bateman, *Environmental Economics*, classify option values as use values: however, this refers to the *possibility* of future use. In the present, option value is a non-use value.
3 Option values may be negative as well as positive. See David W. Pearce and R. Kerry Turner, *Economics of Natural Resources and the Environment*, Harvester Wheatsheaf, Hemel Hempstead, 1990, pp. 132–4.
4 See J.A. Sinden, 'A review of environmental valuation in Australia' *Review of Marketing and Agricultural Economics*, Dec. 1994, pp. 337–68, for a review of techniques and Australian studies. Also see David Pearce et al., *Blueprint for a Green Economy*, Earthscan, London, 1989, Ch. 3 and David Pearce and Dominic Moran, *The Economic Value of Biodiversity*, Earthscan, London, 1994, Ch. 5.
5 Earth Sanctuaries Limited, *Prospectus*, lodged with the Australian Securities Commission, 1 Feb. 1995.
6 For non-technical discussions, see Turner, Pearce and Bateman, *Environmental Economics*, pp. 116–20, or Pearce and Moran, *The Economic Value of Biodiversity*, pp. 65–70.
7 See Turner, Pearce and Bateman, *Environmental Economics*, pp. 120–2 or Pearce and Moran, *The Economic Value of Biodiversity*, pp. 71–7. Hedonic wage studies adopt the same method to estimate the value of environmental attributes at the workplace.
8 David Imber, Gay Stevenson and Leanne Wilks, *A Contingent Valuation Survey of the Kakadu Conservation Zone*, Resource Assessment Commission, RAC Research Paper no. 3, AGPS, Canberra, 1991, Ch. 5.
9 ibid.
10 For example, see Australian Bureau of Agricultural and Resource Economics, *Valuing Conservation in the Kakadu Conservation Zone*, Submission to the RAC, AGPS, Canberra, 1991; Resource Assessment Commission, *Commentaries on the Resource Assessment Commission's Contingent Valuation Survey of the Kakadu Conservation Zone*, Canberra, 1991 and J.W. Bennett and M. Carter, 'Prospects for contingent valuation: Lessons from the South-East Forests' *Australian Journal of Agricultural Economics*, vol. 37, no. 2, Aug. 1993, pp. 79–93.
11 For example, see Leanne C. Wilks, *A Survey of the Contingent Valuation Method*, Resource Assessment Commission, RAC Research Paper no. 2, AGPS, Canberra, 1990; Bennett and Carter, 'Prospects for contingent valuation'; and Pearce and Moran, *The Economic Value of Biodiversity*, pp.49–64.
12 Jeff Bennett, 'The contingent valuation method: A post-Kakadu assessment', *Agenda* vol. 3, no. 2, 1996, pp. 185–94, discusses these issues. Bennett also outlines a new valuation technique, choice modelling, derived from techniques used to measure preferences in marketing and psychology.
13 Also see Bennett and Carter, 'Prospects for contingent valuation: Lessons from the South-East Forests'.
14 Resource Assessment Commission, *Forest and Timber Inquiry—Final Report*, vol. 2B, Appendix U, pp. U16–17.
15 Sinden, 'A review of environmental valuation in Australia', pp. 355–8
16 Bennett and Carter, 'Prospects for contingent valuation: Lessons from the South-East Forests'.
17 Turner, Pearce and Bateman, *Environmental Economics*, p. 109.

12 Monitoring changes in economic–environmental systems

1 Robert Repetto, 'Accounting for environmental assets' *Scientific American*, vol. 266, no. 6, June 1992, p. 94.
2 Michael Common, *Sustainability and Policy*, pp. 193–4.
3 See Charles Perrings, *Economy and environment: A theoretical essay on the interdependence of economic and environmental systems*, Cambridge University Press, London, 1987, section 1.3.
4 For example, Repetto, 'Accounting for environmental assets'; David Pearce et al., *Blueprint for a Green Economy*, Ch. 4; and W.L. Hare (ed.), *Ecologically Sustainable Development*, Australian Conservation Foundation, Fitzroy, 1990, pp. 54–5.
5 Common, *Sustainability and Policy*, pp. 194–5; Pearce et al., *Blueprint for a Green Economy*, pp. 105–9.
6 Common, *Sustainability and Policy*, p. 198.

13 Social coordination in waste disposal and recycling

1 The discussion in this and the following sections draws on Industry Commission, *Recycling*, vol. I, *Recycling in Australia*, Report no. 6, Feb. 1991.
2 Industry Commission, *Interim Report on Paper Recycling*, AGPS, Canberra, 1990.
3 Industry Commission, *Recycling*, vol. I, section 3.4.

14 The economics of pollution control: two parties

1 R.H. Coase, 'The Problem of Social Cost', *The Journal of Law and Economics*, vol. 3, 1960, pp. 1–44.
2 ibid., p. 8.
3 Richard A. Posner, *Economic Analysis of Law*, 4th edn, Little, Brown and Co., Boston, 1992, p. 64.
4 For a discussion of the importance of informal norms, as opposed to the formal law, in averting and resolving disputes between neighbours, see R. Ellickson, *Order Without Law: How Neighbours Settle Disputes*, Harvard University Press, Cambridge, 1991.

15 The economics of pollution control: many parties

1 This section (and parts of Sections 15.3 and 15.7) draws on Ian Wills, 'Information problems in pollution, control: Comparing taxes and marketable permits', *Economic Papers*, vol. 11, no. 3, Sept. 1992, pp. 67–76.
2 Gerry Bates, *Environmental Law in Australia*, 4th edn, Butterworths, Sydney, 1995, pp. 96, 394.
3 For a discussion of political pressures in pollution control, see Paul B. Downing, *Environmental Economics and Policy*, Little, Brown and Co., Boston, 1984, Ch. 6.
4 William J. Baumol and Wallace E. Oates, *The Theory of Environmental Policy*, 2nd edn, Cambridge University Press, Cambridge, 1988, Chs. 11 and 12.
5 Ecologically Sustainable Development Working Group Chairs, *Greenhouse Report*, AGPS, Canberra, 1992, pp. 148–50.

16 Social coordination under certainty

1 For an introduction to issues in making decisions under uncertainty, see, for example, David Pearce et al., *Blueprint for a Green Economy*, pp. 7–19; Edwin Mansfield, *Managerial Economics*, 2nd edn, Norton and Co., New York, 1993, Ch. 13; and Theodore Glickman and Michael Gough (eds), *Readings in Risk*, Resources for the Future, Washington DC, 1990.

2 The following discussion draws on David Lindenmayer, *Wildlife and Woodchips: Leadbeater's Possum, A Test Case for Sustainable Forestry*, University of New South Wales Press, Sydney, 1996.

3 ibid., Chapter 6.

4 ibid., pp. 141–5.

5 ibid., pp. 129–37.

6 Rachel Carson, *Silent Spring*, Penguin, Harmonsworth, UK, 1965.

7 Resource Assessment Commission, *Forest and Timber Inquiry—Final Report*.

8 For a discussion of concern about the outcomes of technologies that pose hazards for people, see Baruch Fischoff et al., 'Defining risk' in *Readings in Risk*, eds Glickman and Gough.

9 See Alonso Plough and Sheldon Krimsky, 'The emergence of risk communication studies: Social and political context' in *Readings in Risk*, eds Glickman and Gough.

10 Industry Commission, *Costs and benefits of reducing greenhouse gas emissions*, vol. 1, AGPS, Canberra, 1992, pp. 10–18.

11 ibid. pp. 34–7.

12 Pearce et. al., *Blueprint for a Green Economy*, pp. 7–9.

13 Commonwealth of Australia, *Intergovernmental Agreement on the Environment*, AGPS, Canberra, 1992, para 3.5.1.

14 The following discussion, and that in the succeeding sections, draws on Ian Wills, 'The environment, information and the precautionary principle' *Agenda*, vol. 4, no. 1, 1997, (forthcoming).

15 See Anthony Chisholm and Harry Clarke, 'Natural resource management and the precautionary principle', Agricultural Economics Discussion Paper 16/92, School of Agriculture, La Trobe University, Victoria, 1992.

16 Aaron Wildavsky, *Searching for Safety*, Transaction Books, New Brunswick, 1988. For Wildavsky, resilience is the capacity to cope with unanticipated dangers after they have become manifest, learning to bounce back. Just postponing action to see what turns up is not enough; society has to equip itself to react quickly to change. See especially Ch. 4.

17 ibid., Ch. 3.

18 ibid. pp. 223–7.

19 Paul Slovic et al., 'Rating the risks' in *Readings in Risk*, eds Glickman and Gough.

17 The economics of global pollution: ozone depletion and climate change

1 For a more detailed explanation, see Kiki Warr, 'The ozone layer' in *Global Environmental Issues*, eds Paul M. Smith and Kiki Warr, Hodder and Stoughton, London, 1991.

2 For a more detailed explanation, see Shelagh Ross, 'Atmospheres and climatic change' in *Global Environmental Issues*, eds Smith and Warr, and Industry Commission, *Costs and benefits of reducing greenhouse gas emissions*, vols 1 and 2, AGPS, Canberra, 1992, Ch. 2 and App. C.2.

3 J.J. Houghton, L.G. Meiro Filho, B.A. Callander, N. Harris, A. Kattenberg and

K. Maskell (eds), *Climate Change 1995: The Science of Climate Change*, Cambridge University Press, Cambridge, 1996, Technical Summary Figure 8, p. 26.

4 ibid. p. 6.

5 For more details, see Alice Enders and Amelia Porges, 'Successful conventions and conventional success: Saving the ozone layer' in *The Greening of World Trade Issues*, eds Kym Anderson and Richard Blackhurst, Harvester Wheatsheaf, Hemel Hempstead, 1992.

6 See Industry Commission, *Costs and benefits of reducing greenhouse gas emissions*, vols 1 & 2, Ch. 6 and App. E.

7 Organisation for Economic Cooperation and Development, *Convention on Climate Change: Economic Aspects of Negotiations*, OECD, 1992, Ch. 1.

8 ibid.; Australian Bureau of Agricultural and Resource Economics and Department of Foreign Affairs and Trade, *Global Climate Change: Economic Dimensions of a Cooperative International Policy Response beyond 2000*, AGPS, Canberra, 1995.

9 Commonwealth of Australia, *National Greenhouse Response Strategy*, AGPS, Canberra 1992.

10 ABARE projections suggest that non-OECD countries' share of global carbon dioxide emissions will grow from 53 per cent in 1990 to 69 per cent in 2020 in the absence of emissions stabilisation in developed countries, and by more if the developed countries successfully stabilise emissions at 1990 levels. See ABARE and Dept of Foreign Affairs and Trade, *Global Climate Change*.

11 OECD, *Convention on Climate Change*, pp. 15–16.

18 Management of common pool resources

1 Geoffrey Blainey, *The Tyranny of Distance*, Sun Books, Melbourne, 1966, Ch. 5.

2 Graeme O'Neill, 'Fishermen's triumph of belief over logic' *Age*, Melbourne, 10 July 1991.

3 ABARE, *Individual Transferable Quotas and the Southern Bluefin Tuna Fishery*, AGPS, Canberra 1989, p. 8.

4 A.E. Caton and K.F. Williams, 'The Australian 1994–95 and 1995–96 southern bluefin tuna seasons', Working Paper, Bureau of Resource Sciences, Canberra, August 1996.

5 ibid. Table 4.

6 Ostrom terms these *appropriation problems* and *provision problems*. See Elinor Ostrom, *Governing the Commons: The Evolution of Institutions for Collective Action*, Cambridge University Press, New York, 1990, Ch. 2.

7 P. Neave (ed.), *The Southern Bluefin Tuna Fishery 1993*, Fisheries Assessment Report, Australian Fisheries Management Authority, Canberra, 1995, pp. 9–10.

8 Ostrom, *Governing the Commons*, Chs 2 and 6, includes a detailed discussion of these information requirements.

9 Industry Commission, *Cost Recovery for Managing Fisheries*, AGPS, Canberra, 1992, pp. 84–9 and E47.

10 Ostrom, *Governing the Commons*, pp. 88–102, discusses the required characteristics.

11 ibid. p. 101.

12 Ronald N. Johnson and Gary D. Libecap, 'Contracting problems and regulation: the case of the fishery' *American Economic Review*, vol. 72, no. 5, Dec. 1982, pp. 1005–22.

13 Ostrom, *Governing the Commons*, Ch. 3.

14 ABARE, *Individual Transferable Quotas and the Southern Bluefin Tuna Fishery*, p. 13.

15 Quoted in Ronald N. Johnson, 'Implications of taxing quota value in an individual transferable quota fishery' *Marine Resource Economics*, vol. 10, 1995, pp. 327–40.

16 ibid. p. 332.

17 ABARE, *Individual Transferable Quotas and the Southern Bluefin Tuna Fishery*, Chs 2 and 5.
18 ABARE, *Foreign Involvement in the Australian Fishing Industry*, AGPS, Canberra, 1993, pp. 36–8.

19 The economic significance of biodiversity

1 Barry Clough, 'Mangrove biodiversity—its value to Australia' *Biolinks*, no. 6, Department of Environment, Sport and Territories, Jan. 1994, pp. 10–12.
2 ibid. p. 10.
3 Charles Perrings, Carl Folke and Karl-Goran Maler, 'The ecology and economics of biodiversity loss' *Ambio*, vol. 21, no. 3, May 1992, pp. 202–3. This whole issue of *Ambio* is devoted to the economics of biodiversity loss.
4 Paul Erlich and Anne Erlich, 'The value of biodiversity' *Ambio*, vol. 21, no. 3, May 1992, pp. 219–26; Department of Environment, Sport and Territories, Biodiversity Unit, *Biodiversity and its Value*, Department of Environment, Sport and Territories, Canberra, 1993.
5 See David W. Pearce and R. Kerry Turner, *Economics of Natural Resources and the Environment*, Harvester Wheatsheaf, Hemel Hempstead, 1990, pp. 129–34.
6 Department of Environment, Sport and Territories, *Biodiversity and its Value*, p. 18.
7 For example, see Flannery, *The Future Eaters*, Pt 2.
8 Erlich and Erlich, 'The value of biodiversity', p. 225.
9 The concern is about the future of humanity, not life in general. Life on Earth has survived mass extinctions in the geological past. See David M. Raup, 'Diversity crises in the geological past' in *Biodiversity*, ed. E.O. Wilson, National Academic Press, Washington DC, 1986.
10 Katrina Brown, David Pearce, Charles Perrings and Timothy Swanson, 'Economics and the Conservation of Global Biological Diversity', Working Paper no. 2, Global Environment Facility, UNDP, UNEP and The World Bank, Washington, 1993, p. 5.
11 Perrings, Folke and Maler, 'The ecology and economics of biodiversity loss', p. 202.

20 The economics of biodiversity loss

1 See Brown, Pearce, Perrings and Swanson, 'Economics and the Conservation of Global Biological Diversity', pp. 25–6. This may well have been the case for the prehistoric megafauna of the Americas (e.g the mammoth), Australia (e.g diprotodonts) and New Zealand (e.g. the moa).
2 Timothy M. Swanson, 'The economics of a biodiversity convention' *Ambio*, vol. 21, no. 3, May 1992, p. 251.
3 Flannery, *The Future Eaters*, Ch. 21.
4 See Brown, Pearce, Perrings and Swanson, 'Economics and the Conservation of Global Biological Diversity', pp. 32–4.
5 World Resources Institute, *World Resources 1992–93*, Oxford University Press, New York, 1992, Table 17.1.
6 Flannery, *The Future Eaters*, Ch. 25; J. Gordon Nelson and Rafal Serafin, 'Assessing biodiversity: a human ecological approach' *Ambio*, vol. 21, no. 3, May 1992, pp. 212–18.
7 'In Sarawak, a clash over land and power' *The Asian Wall Street Journal*, 7 February 1990.
8 Roger Sedjo, 'Property rights for plants' *Resources*, Fall 1989, no. 97, pp. 1–4.
9 Coursey, 'The Demand for Environmental Quality', Working Paper, School of Business, Washington University, St Louis, 1992.

21 Measures to preserve biodiversity

1 For more detail, see Timothy M. Swanson and Edward B. Barbier, *Economics for the Wilds*, Earthscan, London 1992, especially Ch. 3.
2 Roger Sedjo, 'Property rights for plants' *Resources*, Fall 1989, no. 97, pp. 1–4.
3 Brian Dollery, David Schulze and Leigh West 'Swapping Debt for Nature', *Agenda*, vol. 2, no. 3, 1995, pp. 361–5.
4 Swanson and Barbier, *Economics for the Wilds*, p. 215.
5 J. Gordon Nelson and Rafal Serafin, 'Assessing biodiversity: A human ecological approach' *Ambio*, vol. 21, no. 3, May 1992, pp. 212–18.
6 See Paul R. Erlich and Gretchen C. Daily, 'Population extinction and saving biodiversity'; Richard C. Bishop, 'Economic efficiency, sustainability, and biodiversity'; and Norman Myers, 'Biodiversity and the precautionary principle', all in *Ambio*, vol. 22, no. 2–3, May 1993.
7 Myers, 'Biodiversity and the precautionary principle'.
8 Charles C. Mann and Mark L. Plummer, 'The butterfly problem', *The Atlantic Monthly*, vol. 261, no. 1, Jan. 1992, pp. 47–70.
9 Bishop, 'Economic efficiency, sustainability, and biodiversity'.
10 Alan Randall, 'Thinking about the value of biodiversity' in *Biodiversity and Landscape*, eds K.C. Kim and R.D. Weaver, Cambridge University Press, Cambridge, 1993.
11 J.C. Rolfe, 'Ulysses revisted—a closer look at the safe minimum standard rule', *Australian Journal of Agricultural Economics*, vol. 39, no. 1, April 1995, pp. 55–70.
12 Aaron Wildavsky, *Searching for Safety*, pp. 223–7.

22 Economics and environmental problems: important lessons

1 F.A. Hayek, 'The use of knowledge in society' *American Economic Review*, vol. 35, no. 3, 1945, pp. 519–30.

Index

S denotes Section

ECONOMICS AND THE ENVIRONMENT